Food Safety Practices in the Restaurant Industry

Siti Nurhayati Khairatun
Universiti Putra Malaysia, Malaysia

Ainul Zakiah Abu Bakar
Universiti Putra Malaysia, Malaysia

Noor Azira Abdul Mutalib
Universiti Putra Malaysia, Malaysia

Ungku Fatimah Ungku Zainal Abidin
Universiti Putra Malaysia, Malaysia

A volume in the Advances in Hospitality, Tourism, and the Services Industry (AHTSI) Book Series

IGI Global
PUBLISHER of TIMELY KNOWLEDGE

Published in the United States of America by
IGI Global
Business Science Reference (an imprint of IGI Global)
701 E. Chocolate Avenue
Hershey PA, USA 17033
Tel: 717-533-8845
Fax: 717-533-8661
E-mail: cust@igi-global.com
Web site: http://www.igi-global.com

Library of Congress Cataloging-in-Publication Data

Names: Khairatun, Siti Nurhayati, 1975- editor. | Abu Bakar, Ainul Zakiah, 1982- editor. | Abdul Mutalib, Noor Azira, 1985- editor. | Ungku Zainal Abidin, Ungku Fatimah, 1981- editor.
Title: Food safety practices in the restaurant industry / Siti Nurhayati Khairatun, Ainul Zakiah Abu Bakar, Noor Azira Abdul Mutalib, and Ungku Fatimah Ungku Zainal Abidin, editor.
Description: Hershey, PA : Business Science Reference, [2022] | Includes bibliographical references and index. | Summary: "This book provides a series of latest issues relating to food safety management systems as practiced in the restaurant industry around the globe"-- Provided by publisher.
Identifiers: LCCN 2021028571 (print) | LCCN 2021028572 (ebook) | ISBN 9781799874157 (hardcover) | ISBN 9781799874164 (ebook)
Subjects: LCSH: Food service--Sanitation. | Food handling--Safety measures.
Classification: LCC TX911.3.S3 F68 2022 (print) | LCC TX911.3.S3 (ebook) | DDC 363.72/96--dc23
LC record available at https://lccn.loc.gov/2021028571
LC ebook record available at https://lccn.loc.gov/2021028572

This book is published in the IGI Global book series Advances in Hospitality, Tourism, and the Services Industry (AHTSI) (ISSN: 2475-6547; eISSN: 2475-6555)

British Cataloguing in Publication Data
A Cataloguing in Publication record for this book is available from the British Library.

All work contributed to this book is new, previously-unpublished material.
The views expressed in this book are those of the authors, but not necessarily of the publisher.

For electronic access to this publication, please contact: eresources@igi-global.com.

Advances in Hospitality, Tourism, and the Services Industry (AHTSI) Book Series

ISSN:2475-6547
EISSN:2475-6555

Editor-in-Chief: Maximiliano Korstanje, University of Palermo, Argentina

MISSION

Globally, the hospitality, travel, tourism, and services industries generate a significant percentage of revenue and represent a large portion of the business world. Even in tough economic times, these industries thrive as individuals continue to spend on leisure and recreation activities as well as services.

The Advances in Hospitality, Tourism, and the Services Industry (AHTSI) book series offers diverse publications relating to the management, promotion, and profitability of the leisure, recreation, and services industries. Highlighting current research pertaining to various topics within the realm of hospitality, travel, tourism, and services management, the titles found within the AHTSI book series are pertinent to the research and professional needs of managers, business practitioners, researchers, and upper-level students studying in the field.

COVERAGE

- Food and Beverage Management
- Destination Marketing and Management
- Travel Agency Management
- Service Management
- Cruise Marketing and Sales
- Tourism and the Environment
- Hotel Management
- Health and Wellness Tourism
- Service Training
- Leisure & Business Travel

IGI Global is currently accepting manuscripts for publication within this series. To submit a proposal for a volume in this series, please contact our Acquisition Editors at Acquisitions@igi-global.com or visit: http://www.igi-global.com/publish/.

The Advances in Hospitality, Tourism, and the Services Industry (AHTSI) Book Series (ISSN 2475-6547) is published by IGI Global, 701 E. Chocolate Avenue, Hershey, PA 17033-1240, USA, www.igi-global.com. This series is composed of titles available for purchase individually; each title is edited to be contextually exclusive from any other title within the series. For pricing and ordering information please visit http://www.igi-global.com/book-series/advances-hospitality-tourism-services-industry/121014. Postmaster: Send all address changes to above address.

Titles in this Series

For a list of additional titles in this series, please visit: http://www.igi-global.com/book-series/

Prospects and Challenges of Community-Based Tourism and Changing Demographics
Ishmael Mensah (University of Cape Coast, Ghana) and Ewoenam Afenyo-Agbe (University of Cape Coast, Ghana)
Business Science Reference • © 2022 • 300pp • H/C (ISBN: 9781799873358) • US $195.00

Challenges and New Opportunities for Tourism in Inland Territories Ecocultural Resources and Sustainable Initiatives
Gonçalo Poeta Fernandes (CITUR, Polytechnic Institute of Guarda, Portugal)
Business Science Reference • © 2022 • 295pp • H/C (ISBN: 9781799873396) • US $195.00

Rebuilding and Restructuring the Tourism Industry Infusion of Happiness and Quality of Life
André Riani Costa Perinotto (Universidade Federal do Delta do Parnaíba, Brazil) Verônica Feder Mayer (Federal Fluminense University, Brazil) and Jakson Renner Rodrigues Soares (Universidade da Coruña, Spain & Universidade Estadual do Ceará, Brazil)
Business Science Reference • © 2021 • 330pp • H/C (ISBN: 9781799872399) • US $195.00

Handbook of Research on the Impacts and Implications of COVID-19 on the Tourism Industry
Mahmut Demir (Isparta University of Applied Sciences, Turkey) Ali Dalgıç (Isparta University of Applied Sciences, Turkey) and Fatma Doğanay Ergen (Isparta University of Applied Sciences, Turkey)
Business Science Reference • © 2021 • 906pp • H/C (ISBN: 9781799882312) • US $415.00

Socio-Economic Effects and Recovery Efforts for the Rental Industry Post-COVID-19 Strategies
Maximiliano Korstanje (University of Palermo, Argentina)
Business Science Reference • © 2021 • 281pp • H/C (ISBN: 9781799872870) • US $195.00

Risk, Crisis, and Disaster Management in Small and Medium-Sized Tourism Enterprises
Diego R. Toubes (University of Vigo, Spain) and Noelia Araújo-Vila (University of Vigo, Spain)
Business Science Reference • © 2021 • 308pp • H/C (ISBN: 9781799869962) • US $195.00

IGI Global
PUBLISHER of TIMELY KNOWLEDGE

701 East Chocolate Avenue, Hershey, PA 17033, USA
Tel: 717-533-8845 x100 • Fax: 717-533-8661
E-Mail: cust@igi-global.com • www.igi-global.com

Table of Contents

Detailed Table of Contents

Food poisoning cases in Malaysia showed an increasing trend every year where 496 episodes were reported in 2018 as compared to 401 episodes in the same week of the year 2017. Bacillus cereus is one of the foodborne pathogens related to food poisoning cases in Malaysia. The main cause for the outbreak of B. cereus is the unregulated temperature during holding time. This study was conducted to detect the presence of aerobic bacteria and B. cereus present in ready-to-eat food in Northern Perak. A total of 83 food samples were collected and analyzed for the microbial count. The result shows that aerobic bacteria and B. cereus were detected in 28% of the samples. B. cereus count in food samples tested ranged from 100 cfu/g to 42000 cfu/g, whereas the aerobic bacteria recorded a range of 500 cfu/g to 2100000 cfu/g. The highest percentage of B. cereus was found in rice-based food, followed by meat, poultry, and gravy dishes. Positive colonies of B. cereus were further tested for anti-microbial resistance profile. Most B. cereus isolates showed resistance to tetracycline and clindamycin.

In this modern era of digitalization and consumer awareness regarding food safety issues, it has become important to build proper strategies that can ensure the quality and safety of the food items from farm to forks. People love to eat at restaurants not only during business meetings but also with their family for fun and entertainment. The choice and safety of the food is vital to attract the consumer in this competitive

Polystyrene is commonly used on a daily basis for the packaging of takeaways as if people assume that it is safe. Although studies show the migration of styrene into food could cause adverse health impacts such as cancer, neurotoxicity, and hormone-related problems, less is known about public awareness of its risk. This chapter presents a study conducted to assess the level of knowledge, attitude, and practice (KAP) of polystyrene food packaging usage among food operators (i.e., hawkers, restaurants, and night market). Data were collected using self-administered to survey 115 food operators. Results show that the food operators have a moderate knowledge and attitude in polystyrene usage. Their practice in polystyrene usage is still poor. Food operators from restaurants had the highest level of KAP compared to hawker and night markets. The finding could provide some guides in developing interventions to educate food operators about food safety issues related to polystyrene.

Transmission of hazardous materials could be aggravated by inappropriate handling and storage practices. This results in cross-contamination to foodstuff or cooking utensils. The introduced hazards in the food supply chain might lead to client and reputation loss. The implementation of food safety is necessary to secure safety concerns. All employees should take initiative to be aware and have good attitudes regarding proper hygiene and sanitary practices to assure their product integrity and safety for human consumption. Therefore, this chapter delivered the appropriate and standard food safety protocols to all individuals involved in food storage, preparation, and serving. The scope was structured into (1) identification of hazardous ingredients, (2) purchasing and receiving raw materials, (3) transporting and storage, (4) cooking and reheating, (5) food serving and displaying, (6) leftover storage, and (7) cleaning and sanitation.

Being knowledgeable about food safety is one of the strategies to address food-borne diseases (FBD). The systematic review was focuses on food safety knowledge and the respective interventions. Generally, numerous relevant studies have been done to determine the level of food safety knowledge among food handlers, but studies from a Malaysian perspective were limited. Therefore, the present study reviewed a number of previous studies regarding level of food safety knowledge and type of interventions that have been done among various categories of food handlers in Malaysia. For the review purpose, preferred reporting items for systematic reviews and meta-analyses (PRISMA) was adopted based on Science Direct, Scopus, and Google Scholar databases. A total of 22 resulted from the searching and were analyzed systematically. The review of food safety knowledge was divided into three themes consisting of food handlers at premises, consumers, and students. The results of this review have identified the knowledge gap of food handlers, and the authors provide recommendations for future food safety education.

The Maldives relies completely on imported food for its staples such as rice, flour, and sugar. It is reported that the Maldives produces less than one-tenth of its overall food requirements. Due to its huge dependence on imported food products, Maldives is exposed to a high risk of contaminated food and foodborne illnesses caused within the supply chain. This chapter aimed to investigate the level of awareness among the public of the Maldives relating to food safety practices in the food industry. An online survey was developed and administered to the consumers in the Maldives. A total number of 369 usable responses were analyzed statistically. The findings highlight that even though there was a high level of awareness of the participants towards food safety in general, participants were poorly aware of the activities and the consumers' roles in the food safety practices. This study serves as a baseline study for future research in this area, particularly in the Maldives.

Section 3
Food Adulteration and Defense

This section presents two topics covering the level of awareness of food terrorism activities among consumers and food adulteration.

Food terrorism issues have gained attention from the food industry globally. Food terrorism is a threat, sabotage, or contamination act to the food supply chain committed intentionally by people with a purpose to harm the public, jeopardize the economy, and disrupt the social and political system. More seriously, injury and

reviewed. Various laws and regulations are in effect around the world in order to prevent adulteration and ensure food safety to protect the consumers. Regulatory agencies play an important role in putting a check to food adulteration by monitoring the quality of food and penalizing defaulters.

Foreword

Food is one of the most important commodities in the world to ensure the survival of human beings. Enjoying safe and nutritious food is the right of human beings. This handbook exists to provide the extensive research works relating to numerous food safety current trends, practices, and issues, particularly in restaurant establishments, to the readers. Ten book chapters are all-inclusive of three major topics namely 1) Food Safety, Control, and System 2) Food Safety: Perspectives of Cognitive and Behavior, and 3) Food Adulteration and Defense. The diversity of authors from countries around the globe will offer an exclusive knowledge spectrum, both theoretical and practical, that could be useful to comprehend the food safety practices of a specific population. It is hoped that all stakeholders in the food service industry as well as in academia will benefit from the publication of this handbook.

Eman A. Alam
Biosciences Research Support Foundation (BRSF), Egypt & Al-Azhar University, Egypt

Preface

The restaurant industry is often linked with foodborne illness outbreaks; particularly due to the unsafe food handling by workers. Seven safe food handling practices in the restaurant industry include washing hands, preventing contamination of food, using gloves, cooking food to the right temperature, keeping heated food hot, keeping cooled food cold, and reheating food to the right temperature. Time pressure, lack of equipment and resources, management and co-workers, negative consequences, education and training, and restaurant procedures are identified as factors linked to the seven safe food handling practices. Food safety systems in the restaurant industry should cover the entire flow of food from receiving to cleaning for the operation. As the global borders continue to be redefined with increased opportunities to supply food globally through international trade, it is becoming more and more important to emphasize food safety to ensure the safety and quality of food supplied throughout the food supply chain. Food safety continues to be a concern globally and especially in developing nations that are estimated to have the highest incidence rate of foodborne illness globally; as evidenced by the increased annual cases in many developing countries.

Food Safety Practices in the Restaurant Industry consists of 10 chapters divided into three sections: 1) Food Safety, Control, and System; 2) Food Safety: Perspectives of Cognitive and Behavior; and 3) Food Adulteration and Defense. Those chapters discuss multidimensional aspects of ensuring the safety and quality of food throughout the food supply chain; including various hazards affecting safety quality of food, food handler practices, and consumers knowledge, attitude, behavior, awareness, and opinion of food safety, challenges of food adulteration and terrorism, various techniques to authenticate food safety, food safety standards, and management systems.

All types of hazards and allergens can lead to deleterious effects on the human body; strategies should be devised to avoid them. Foodborne diseases exert a significant effect on the health, morbidity, mortality, and economic development of a country. A food safety management system is concerned about contamination hazards that might occur accidentally within the food supply chain. In addition to food safety, more recently food defense has been suggested to be of concern by regulatory authorities, industry, and consumers alike that is also discussed in this book. The food defense system deals with intentional contamination done to the food supply chain which causes public harm or economic disruption. The key difference between these two systems is the motivation for committing adulteration. Outbreaks of both unintentional and deliberate foodborne diseases can be managed by the same mechanism.

The target audience of this book is academicians and researchers in the food safety area, government enforcement agencies, and policy makers, foodservice operators, including workers and managers, students, and the general public that should be concerned about the safety and quality of food they consume daily.

Food Safety Practices in the Restaurant Industry covers three important elements in the food industry. Each section encapsulates the current issues, problems, and solutions related to food safety practices and management systems within the industry. Those sections are categorized as follows:

Section 1: Food Safety, Control, and System

Section 2: Food Safety: Perspectives of Cognitive and Behavior

Section 3: Food Adulteration and Defense

In the first section, four chapters highlighting issues with food safety audits, a study of the prevalence of bacteria in ready-to-eat foods, opportunities, and challenges to authenticate food safety issues, and a strategic approach in handling food microbial hazards in restaurants. The authors discussed the raising issue of foodborne illnesses, adulteration, and increasing awareness on food safety globally. The first chapter discusses food safety auditing procedures to ensure the food that is manufactured, stored, and sold to the consumers is safe and complies with the national and international standards. The next chapter covers the occurrence of pathogenic bacteria such as *Bacillus cereus* in ready-to-eat food. The authors indicate that due to temperature abuse, pathogenic bacteria can easily grow and contaminate food, and might lead to serious illness. Next, techniques to authenticate food were covered in the third chapter. The authors suggest several strategies to ensure the food intended to be sold is safe and at its highest quality without damaging the food with rapid and non-destructive techniques such as fluorescence, NIRS, MIR, and Raman spectroscopy. In another chapter, the authors discuss various food safety management systems that are essential to control hazards in the food industries such as HACCP, GMP, GHP, and others.

The second section is dedicated to food safety practices from the perspectives of human cognition and behavior. Four topics delve into food safety knowledge and practices related to consumers and employees of restaurants include issues relating to knowledge, attitude, and practices of the use of polystyrene food packaging among food operators, standard food safety practices in a restaurant, a literature review on the level of food safety knowledge among food handlers, and a study of consumers awareness and opinion toward food safety practices and policy. In the first chapter of this section, the authors examine issues related to knowledge, attitude, and practice (KAP) of the operators and consumers relating to food safety. The authors identify the KAP of the food operators on the usage of polystyrene food packaging. The results show that the food operators have moderate knowledge and poor practices while using this food packaging that can lead to health and environmental issues that warrants intervention programs. The next chapter presents the emphasis made by the authors that the food safety behavior related to identification of hazards, purchasing and receiving raw materials, transporting and storage, cooking and reheating, serving and displaying, leftover storage, cleaning, and sanitation. In the next chapter, the authors suggest that food operators should undergo food safety education due to some knowledge gaps identified in the study. In the last chapter, the authors evaluate consumers' awareness in the Maldives related to food safety practices and policy. The data show that there is a high level of food safety awareness among the consumers. However, due to the geographically dispersed islands, effective awareness and surveillance programs are often difficult to commence.

In the final section, the chapters deal with the latest research trends in food safety management systems namely food adulteration and defense. In the first chapter, the authors highlight two important issues. Food terrorism is an act intended to harm consumers and paralyze the economy. The authors report that the level of awareness related to food terrorism among Malaysians residing in Klang Valley is average. Therefore, education relating to food terrorism might increase the awareness level among the citizens and future research should be done to further investigate the situation. Finally, in the next chapter, the author emphasizes the importance of the regulatory agencies to combat food adulteration and constantly monitor the quality of the food. This is to ensure that the consumers receive safe and quality food, and this could prevent harmful substances that might affect their health.

In conclusion, all 10 chapters published in this book showcase the current matters associated with food safety practices and management systems in the food industry, particularly in the restaurant sector. The authors of this book have done extensive research, thus the data shared is a significant added value in the body of knowledge and also will benefit the readers as well as the stakeholders in the foodservice industry.

Acknowledgment

We, the editorial team, would like to express our gratitude to all authors and co-authors for their dedication and cooperation in contributing their excellent research to this handbook. All authors are, either food experts, food scientists, researchers, faculty members, graduates, postgraduates, or government officials, who teach, involve, and deal with food safety education and research.

We also thank the IGI Global Publisher for providing us with this opportunity to publish this book.

Siti Nurhayati Khairatun

Ainul Zakiah Abu Bakar

Noor Azira Abdul Mutalib

Ungku Fatimah Ungku Zainal Abidin

Section 1

Food Safety, Control, and System

This section presents four chapters dealing with food safety audits, a study of the prevalence of bacteria in ready-to-eat foods, opportunities and challenges to authenticate food safety issues, and a strategic approach in handling food microbial hazards in restaurants.

Chapter 1
Food Safety Audits:
Ensuring Quality and Safe Food to Our Plate and Palate

Ruby Siwach
College of Dairy Science and Technology, Lala Lajpat Rai University of Veterinary and Animal Sciences, India

ABSTRACT

Food safety has gained global attention due to rising issues of food-borne illnesses, adulteration, and increased consumer awareness about food safety worldwide. It is a challenge for the governments and the food industry itself to maintain food safety throughout the food and supply chain. There are several systems and processes adopted by various countries to ensure food safety, and the food safety audits are one of the indispensable tools to achieve the goals of food quality and safety. Rising trends of consuming processed foods, eating out in restaurants and cafes, home deliveries of food from outside worldwide have made the auditing process very essential to ensure that the food products are being manufactured, stored, and sold in compliance with national and international standards. This chapter aims at providing an overview of the food audit processes, scope, importance, challenges, and future trends.

INTRODUCTION

The global food industry has witnessed large expansion in productions and markets due to globalization. Besides that, the increase in the education level of consumers; the health and awareness campaigns run by various national governments and

DOI: 10.4018/978-1-7998-7415-7.ch001

international agencies like WHO; and the top of it the advertising industry have made the consumers worldwide very health conscious. These days consumers consider it essential to have a look at the label of the food product not only to know its cost but also to know about its nutritional value as well as whether the product quality and safety are being audited/certified by any national or international agency or not. Therefore, consumers want food packed, high nutrition, ready to eat, longer shelf life, but not at the cost of food quality and safety.

Food quality is a holistic concept that includes all the factors that will influence consumers' acceptability of any product including positive attributes such as nutritional value, cultural value, color, taste, flavor, texture as well as negative attributes such as spoilage, contamination, adulteration, and food safety hazards. On the other hand, food safety means the assurance that a given food will not cause any adverse health effect, toxicity, or injury to the consumers as it has been handled, prepared/manufactured, and stored in compliance with the food standards laid down by various governments and international bodies and it is free from any contaminants and adulterants or any non-permitted additives (The National Council of Educational Research and Training [NCERT], 2021-22).

As food safety has gained utmost importance, the food manufacturers, processors, retailers, and regulatory agencies across the globe are increasingly turning to the food safety auditing process to assure their consumers base about the safety and quality of their products. The auditing process involves the inspection of the food business unit, starting from the raw material to the manufacturing processes and systems; the manufactured products and throughout the supply chain till it reaches the end consumer, in a very detailed and systematic manner by a trained individual (auditor) possessing the experience, education, and knowledge of the above disciplines, but they must also have skills in conducting auditing in a systematic, objective and professional way (Food Safety and Standards Authority of India [FSSAI], 2006). For the food industry the auditing process has become indispensable as it effectively ensures food quality and food safety in several ways:

- The given food products, processes employed and the systems involved comply with national and international standards.
- The audit programs find deficiencies and non-compliances in the audited elements thereby create room for further improvement in the operations, processes, and systems of a food manufacturing and processing unit.
- Food safety and quality audits are used to evaluate management systems in the food industry followed by prioritization of management actions.
- Audits help a business unit to obtain certifications to certain food safety and quality standards.

- Audits increase the credibility of the auditees (the business unit on which audit is to be performed) for consumers in the global marketplace and thus help them to stay and compete for the betterment of food safety and quality.
- Audits give the consumers the necessary satisfaction and trust in a given product.

Therefore, this book chapter has been written to impart readers a detailed overview of the food audit processes, basic principles of auditing in the food industry, the scope of auditing in food systems, the importance of conducting audits, skills and knowledge required for auditors, challenges in auditing processes and systems, recent and future trends in food audit systems.

BACKGROUND

The word audit has been derived from the Latin word "auditus" which means "hearing" and the history of auditing dates back to medieval times when auditing was done by hearing the accounts and was aimed at checking any negligence or fraudulence (Matthews, 2006). Earlier the auditing process was mainly restricted only to the financial accounts of any business unit, however, later on with the rapid growth of industrialization in the 19th century and dawn and spread of globalization in the 20th century, and due to the focus of industry on consumers' trust and confidence owing to cut-throat competition in the marketplace in the 21st century, the auditing has become an integral part of every process associated with any business or industry whether primary, secondary, or tertiary. In recent times the audit covers not only financial accounts of a business unit or industry but covers the entire process involved in the procurement, manufacturing, and distribution of a given product starting from the raw material till the end product reaches the retailer and the food industry is not an exception to it.

The auditing process has evolved itself over the centuries from a mere hearing process to a systematic, well-planned, and documented process conducted by a certified auditor for getting audit evidence and thereafter evaluating it objectively to determine the extent to which the audit criteria i.e. regulations and standards specified by national and international agencies for a given food product are fulfilled (International Organization for Standardization [ISO], 2018a).

The need for a comprehensive and robust audit system to ensure food safety was felt in the food industry due to several incidences of food poisoning outbreaks across the world and the 1993 Jack in the Box restaurant E. coli outbreak in the U.S. was one of them, which occurred due to consuming beef patties contaminated with *Escherichia coli O157:H7* bacterium, caused the death of four children and infected

732 people across four states in the United States (Millstone & Van Zwanenberg, 2003).

In the United Kingdom, the food safety regulations had to be revised and a ban was put on beef exports by European Union (EU) in 1996 due to Bovine Spongiform Encephalopathy (BSE) crisis to gain credibility and consumer confidence. Bovine spongiform encephalopathy is also known as mad cow disease and when spreads to humans by eating contaminated food, is known as Creutzfeldt–Jacobs disease (Ramasamy et al., 2003).

Several incidences and changes across the globe along with the release of the General Principles of Food Hygiene by the Codex Alimentarius in 1969 demanded a system that could prevent any such outbreaks and can maintain consumers' trust; and thus, the food safety and quality audit systems came into existence.

In the 1990s and the early 2000s, many food manufacturers themselves developed and introduced many food safety audit schemes as a tool to ensure the quality and safety of the raw material they were buying from their suppliers. This was a significant step towards food quality and safety, but it gave birth to the problem of multiple audits every year—sometimes 20 or more to which suppliers were subjected, and to solve this problem the Global Food Safety Initiative (GFSI) was launched in 2000 to attain the objective of "certified once, accepted everywhere" to reduce the costs and multiple audits; and also to pursuit for food safety culture in the whole supply chain across the globe.

The GFSI launched many audit schemes such as Food Safety System Certification 22000, International Featured Standards (IFT), and many more under one umbrella. Although these audit schemes served well, these were privately owned schemes and there was a global demand for an international food safety standard and after detailed meetings and discussions on international platforms and thorough examination of food laws and regulations from around the world the ISO 22000 standard was developed as an international food safety standard by the International Organization for Standardization (ISO) using the ISO 9000 standard for quality and the Codex Food Hygiene document. The ISO 22000 standard is a food safety-specific standard that integrates the HACCP system as described by Codex Alimentarius, along with three other elements: interactive communication, prerequisite programs, and system management (Stier, 2020). The audit in the food industry can be of three types:

First-party audit: This is also referred to as an internal audit where the audit is being conducted by the company or organization by itself or by an external party on its behalf.

Second-party audit: This is also referred to as an external audit where the audit is conducted by another company or organization i.e. when the auditee and the auditor are different but there is a direct commercial relationship between the two. For example, the audit conducted by a food company on one of its suppliers or the audit conducted by a retailer on the food business unit.

Third-party audit: This is also referred to as an external audit where the audit is conducted on a food business by an organization, body, or agency on behalf of another organization or scheme which has no direct commercial relationship with the food company. For example, the certification audits under the Global Food Safety Initiative (GFSI).

With this background, this chapter aims to enrich the knowledge and concept of the readers about the safety and quality auditing system in the food industry across the globe.

PRINCIPLES OF AUDITING

There are several underlying principles behind a successful audit program and only strict adherence to these principles can make an audit program efficient, effective, and reliable:

Integrity: Integrity refers to the personality attributes of the auditor and the auditing team managing an audit program such as professionalism and works ethics (honesty, responsibility, impartiality, upright character, and ability to be incorruptible, ability to resist influences on the judgment while carrying out an audit and more).

Fair presentation: The audit evidence, audit conclusions, and reports should be a truthful, objective, and accurate reflection of the audit activities. Any difference of opinions between the audit team and the auditee observed while conducting the audit should be reported and consensus-based. Fair representation of the entire audit program includes the truthful and accurate representation of the findings which are observed or heard for example the Lactometer no. 11 is out of calibration since the due date of calibration has passed; then arriving at inferences from such facts on basis of evidence and in consensus with the auditee, for example, the auditee has no calibration program for calibration of Lactometer as required by Standard I.S. 9585; and the truthful statement of any non-compliances encountered such as the Hazard Analysis Critical Control Point (HACCP) plan has not been documented well as per prescribed standards.

Due diligence: The ability of the auditors to apply professionalism and to make reasoned judgments in all the audit situations is a must-needed principle while conducting an audit. Proper professional conduct helps in building up the relationship of trust and mutual respect between the auditor and the auditee which is essential for the exchange of basic data and information between the two to conduct an audit. This principle if followed creates a working environment where both the parties are willing and can work together to achieve both business improvement and food safety and quality audit objectives.

Confidentiality: The auditors should not use the confidential information available to them while conducting the audit, inappropriately for personal gain or against the legitimate interests of the auditee. The confidential and sensitive information should be handled securely by the auditor.

Independence: The auditors should maintain the objectivity of the audit evidence and impartiality of the entire audit process by avoiding biases and conflicts of interest. For internal audits, the assigned auditors should not be chosen from the system or activity being audited.

Evidence-based approach: The audit evidence should be collected objectively using a rational method so that these remain verifiable and the reproducible audit conclusions can be reached out every time for a given set of audit criteria and scope, even when performed by different auditors.

Risk-based approach: The audit process should take into consideration the risks and opportunities that are important for the audit client, and for achieving the audit program objectives (ISO, 2018b).

AUDIT PROCESSES

Food safety and quality audit are conducted to identify any areas of potential improvement in food safety processes and systems of a food business unit. It also identifies areas having deficiencies so that appropriate corrective action may be taken to restore compliances (Food Standards Australia New Zealand [FSANZ], 2001). Audits are of three types- first party, second party, and the third party depending upon the conducting organization/agency and can be conducted by private organizations (such as certification bodies for example GFSI, contractors) or government bodies or by customers of a business's product or services. Audits are an integral part of the food business units covering both products and systems starting from the procurement of raw materials to manufacturing of product and throughout the supply chain to the retailer to ensure that products and systems comply with the regulatory requirements and standards. Regardless of the audit criteria and scope the food safety and quality audits are conducted by following steps:

- Planning
- Execution
- Corrective and preventive action
- Verification
- Audit Evaluation

(Global Food Safety Resource [GFSR], 2016).

Planning: The audit program planning covers some areas that need to be carefully planned before commencing an off-site audit program. The planning for conducting a specific audit, by the audit team leader should be done on a risk-based approach and should include the preparation of an audit program according to the auditee and the processes and systems to be audited to achieve the objectives effectively. The audit team leader communicates with the auditee to provide necessary information related to the audit objectives, scope, criteria, methods, and audit team composition, including any technical experts; and also asks for the information related to risks and opportunities identified by the organization and how they are being addressed. The auditor plans for arranging the applicable statutory and regulatory standards and other requirements relevant to the systems, processes, products, and services of the auditee, so that the documented food safety program of the food business unit can be reviewed for compliance before an on-site assessment or during off-site assessment (FSANZ, 2001; ISO, 2018a).

Execution: The audit program is initiated by appointing and assigning the auditors and formally notifying them. For the execution of the audit plan, the audit team members should collect and review the information related to their audit assignments and prepare documented information for the audit, such as physical or digital checklists; audit sampling details, audiovisual information, and more. A process audit is carried out through performing a review of procedures and documentation, and interviewing persons directly involved in the process being subjected to audit. The first step while conducting an on-site audit is - appointing and assigning roles and responsibilities of guides and observers with approval from the audit team leader, audit client, and/or auditee. The main job of the guides or observers is to assist the auditors throughout the audit process. While conducting an audit, the audit team exchange information, assess audit progress and reassign work between the audit team members, as needed and also maintains communication with the auditee, the audit client and regulators when statutory and regulatory requirements require mandatory reporting of nonconformities.

Corrective and preventive action: In the next step the auditor evaluates the data and evidence collected to decide whether the audited products and systems are consistent and following documented policies, objectives, procedures, and records. The job of the auditor is to search for the reasons behind non-compliance and then correlate it with the management system and a standard, giving the food company the tools, they need to identify and address the inconsistency (Surak & Lorca, 2007). Corrective action is performed for a nonconformity that has already occurred and involves locating and documenting the root cause behind the non-conformities, then scanning the entire system to ensure that no other similar nonconformities could occur, and thereafter analyzing the effect that such nonconformity may have had on a product or service produced before the nonconformity was identified and dealt with. The corrective action is taken according to the severity of the situation either by recalling the product, notifying the consumer, or downgrading or scrapping the product. On the other hand, a preventive action addresses the chances and potential for nonconformity to occur and involves the establishment of proactive measures to prevent a potential nonconformity from occurring, and thereafter analyzing the entire process and system to determine what actions are required and what controls should be in place to prevent non-conformity from occurring (Westcott, 2005).

Verification: The verification stage of an audit involves the conduction of the process of assessment to evaluate how effective the corrective and preventative actions are in achieving their objectives. The individual responsible for conducting the food safety audit should not be the same individual who determined the corrective action, to add a degree of impartiality and a 2nd viewpoint.

Evaluation: All the audit evidence collected using appropriate sampling is evaluated against the audit criteria to reach audit findings which are further reviewed to reach audit conclusions. As per the audit plan, the audit findings should include conformities and good practices along with their supporting evidence, nonconformities along with their supporting audit evidence, opportunities for improvement, and any other recommendations to the auditee. The audit conclusions should be reported by the audit team leader to provide a complete, accurate, concise, and clear record of the audit.

Therefore, an on-site audit involves on-site audit management according to the audit plan, an opening meeting with the representatives of the company for understanding the processes, products, and systems to be audited, verification that the audited elements are under compliance, and thereafter communication of results and observations related to compliances as well as non-compliances among team members and with the auditee (ISO, 2018a).

GLOBAL AUDITING AGENCIES, SCHEMES, AND STANDARDS

An auditing agency means an agency recognized by the Food Authority of a country for conducting food safety and quality audits by the prescribed regulations and standards. For example, the Food Safety and Standards Authority of India (FSSAI) grants recognition to private auditing agencies for carrying out food safety audits after fulfillment of certain criteria such as the agency is established or registered and holds a valid accreditation on ISO/IEC 17020 or ISO/IEC 17021 or ISO/TS 22003 from National Accreditation Board for Certification Bodies for Management System Certification, for the required food categories specified by the Food Authority; and it has at least three auditors with the prescribed qualifications and requirements.

The food manufacturing and processing units are different concerning the products being manufactured and processed at their sites and processes and systems involved in the production, therefore no auditing agency can use the "one size fits all" approach for conducting the auditing and usually consists of different schemes and the auditee can choose one of them as per its requirements (Safe Food Alliance, n.d.). Food safety and quality audit are conducted to determine compliances and non-compliances in the products, processes, and systems of a food business unit, and to have a standard against which these can be determined is an essential requirement for any auditing agency. The standard refers to the criteria such as specific legislation/regulation, technical requirements, standard operating procedures, or codes of practice against which the audit is being conducted. In absence of a standard, the auditor cannot conduct audits objectively and there would be no uniformity in the audit findings and conclusions when the audit is conducted on a given auditee by different auditors. Without standards, the subjective assessment of the elements to be audited will result in a breakdown of the audit program. The standard or audit criteria provide unambiguous requirements which the company can aspire to meet and the auditor can audit against. By referring to standards all the stakeholders involved in the audit process can understand what is expected from them, side by side the scope for arbitrary interpretation of the circumstances and findings is minimized. Various global auditing agencies, schemes, and standards are as follows:

Global Food Safety Initiative (GFSI): The GFSI is a private organization that approves different auditing platforms once the auditing platform meets the criteria as specified by GFSI thus providing recognition to specific food safety audits which are conducted to ensure that the food facilities are manufacturing and processing safe and quality food for the consumers. Therefore, if a food processor or manufacturer has the GFSI certification, it can effectively gain credibility by showing its current and potential customers that its plant is operating with a structured, comprehensive, and effective food safety program. GFSI over the years has established benchmarked schemes that focus on food safety and the essential prerequisite programs. The GFSI benchmarked schemes are:

- PrimusGFS
- Global Aquaculture Alliance (GAA)
- GLOBALG.A.P.
- Food Safety System Certification 22000 (FSSC 22000)
- Global Red Meat Standards (GRMS)
- CanadaGAP
- Safe Quality Food (SQF)
- BRCGS
- International Featured Standards (IFS)

- Japan Food Safety Management Association (JFSM)

Out of these, the FSSC 22000—Food Safety System Certification 22000, BRC Global Standard, IFS—International Featured Standards, and SQF Food Safety Code are approved process-related audit schemes.

PrimusGFS: PrimusGFS is a GFSI recognized audit certification program, having a presence in 22 countries, and is the only GFSI audit that is fully integrated into an online platform. It is working with a large community of stakeholders such as growers, shippers, retailers, food service providers, restaurants, scientists, consultants, academics, etc. to provide for the certification of produce sector products — from growing operations to minimally-processed produce (fresh-cut) products. Depending on the operation being audited, PrimusGFS audits cover Food Safety Management Systems (FSMS), Good Agricultural Practices (GAP's), Good Manufacturing Practices (GMP's), and Hazard Analysis Critical Control Points (HACCP). PrimusGFS audits are performed by certification bodies (auditing companies) that are approved by the PrimusGFS Certification Program and accredited under ISO/IEC 17065 or ISO/IEC 17021 to perform PrimusGFS audits. PrimusGFS also licenses training centers to provide required training for auditors, as well as optional education for auditees which are very crucial for conducting robust and efficient audits.

Global Aquaculture Alliance: Global Aquaculture Alliance changed its name to Global Seafood Alliance in 2021. It is a non-governmental organization and through the development of its Best Aquaculture Practices (BAP) certification standards, it has become the leading standards-setting organization for farmed seafood. It ensures the robustness and credibility of the BAP standards through quality control, compliance monitoring, and technical expertise. Implementation of BAP standards can meet the demands of the growing global market for wholesome seafood produced in an environmentally and socially responsible manner.

GLOBALG.A.P.: The scheme is a trademark and a set of standards for good agricultural practices. It is a global organization to bring safe and sustainable agriculture worldwide with an understanding that every generation has a right to safe food and extends support to farmers enabling them to connect to markets where they can sell their safely and sustainably produced agricultural products by developing and implementing farm assurance systems that are based on facts and recognized across the supply chain. GLOBALG.A.P. certification is available for three scopes of production (i.e., crops, livestock, and aquaculture) and consists of a total of more than 40 standards.

FSSC 22000: The Food Safety System Certification 22000 (FSSC 22000) offers a complete certification Scheme for the auditing and certification of Food Safety Management Systems (FSMS) of an organization. It also offers certification for combined FSMS and Quality Management Systems (FSSC 22000-Quality). The FSSC

22000 Scheme is managed by Foundation FSSC and governed by an independent Board of Stakeholders, consisting of representatives from several sectors in the food industry. This scheme uses international and independent standards such as ISO 22000, ISO 9001, ISO/TS 22003, and technical specifications for sector-specific Pre-Requisite Programs (PRPs), such as ISO/TS standards and BSI PAS. Besides these standards, this scheme contains additional requirements to ensure consistency, integrity and to provide governance and management. FSSC 22000 supports all the sectors of the food supply chain starting from animal farming, to animal feed production, food packaging manufacturing, food manufacturing, transport and storage, catering, and wholesale and retail.

Global Red Meat Standards (GRMS): The Global Red Meat Standard has been developed by the Danish Agriculture & Food Council in partnership with its abattoir members and the Danish Meat Research Institute. It is a certification program customized to the specific requirements of the red meat industry. GRMS was first published in 2006 – and since 2009 it has been benchmarked against GFSI benchmarking requirements. GRMS aims to deliver transparency within animal welfare, quality, food safety, and hygiene in factories that slaughter, cut, debone, process, and handle meat and meat products from pork, beef, lamb/sheep, goat, and horse, through an independent certification process based on ISO/IEC 17065.

CanadaGAP: The CanadaGAP certification program was launched in 2008 by the Canadian Horticultural Council (CHC), and currently has six generic HACCP models and two on-Farm-Food-Safety (OFFS) manuals: one specifically for greenhouse operations and a second for other fruit and vegetable operations. It is a voluntary food safety program for fruit and vegetable businesses operating in Canada and is owned and operated by CanAgPlus, a not-for-profit corporation. CanadaGAP focuses on good agricultural practices, GMP, and HACCP programs (commodity-specific HACCP plans) and was GFSI benchmarked in 2010. The program is designed to be realistic, cost-effective, voluntary, and market-driven. It is based on industry input and needs, technically sound, and credible as well as consistent between regions and commodities and buyer recognition.

Safe Quality Food (SQF): The SQF Institute is a Division of the Food Marketing Institute in the USA and manages the SQF supplier assurance programs that are used by food retailers, manufacturers, and primary producers internationally to assure the safety and quality of their food supply. The food safety codes of SQF are designed for all parts of the food supply chain from primary production through manufacturing, to storage and retail; and are internationally accredited and benchmarked by the GFSI. Various food safety codes under SQF are:

- SQF Food Safety Fundamentals (for small farm businesses)
- The SQF Food Safety Code for Primary Production

- The SQF Food Safety Code for Manufacturing
- The SQF Food Safety Code for Storage and Distribution
- The SQF Food Safety Code for Manufacture of Food Packaging
- The SQF Food Safety Code for Retail
- The SQF Food Safety Code for Foodservice
- The SQF Quality Code

British Retail Consortium Global Standards (BRCGS): BRCGS was founded in 1996 by retailers in pursuit of harmonization of food safety standards across the supply chain. BRCGS also has several standards such as Packaging Standard, Consumer Products Standard, Storage and Distribution Standard, Agents and Brokers Standard, Retail Standard, and Ethical Trade, and Responsible Sourcing Standard. Each Global Standard is regularly reviewed, revised, and updated at least every three years after extensive consultation with a wide range of stakeholders. BRCGS focuses on the post-farmgate (starting with packhouses), through to retail.

International Featured Standards (IFS) Food: International Featured Standard (IFS) Food is a GFSI recognized standard for certifying the safety and quality of food products and production processes. Intertek is an accredited certification body of International Featured Standard with accreditation for globally recognized programs including the Food Safety System Certification (FSSC 22000), British Retail Consortium Food Standard (BRC), and Safe Quality Food (SQF). Intertek goes beyond testing, inspecting, and certifying products and is a Total Quality Assurance provider to industries worldwide. It has its presence in more than 100 countries and delivers innovative assurance, testing, inspection, and certification solutions for food business units' operations and supply chains.

Japan Food Safety Management Association (JFSM): The JFSM was established in 2016 and it contributes to food safety and strengthens consumer's confidence by establishing international food safety management standards in Japan and related schemes, by managing an accreditation and certification scheme to the Japanese food safety management standards, by training and developing human resources among food industries and by conducting surveys related to food safety and providing information to the food industries.

The GFSI audit schemes have gained incredible popularity and acceptance around the world. Though these schemes are not mandatory, most buyers of foods, beverages, packaging, and ingredients expect that their suppliers have an established GFSI audit scheme in place.

NSF International: This international body is a leading American organization involved in certification, testing, and auditing to support food safety worldwide. National Sanitation Foundation (NSF) is accredited by the American National Standards Institute (ANSI) and has developed over 80 currently active voluntary

American National Standards under the scope of public health, safety, environment, and sustainability assessment. Besides certification, it also offers online, on-site, and on-demand training and education and world-class strategic and technical consulting services to the companies. The NSF mark is valued by consumers, manufacturers, retailers, and regulatory agencies worldwide and assures consumers that a product has been tested by a trusted independent certification organization.

British Standards Institution (BSI): BSI is the national standards body of the United Kingdom and reaches across the international standards community. It provides auditing (i.e. supplier audits, custom audits, and internal audits), certification (ISO certification, and others such as IATF, FSSC), product testing, certification (BSI Kitemark, CE marking, and verification, market access solutions), and training services to various stakeholders.

Australian National Food Safety Auditor (NFSA): The NFSA Auditor certification program has been developed by Exemplar Global in conjunction with the Australian Government, primary producers, food manufacturers, retailers, trainers, and certification bodies. The Australian Government Department of Agriculture, Fisheries, and Forestry initiated the project to develop a single national certification scheme that encompasses first-, second-, and third-party auditors.

China FSMS: China Quality Mark Certification Group Co., Ltd. is a core enterprise, approved registered in China which has an independent legal entity of the third-party certification and training institutions. The enterprise implements the GB/T22000-2006 standard to establish and improve the food safety management system, to improve the ability to ensure food security, to reduce the risk of food insecurity, and to increase consumers' satisfaction.

GETTING READY FOR AUDITS

Audit readiness may be defined as the awareness of a company and its staff about audit principles, processes, programs, and the preparedness to undergo a food safety and quality audit at any time. Due to the presence of several national and international food safety standards, certification bodies such as FDA, USDA, GFSI, SQF, BRC, FSSC, IFS, and more, and different types of audits (first party, second party, and third party), it can be very challenging for a company and its staff to prepare itself for food safety and quality audit. Depending on the nature of the food business unit and the standard under which it is being audited, the audit plan, scope, criteria, and management will vary. However, manufacturers who have strong food safety and quality programs in place take audits as an opportunity to identify the areas of deficiency and improvement in their systems, products, and processes (Ecolab, 2011). Any organization can make itself audit-ready by implementing the following steps:

1. **Awareness:** The documented food safety procedures related to food products, processes, and systems in any food business unit to be audited should be followed thoroughly and consistently as these are the first ones to be reviewed by the audit team leader while conducting the audit. The organization and staff should be fully aware of potential risks and corrective and preventive actions should be part of everyday activities. For example, if the staff understands that by following good hygienic practices, they are preventing the contamination of a food product or any tool that might come in contact with food or food product then the cross-contamination and occurrence of foodborne illnesses can be prevented.
2. **Risk-based prioritization:** All tasks, actions, and timelines in any food business unit should be prioritized based on their risks to food safety and quality rather than audit date. The understanding by the management and staff of any organization of the fact that the purpose behind audit readiness is not to pass the audit only but to maintain a culture of food safety in every area of the organization.
3. **Establishing a culture of peer and self-review:** Companies should empower those at the top of the hierarchy to coach their subordinate employees and also the employees should conduct peer observations, not only to bring compliance but also to keep the entire workforce committed to upholding the culture of food safety. If the tools such as self-review and daily checks are imbibed in the everyday activities of an organization then every employee will become his/her auditor.
4. **Sense of belongingness:** Establishing and maintaining a culture of communication with all the employees and a conducive working environment can make them more efficient and transparent in their assigned duties. This can earn their trust and create an environment in which employees feel like they belong to the organization.

If a company is not feared of visits by auditing agencies and rather waits for the agency to make an impromptu visit that means the company and its staff are audit-ready (Chegeni, n.d.).

DIGITIZATION IN FOOD SAFETY AND QUALITY AUDITS

History has witnessed significant changes in the processes of conducting an audit and collecting audit evidence, starting from a mere hearing process to a systematic and well-documented process. In recent times, the audit processes have witnessed another revolutionary change (i.e. digitization) by shifting from paper-based auditing and reporting to auditing based on tools such as the Internet of Things (IoT), blockchain, mobile solutions, big data analytics, full end-to-end traceability, and recall management and even artificial intelligence (Sebastian, n.d.). This modernization and digitization in all aspects of the food safety and quality audits infers many advantages to the food safety management systems and the food industry as a whole:

Better decision and policymaking: Digitization of the entire auditing process generates enormous digital data at the national and global level, which can be analyzed by advanced statistical tools easily, to reach various conclusions and inferences significant to the policymakers and can be used for better public health decision making, for example, what adulterants are common in a given food product, supplies from which region are showing highest non-compliances and which process steps are at high risk of getting a non-compliance and what are the lacunae in the supply chain.

Rapid data exchange: Digitization facilitates audit data exchange between all the stakeholders and parties involved in the auditing process, thus making it easier to correspond and communicate regarding corrective actions and non-compliance.

Real-time monitoring: The monitoring data obtained from digitizing the audit process allows different Food Safety Management Systems of an organization to check the entire procurement, production, and supply chain systems for compliance and non-compliance on a real-time basis.

Training of the auditors: Training of the auditors is an essential component of a robust audit process and digitization has made online training on principles and methodology of the audit processes, preventative controls such as detecting and preventing adulteration, safe food handling, food handler health awareness and more, available to auditors, enabling them to gain knowledge, and competence without traveling. Such common digital platforms providing online training can be accessed virtually by auditing agencies for their employees working as external auditors, food industry for their internal auditors, and foodservice operators (e.g. restaurants, hotels, canteens) for their employees across the globe covering a very large workforce compared to off-line training and simultaneously saving costs for each stakeholder.

Tracking of the supply chain: Digitization enables real-time tracking and monitoring of vehicles used for transportation and delivery of food products and quickly identifies vehicles having deviation from the prescribed temperatures thus preventing temperature shock to the product, maintaining product quality and safety, and also the deviations can be reported and corrected immediately.

No multiple audits: Digitization has transformed supplier management in the food industry. All the relevant nutritional information and certifications are verified and digitized on a common digital platform and are available for the buyers to view and thereby reducing the need for multiple supplier audits thus saving time and costs for all the stakeholders.

Addressing the shortage of auditors: The ratio of auditors to the food manufacturing facilities to be audited across the globe is not satisfactory due to the shortage of competent auditors and better use of digital tools and technology can solve the problem of shortage of auditors in the auditing field and thus improving the efficiency and effectiveness of the auditing process (Harris, 2019).

Traceability and food safety: Digitization has transformed the traceability and recall processes in the food industry making auditing and the entire food safety system more robust. Technology has enabled manufacturers and processors to trace their food right from their suppliers to retailers. These days many larger buyers are expecting their suppliers to have high-end tracking technologies. For example, after the 2018 Romaine lettuce, *E.coli* outbreak the Wal-Mart demanded that all of the company's leafy green suppliers should have blockchain technology which allows for transparency in terms of recording, storing, and securing data (Keogh, n.d.). Blockchain is a system of linked chain that stores auditable data and information in units called blocks across a network of users in a virtual open space, allowing all users to simultaneously look at records, reports, audit evidence, etc. in real-time and since the data and information are not stored at any particular single location, it is almost impossible to hack the information. Blockchain has immense potential to revolutionize traceability and recalls, thus food safety in the food industry. Walmart has employed this technology for conducting traceability tests on mangoes and the blockchain remarkably increased the efficiency of traceability by reducing the time required to trace the product's origin. By using blockchain, the time required to trace mangoes back to their original farm was reduced from six days, 18 hours, and 26 minutes to only 2.2 seconds. During an outbreak of disease or contamination, tracing and recall in few seconds by using blockchain technologies can save many lives. Blockchain is also being used by some certification bodies in audits (Food and Agricultural Organization [FAO], 2019). Another recent development in traceability is 2DBarcodes storing a large amount of product information at the Point-Of-Sale, for example, batch number, lot number, best-before date, use-by date, pack date, product weight, product price, and more.

Predictive modeling enabled risk reduction: Predictive modeling has emerged as another innovation in the digitized audit system. In financial auditing, predictive modeling provides the auditors with early warning indicators in real-time (throughout the audit year) before the year-end annual audit. Unilever used predictive modeling to develop a probabilistic exposure assessment to assess the safety of REPFEDs (refrigerated processed foods of extended durability) about *Bacillus cereus* by taking into consideration many factors including raw material contamination, heat treatment, bacterial spore heat resistance, injured spore lag time and growth potential, chilled supply-chain market (split into the warehouse, retail, consumers steps) and consumers' habits (e.g. time for shopping, a time before consuming) to predict the growth of *Bacillus cereus* (Membré et al., 2007). Food safety predictive modeling would help in conducting the audits which follow a risk-based approach and aim at maximizing risk reduction, rather than focusing on the traditional model of interval-based audits followed by the remedial intervention (Martin, 2018).

Benefits of smart glass technology: The smart glass technology coupled with intelligence allows real-time, hands-free two-way audio-visual communication over any distance across the globe. The person wearing smart glasses can talk to a remote viewer and view images or video messages on the device's visual display. The remote viewer – via computer, smartphone, or tablet – can see the other person's point of view through the lens of the wearable device. The use of smart glasses by many companies in training auditors and conducting shadow audits by allowing experts to be "virtually" on-site anywhere in the world at any time and has significantly reduced the traveling cost and other expenditures compared to on-site training and audit programs.

LESSONS TO BE LEARNT

Various food manufacturing and processing units, national and international bodies, retail chains, and global auditing agencies are setting examples by being pioneers in adopting new techniques, methodologies, and processes in the auditing system.

With the digitization of their entire food safety management system, Dubai has taken a global leader in protecting the health and well-being of residents of Dubai and the millions of visitors. Dubai has taken purposeful steps in advancing towards an integrated and digitized approach and announced its open data and blockchain policy in 2016.

Freshchain, an Australian-owned and operated company is using blockchain and artificial intelligence to improve traceability throughout the supply chain.

National Sanitation Foundation (NSF) international has begun to trial smart glasses in actual food safety audits of foodservice operations, to augment auditor knowledge and access to data. For NSF, this brings greater accuracy and efficiency when executing over 150,000 food safety and quality audits around the globe each year. At a time when there is a much greater need for specialized expertise, and an aging workforce of auditors that is no longer willing to travel to conduct on-site audits, the smart glasses technology introduces the capability to bring these experts to the audit without them ever having to leave their home or office.

In Australia and New Zealand, the Joint Accreditation System of Australia and New Zealand (JAS-ANZ) is conducting a trial on using smart glasses. The trial is aiming to evaluate smart glass technology for effectiveness by comparison with having JAS-ANZ staff, technical experts, or mentors on-site, including evaluating the costs of the glasses (including associated costs such as couriering, mobile phone costs, etc.) compared to the costs of having JAS-ANZ staff on-site.

In Europe, risk-based approaches have been planned where the businesses with good audit performance and low-risk may be audited less frequently and replaced with remote audits in some years. BRCGS has brought many changes in its new version from the expansion of the requirements for environmental monitoring (sanitary design, personnel practices, and operational methods) to an increased number of unannounced audits to further strengthen food safety systems (Solar, n.d.).

CHALLENGES, ISSUES, AND PROBLEMS IN AUDITING

The food safety and quality audit processes have evolved over the years to keep pace with changing consumers' requirements, the regularly revised national and international standards, and advanced technology. However, any changes in existing audit processes and techniques though can solve some existing problems and can bring improvements in operations but side by side might pose new challenges in its adoption and execution. Also, the dynamic nature of the global food markets keeps bringing new issues and problems in the auditing area which further demand improvements in audit processes and the cycle continues to go on. The challenges and problems faced by different stakeholders of the auditing process are as follows:

Multiple audits: The problem of multiple audits is as old as the audit process itself. The food manufacturers and processors usually complain that despite the launch of commonly recognized private schemes such as GFSI and international standards such as ISO 19011: 2018, they are still subjected to multiple audits, sometimes 10 or even more. Most of these audits are conducted by different buyers using their audit schemes and demands. There are several reasons behind the multiple audits and a lack of trust in audit findings of the private schemes is one of them. Some buyers are of the view that many existing audit schemes have become mere checklists that do not provide the details that they want to see in the product and services and therefore they must conduct their audit as per their requirement. Many buyers complain that in many cases the food poisoning outbreaks have happened even after the third-party audits thus questioning the efficiency of third-party audits and they have to conduct an audit on their own to ensure that food quality and safety are maintained as per consumers' satisfaction. Another issue that concerns buyers is that an auditor might conduct an audit on various elements of products, processes, and systems in a food business unit as per the audit plan, yet the auditor, when evaluating these elements, may only rate them either as fully compliant, or partially compliant or non-compliant and thorough details are provided only when the given point is fully non-compliant. The detailed information about the elements which are partially compliant is equally important as these are the sources for future non-compliances. Besides many food companies, want additional details regarding the operations of a supplier and want to rely on this detailed information when evaluating a supplier as a potential partner, so they are of the view that they have to do their audit even though the company has passed a food safety audit (Stier, 2020).

Old technology: The continuous use of paper-based systems in auditing, and the failure to communicate information across the stakeholders on a real-time basis is another challenge.

Challenges in adopting digitization in the audit process: Although the benefit of digitizing food safety audits outweighs the traditional paper-based approach, the former can prove to be a challenge if not implemented and managed correctly. Digitization needs to be introduced and implemented sustainably to minimize the loss of jobs. Both experts agreed that while advanced technology seems to hold promise in helping the audit situation, food manufacturers are not generally buying them because they are expensive, company owners do not see a high return on the investment, and some are even wary of making it easier to be prosecuted should something go wrong.

Problems faced by the auditors: There are several problems faced by the auditors such as advanced age, the physical and time demands of the job and inadequate compensation, traveling nature of the job, contractual work so no working days - no money, no compensation for travel time to facilities, or for the time they need to undergo training. Besides that, auditing is a never-ending job for the auditor as the audit reports are to be written along with follow-up on any corrections that need to be made.

Challenges faced by auditing schemes: The organizations running auditing schemes face the problem of the auditee and their perception of audit only as compliance and as an unavoidable necessary activity to an improved food safety culture across the business at any point in time. Organizations across the globe are trying to address this issue to make the audit process a sought activity rather than a burden.

RECOMMENDATIONS AND SOLUTIONS

Focus on auditors: Auditors are the backbone of any robust and efficient audit program and to carry out efficient audits, it is essential to pay attention to their work environment, monetary benefits, and other incentives. The creation of a lucrative career path having well-paid entry-level positions and frequent promotions to move up in the profession are need of the hour to attract the food safety auditors. Also, the introduction of blended and off-site online audits with the use of modern technology can ease the burden of traveling on the auditors.

Unscheduled or unannounced audits: Unscheduled or unannounced audits can solve the problem of the one-day-per-year mentality of the auditee, faced by the auditing schemes, and are being used as a tool for ensuring year-round compliance. All the stakeholders should understand the importance of unscheduled or unannounced audits in reducing costs and bringing the culture of food safety in the longer term. The governments and the certification bodies across the globe should focus on incentivizing food businesses to move from a compliance/ market access mindset to the opportunity of a robust food safety culture with unannounced audits.

Adoption of risk-based auditing: Risk-based auditing has emerged as a new development while conducting food safety and quality audits. It is based on more frequent audit visits for poor performers/higher-risk products, processes, and systems thereby giving them ample opportunities to improve, and, very few and less frequent audit visits for good performers/lower-risk products, processes, and systems to incentivize them to maintain their performance by reducing their costs. However, it has not been adopted by all major auditing schemes like GFSI and others.

Regular revisions in standards: All ISO standards are reviewed and revised regularly to make sure that they remain relevant to the changing needs of all the stakeholders and the ever-changing marketplace. Over the years, ISO has developed additional standards for supporting the certification of food safety management systems and for defining prerequisite programs and auditor competencies (Stier, 2020). ISO 22000:2018 encompasses the latest trends and food safety requirements, and is a timely response to the rising global demand for food safety and to overcome the challenges faced by the food industry as it includes improvements in definitions, a better understanding of the concept of risk, distinguishing between risk at the operational level and the strategic level of a management system and more. The third edition (ISO 19011:2018) has added new approaches in the principles of auditing, expanded the guidance on conducting, and managing an audit program, expanded guidance on new concepts such as leadership and commitments, virtual audits, compliance, and supply chain. Therefore, it will help to address the growing need for robust and sustainable food security systems (ISO, 2018a).

Regular revisions of audit programs: The audit programs should be reviewed to identify needs for changes in planning, methodologies, and execution and possible opportunities for improvements.

Digitization in auditing: Many private auditing agencies, schemes, and big players like Walmart are adopting new technologies like blockchain, the internet of things, smart glasses, and digital platforms for real-time communication of their operations starting from procurement to production, processes, tracing, recalls, and audit systems amongst all stakeholders. For example, the Foodwatch is a cutting-edge digital platform for food safety and nutrition functional in Dubai. However, digitization is in the budding stage and there is no clear trend at present on which digital technology will be most useful for the industry to pursue and further research is required. Moreover, the digital platforms need to be designed at the international level by organizations like ISO for uniform and rapid adoption of digital technology.

Regular update of auditing schemes: GFSI fully recognizes the option of partially remote audit, however fully remote audits are yet to be recognized by GFSI. Therefore, GFSI should update itself by recognizing fully remote audits so that all the stakeholders across the globe can maintain the robustness of the food safety systems during situations like the COVID-19 pandemic which does not allow direct physical audits.

FUTURE AND EMERGING TRENDS

Remote or blended auditing: Remote auditing or blended (mix of on-site and remote) auditing is a trend on the global audit horizon, brought into sharp focus through the COVID-19 pandemic. Digitization in the auditing sector has made it possible to conduct full or partial remote auditing which has emerged as a future trend in auditing. Since October 2020, all FSSC 22000 licensed Certification Bodies (CBs) can deliver FSSC 22000 auditing and certification for food safety and quality by utilizing the full remote audit option. Similarly, since July 2020 CanadaGAP has introduced the option of a partial remote audit in its auditing process.

Risk-based auditing: Risk-based auditing has emerged as a new development while conducting audits, which includes more frequent audit visits for poor performers/higher-risk products, processes, and systems thereby giving them ample opportunities to improve, and, very few and less frequent audit visits for good performers/lower-risk products, processes, and systems to incentivize them to maintain their performance by reducing their costs.

Self-assessments: Strengthened self-assessments are also growing in prevalence, with self-assessments forming part of some schemes' processes to help in reducing costs.

CONCLUSION

Food safety and quality audits have gained utmost importance in recent times for every stakeholder in the food industry such as suppliers, manufacturers, processors, and end consumers by offering benefits to each one of these. The suppliers benefit from reducing unnecessary multiple audits thereby reducing their cost, the manufacturers and processors benefit from gaining the credibility and confidence of the consumer which is essential for staying in the marketplace and the consumers benefit from the assurance that they are consuming safe and healthy food and are getting value for their money. When the audit is performed in any food industry or food service providers (restaurants, hotels, etc.) by following the principles and detailed methodology is described in the ISO 19011 (2018), one can surely achieve the objectives of ensured quality and safety in the given food product. However, the auditing process is still evolving and digitization of auditing has opened up new avenues for the auditing sector of the food industry and has the potential to deliver real-time assurance of food quality and safety on basis of predictive modeling which gives the insight into what went wrong to what is likely to go wrong. Though digitization seems promising the global food industry needs development and implementation of common digital platforms at the global level by organizations like ISO for the benefit of one and all. Moreover, the key to improving food safety is the generation of a robust food safety culture across the business system and processes which can be established by overcoming the 'one-day one-year' mentality about the audit programs.

REFERENCES

Chegeni, B. (n.d.). How to be audit ready. *Global Food Safety Resource.* https://globalfoodsafetyresource.com/how-to-be-audit-ready/

Ecolab. (2011). Food safety audits 101: A brief history and preparation essentials. *Food Safety Magazine.* https://www.food-safety.com/articles/3855-food-safety-audits-101-a-brief-history-and-preparation-essentials

Food safety and standards authority of India (FSSAI). (2006). *Food Safety and Standards Act 2006.* Ministry of Law and Justice, New Delhi. https://fssai.gov.in/upload/uploadfiles/files/FOOD-ACT.pdf

Food Standards Australia New Zealand (FSANZ). (2001). *Food safety: An audit system.* https://www.foodstandards.gov.au/publications/Pages/anauditsystem.aspx

Global Food Safety Resource (GFSR). (2016). *Food Safety Audits.* https://globalfoodsafetyresource.com/food-safety-audits/#

Harris, L. (2019). *Auditor shortage slowly being addressed.* Global Food Safety Resource. https://globalfoodsafetyresource.com/auditor-shortage-slowly-addressed/

Herron, M. (2018). Is predictive analytics the end of the annual audit? *AccountancyAge*. https://www.accountancyage.com/2018/05/21/is-predictive-analytics-the-end-of-the-annual-audit/

International Organization for Standardization (ISO). (2018a). *International standard ISO 19011 - Guidelines for auditing management systems*. https://www.borhanjooyan.com/DL/ISO-19011-2018.pdf

International Organization for Standardization (ISO). (2018b). *Management systems for food safety*. https:// committee.iso.org/home/tc34sc17

Keogh, J. G. (n.d.). Dubai leadership: The digitization of food safety management. *Global Food Safety Resource*. https://globalfoodsafetyresource.com/dubai-leadership-digitization-food-safety-management/

Matthews, D. (2006). *A history of auditing: The changing audit process from the 19th century till date*. Routledge, Taylor & Francis Group.

Membré, J. M., Amézquita, A., Bassett, J., Giavedoni, P., Blackburn, C. D. W., & Gorris, L. G. M. (2006). A probabilistic modeling approach in thermal inactivation: Estimation of postprocess Bacillus cereus spore prevalence and concentration. *Journal of Food Protection*, *69*(1), 118–129. doi:10.4315/0362-028X-69.1.118 PMID:16416909

Millstone, E., & Van Zwanenberg, P. (2003). The evolution of food safety policy–making institutions in the UK, EU and Codex Alimentarius. *Social Policy and Administration*, *36*(6), 593–609. doi:10.1111/1467-9515.t01-1-00306

Ramasamy, I., Law, M., Collins, S., & Brooke, F. (2003). Organ distribution of prion proteins in variant Creutzfeldt-Jakob disease. *The Lancet. Infectious Diseases*, *3*(4), 214–222. doi:10.1016/S1473-3099(03)00578-4 PMID:12679264

Safe Food Alliance. (n.d.). *What is GFSI? Getting started with the Global Food Safety Initiative (GFSI)*. https://safefoodalliance.com/food-safety-resources/what-is-gfsi/

Sebastian, J. (n.d.). Effectively digitizing food safety audits. *Global Food Safety Resource*. https://globalfoodsafetyresource.com/effectively-digitizing-food-safety-audits/

Solar, L. (n.d.). BRC Global Standards V.8: New changes for improving compliance. *Global Food Safety Resource*. https://globalfoodsafetyresource.com/brc-global-standards-v-8-new-changes-improving-compliance/

Stier, R. F. (2020). The evolution of food safety management systems—from early 1900s to audits today. *Food Engineering*. https://www.foodengineeringmag.com/articles/98678-the-evolution-of-food-safety-management-systemsfrom-early-1900s-to-audits-today

Surak, J. G., & Lorca, T. A. (2007). Process auditing for food safety. *Food Safety Magazine*. http://www.foodsafetymagazine.com/magazinearchive1/augustseptember2007/process-auditing-for-food-safety/

The National Council of Educational Research and Training (NCERT). (2021-22). *Food safety and food quality*. https://ncert.nic.in/textbook/pdf/lehe106.pdf

KEY TERMS AND DEFINITIONS

Adulteration: Food adulteration refers to the process of lowering the quality of the food either by substituting the main product wholly or partly and either by the addition of inferior quality material or by extraction of valuable ingredient and the added substances may be injurious to health, e.g., removal of fat from milk and selling it as full cream milk is adulteration.

Audit: A food safety audit is a structured activity performed by an auditor who checks and collects the evidence to ensure that the given food product, processes involved, and the supply chain points comply with the requirements, criteria, specifications, and procedures set out in the food safety standards, legislation, and internal standards of operation.

Audit Agenda: The audit agenda is the time-table schedule for conducting the actual audit and sent to the auditee (business unit being audited) well in advance of the audit to enable the auditee to make sure that the required personnel and documentation of each section/unit (purchase, production, quality) to be audited are ready when the auditor conducts the audit.

Audit Conclusions: Audit conclusions refer to the overall outcome of an audit program provided by the audit leader or the audit team after a thorough evaluation of audit results. The checklist findings and all the data and information gathered during the process of auditing should be reviewed by the auditor against the reference standard.

Audit Criteria: Audit criteria refer to the range of requirements or standards against which the audit is conducted. These requirements can be related to one or more management system standards, technical standards for a given product, statutory and regulatory requirements, or any other requirements specified by relevant interested parties. The food safety and quality audit may be conducted against the above-mentioned audit criteria, separately or in combination, including but not limited to these.

Audit Scope: The scope of food audit refers to the details regarding coverage of the audit activity such as the physical and virtual location of the unit, products, and processes, time, etc.

Food Auditor: A food auditor is a trained person having multi-disciplinary knowledge about the manufacturing processes, standard operating procedures, risk assessment, monitoring, human resources, food quality and safety, microbiology, chemistry, toxicology, engineering, and so on with the competence to conduct an audit in a systematic, independent, and documented manner for obtaining the audit evidence, and thereafter evaluating it objectively to assess the extent of compliance to national and international food safety standards.

Food Quality: Food quality is a holistic concept that includes all the factors that will influence consumers' acceptability of any product including positive attributes such as nutritional value, cultural value, color, taste, flavor, texture as well as negative attributes such as spoilage, contamination, adulteration, food safety hazards.

Food Safety: Food safety means the assurance that a given food will not cause any adverse health effect, toxicity, or injury to the consumers as it has been handled, prepared/manufactured, and stored in compliance with the food standards laid down by various governments and international bodies and it is free from any contaminants and adulterants or any non-permitted additives.

Food Standard: The food standards are laid down by various governments and international bodies for manufacturers/suppliers in the food industry to ensure that safe and quality foods are reaching the consumers. These consist of a set of criteria and specifications regarding various physical, chemical, and microbiological parameters such as raw materials used during manufacturing, chemical composition, color, permissible food additives, and maximum bacterial content, etc. that a particular food product must meet to ensure that consumers get safe food and value for money.

Management System: It refers to the set of interrelated, interconnected, or interacting elements of an organization that are involved in establishing the organization's structure, roles and responsibilities, planning, operation, policies, rules, objectives, and the processes to achieve those objectives addressing single or multiple disciplines, e.g., quality management, financial management, or environmental management or the entire organization.

Chapter 2
Prevalence of Bacillus cereus in Ready-to-Eat Foods in Northern Perak, Malaysia

Noor Azira Abdul Mutalib
Universiti Putra Malaysia, Malaysia

Noor Aniza Abdul Rahim
Universiti Putra Malaysia, Malaysia

Ungku Fatimah Ungku Zainal Abidin
Universiti Putra Malaysia, Malaysia

ABSTRACT

Food poisoning cases in Malaysia showed an increasing trend every year where 496 episodes were reported in 2018 as compared to 401 episodes in the same week of the year 2017. Bacillus cereus is one of the foodborne pathogens related to food poisoning cases in Malaysia. The main cause for the outbreak of B. cereus is the unregulated temperature during holding time. This study was conducted to detect the presence of aerobic bacteria and B. cereus present in ready-to-eat food in Northern Perak. A total of 83 food samples were collected and analyzed for the microbial count. The result shows that aerobic bacteria and B. cereus were detected in 28% of the samples. B. cereus count in food samples tested ranged from 100 cfu/g to 42000 cfu/g, whereas the aerobic bacteria recorded a range of 500 cfu/g to 2100000 cfu/g. The highest percentage of B. cereus was found in rice-based food, followed by meat, poultry, and gravy dishes. Positive colonies of B. cereus were further tested for anti-microbial resistance profile. Most B. cereus isolates showed resistance to tetracycline and clindamycin.

DOI: 10.4018/978-1-7998-7415-7.ch002

INTRODUCTION

Food poisoning is the result of eating contaminated, spoiled, or toxic food. According to the Centers for Disease Control and Prevention (CDC) Trusted Source, 1 in 6 Americans will contract some form of food poisoning every year. In Malaysia, 50.90 incidence rate per 100,000 population of food and waterborne disease involved food poisoning, followed by dysentery (0.67), typhoid and paratyphoid (0.55), hepatitis A (0.29) and cholera (0.25) (Ministry of Health Malaysia, 2020). Symptoms can differ depending on the source of the infection. The incubation time for symptoms to appear also depends on the source of the infection, but it can range from 30 minutes to 15 hours (Food & Drug Administration [FDA], 2012).

Common causes of food poisoning typically include at least three of the following symptoms: abdominal cramps, diarrhea, vomiting, loss of appetite, mild fever, weakness, nausea, headaches. *Staphylococcus aureus, Escherichia coli, and Bacillus cereus* have been identified as the most common pathogenic microorganism from food and hand swab samples (Okareh & Erhahon, 2015) and the isolates of *Bacillus cereus* has been shown to demonstrate resistance to various common antibiotics (Penido et al, 2013). In addition, a significant percentage of outbreaks of foodborne diseases are caused by foods that are poorly processed (Fang et al., 2003). The objectives of this study were to determine the prevalence of *B. cereus* in ready-to-eat (RTE) food and the antimicrobial resistance of detected colonies towards common antibiotics.

BACKGROUND

Food Poisoning Cases in Restaurants

Almost 1 in 10 people in the world fall ill every year after consuming spoiled food, causing 420,000 deaths (FDA, 2018). From 1992 to 2009, 677 restaurant outbreaks reported in England and Wales affected 11,795 people with 491 hospitalizations (Gormley et al., 2012). Department of Standard Malaysia (2020) reported that food and water-borne diseases increased by 12% in 2019 where 97% are food poisoning cases.

Food handling and preparation practices in the restaurant followed by food handlers are the most common contributing factors of food poisoning reported in restaurants in the United State from 1998 to 2013 (Angelo et al., 2017). Numerous studies on local ready-to-eat (RTE) food have linked hygiene practices to the cross-contamination from hands to food (Okareh & Erhahon, 2015 and Lee et al., 2017). Apart from cross-contamination, environmental contamination, and under-cooked dishes were identified as contributing factors of food poisoning in the restaurant in the United Kingdom and Pittsburgh which caused a large outbreak in 2009 and 2013, respectively (Severi et. al., 2012; Torso et al., 2015).

Bacillus cereus as Food Spoilage

Bacillus cereus is a Gram-positive, spore-forming bacteria, capable to grow in a wide range of temperatures which is often related to causing foodborne disease (Christiansson, 2002; Guinebretiere et al., 2003 and Tirloni et al., 2019). Even though *B. cereus* is commonly related to rice, there are many outbreaks from a variety of foods in Europe since 1950. It naturally exists everywhere and the heat resistance of its endospores, joint with psychrotrophic properties, make it a likely contaminant of cooked and chilled foods (Christiansson, 2002 and Guinebretiere et al., 2003).

This bacterium is divided into two types of *B. cereus*-related illness which are gastrointestinal (emetic) diseases and diarrhoeal disease. The gastrointestinal type is most likely the result of eating contaminated rice or other starchy product which produce toxins between 30 minutes and 5 hours after ingestion. On the other hand, the diarrhoeal form caused by the presence of *B. cereus* in the intestine resulted from eating infected food. In some outbreaks both diarrhoeal and the emetic types of illness had been reported together (Christiansson, 2002).

B. cereus does not require a complex nutrient and is commonly found in soils with a low nutrient amount (Kotiranta et al., 2000). Due to this characteristic and its ability to form spores, *B. cereus is* able to spread easily and is related to a wide variety of infection-related foods, including cooked meat and vegetables, boiled or fried rice, vanilla sauce, custards, soups, ice cream, herbs, and spices. Since this organism's spores are heat-resistant and can withstand cooking temperature, this bacterium can contaminate cooked foods (Schneider et al., 2017). Previous studies by Gadaga et. al., (2008) and Fang et. al, (2003) also found the prevalence of *B. cereus* in the stew, vegetables, and salads, chilled ready-to-eat (RTE) food, sandwiches, cold noodles, and rice balls.

Antibiotic Resistance of *Bacillus cereus*

In 2013, the first Antibiotic Resistance Threats Study in the United State (US) had alarmed about antibiotic resistance where at least 2 million people reported getting an antibiotic-resistant infection every year in the US resulted in 23,000 people dying (US Department of Health & Human Services, 2019). Antimicrobial agents such as antibiotics are necessary to treat infections caused by bacteria. However, their overuse and misuse in veterinary and human medicine has been related to the development and spread of resistant bacteria. This condition results in making the treatment of infectious diseases ineffective in animals and humans.

Earlier studies found that *B. cereus* isolates were highly resistant to beta-lactam antibiotics (Shilla, 2011; Bilung et. al., 2018 and Torkar & Bedenić, 2018). Since beta-lactam antibiotics are frequently used to treat food poisoning, the incidence of resistant strains to these antibiotics may be a matter of concern as it may result in ineffective medications and useless efforts to monitor and prevent them. Another research by Jawad et. al., (2016) also found some of these bacterial strains be resistant to most commercial antibiotics, including tetracycline, gentamicin, vancomycin, penicillin, nalidixic acid, and nitrofurantoin.

METHODOLOGY

Sample Collection and Microbial Analysis

Eighty-three food samples were collected from twenty-five food facilities in Northern Perak, Malaysia from June to August 2020. Ethical approval (JKEUPM-2019-043) was obtained from the Ethics Committee of Universiti Putra Malaysia on 11 June 2020. Food samples were collected and kept in a cold box occupied with ice and were taken to the laboratory for microbial analysis. For this study, food samples were analyzed for aerobic bacteria using APC agar and *Bacillus cereus* (BC). Enumeration steps for each food sample were done as in Figure 1. This method referred to Bacteriological Analytical Manual (2012) and Biokar Diagnostic (2019).

Antibiotic-Susceptibility Testing of Positive Colonies

Antibiotic-susceptibility testing (AST) for each *Bacillus cereus* colony detected was done as below. Five types of antibiotics were used to test the antibiotic resistance of the colonies. This method (Figure 2) referred to Cockerill et al, (2013).

FINDINGS AND DISCUSSION

Microbial Count in Food Sample

The microbial analysis tested were aerobic bacteria and *Bacillus cereus* (Figure 3). From 83 food samples, 28% were detected with *Bacillus cereus,* while 61% were detected with aerobic bacteria only. *B. cereus* count in food samples tested, ranged from 1.0×10^2 cfu/g to 4.2×10^4 cfu/g, whereas the aerobic bacteria recorded a range of $5.0\text{x}10^2$ cfu/g to $2.1\text{x}10^6$ cfu/g.

Figure 1. Steps for APC and BC microbial analysis of food sample

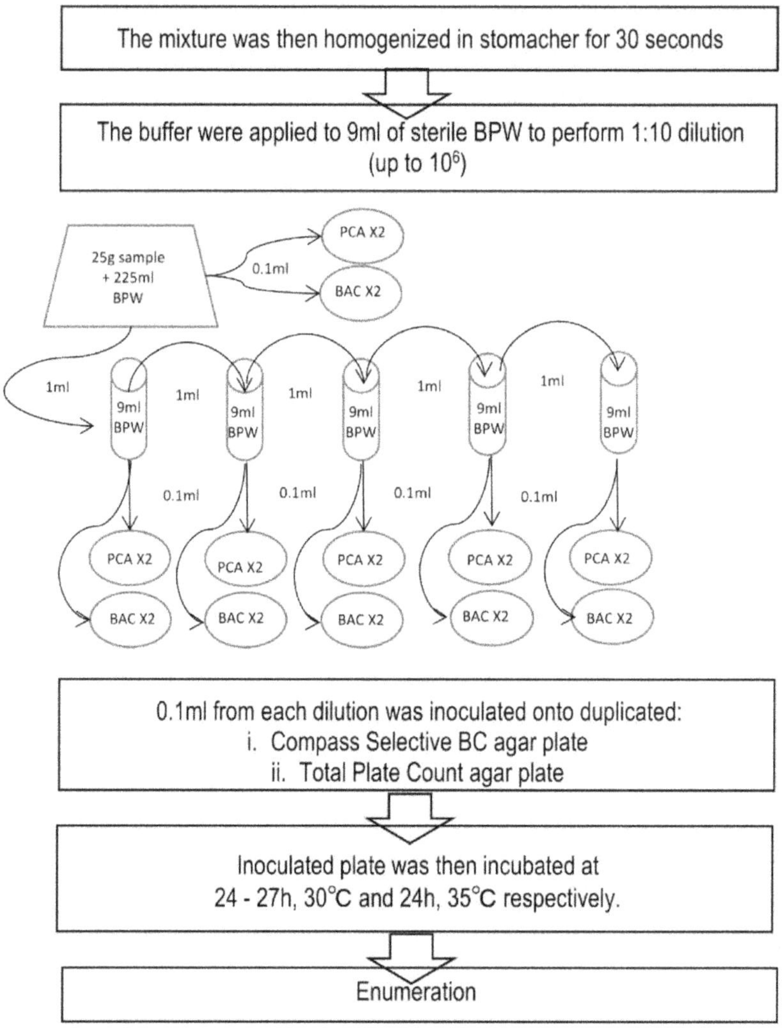

There are guidelines for unsatisfactory ready-to-eat food in Australia and Europe which limit to less than 10^5 CFU/g for total plate count and less than 10^4 CFU/g for *B. cereus*. These microbiological limits are also used worldwide by other countries. More than 10^4 CFU/g *B. cereus* in food are reported to cause food poisoning (Bolton, 2009 and Food Standard Australia New Zealand [FSANZ], 2018). Malaysia does not have a specific microbiological limit for ready-to-eat food but in Part VII Incidental Constituent: 39(2), there is a statement stated that "no person shall import, prepare or advertise for sale any food ready for consumption that is contaminated with pathogenic microorganism" (Malaysia Food Regulation,1985).

Figure 2. Steps for AST analysis of positive B.cereus colony

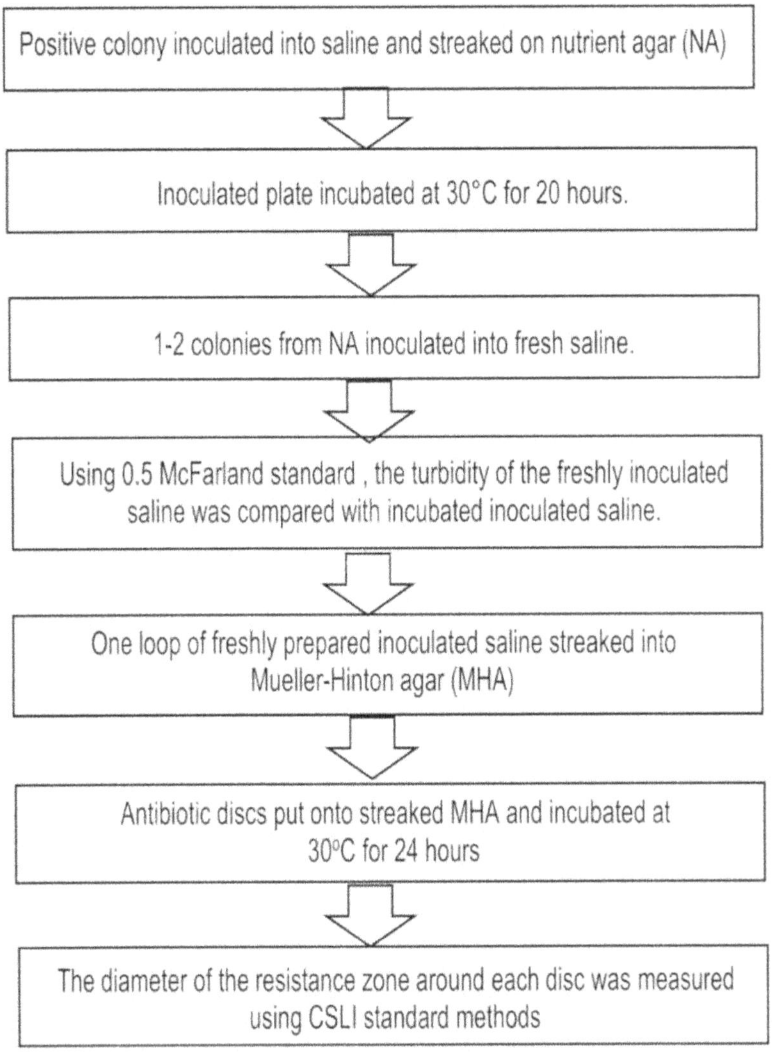

The type of food tested in this study were rice-based, poultry, meat and gravy dishes as shown in Table 1.

The highest percentage of *Bacillus cereus* was detected in meat and rice-based food, followed by poultry, and gravy dishes. However, no significant difference between the types of food to microbial count was found. Previous studies also showed the same type of food reported incidences of *B. cereus* which 50-73% in cooked rice, 27-33% in curry and 17-34% in cooked meat (Sandra et al., 2012; Erihun et al., 2015; Ali et al., 2017; Yu et al., 2020; Sornchuer & Tiengtip, 2021).

Figure 3. (left) Colonies of Bacillus cereus on COMPASS selective media for Bacillus cereus agar, (right) Aerobic colonies on Total Plate Count agar

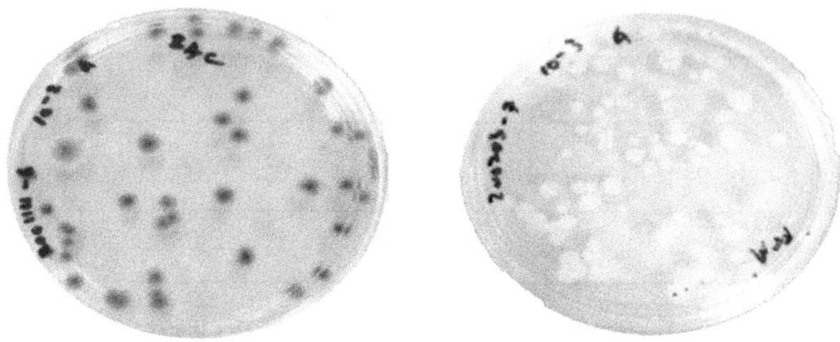

Table 1. Percentage of microbial detected based on type of food

Type of RTE food	No. of food	APC (% detected)	BC (% detected)
Rice	40	37 (93%)	15 (38%)
Poultry	11	9 (82%)	2 (18%)
Meat	7	7 (100%)	3 (43%)
Gravy	25	21 (84%)	3 (12%)
Total	83	74	23

RTE - ready-to-eat, APC - aerobic plate count, BC - Bacillus cereus

Rice had been the most common food contaminated with *B. cereus*. Additionally, even though precooked rice is contaminated at very low levels of *B. cereus* due to its natural presence in different substrates, storing at room temperature would allow rapid spore germination and toxin-producing food contamination (Choi et al., 2014 and Tirloni et al., 2019). A study done by Bilung et al., (2018) found that 100% of samples of local unhusked rice collected in Sarawak, Malaysia were detected with *B.cereus*. However, the percentage of *B. cereus* detected in cooked food was considered low compared to food detected with aerobic plate count. A study was done by Tirloni et al., (2019) reported a low percentage of which only 3 out of 38 total samples (7.89%) were detected with *B. cereus* in fried rice meals.

Other studies towards various cooked foods in China and Thailand reported 35% and 21% samples respectively, were positive with *B.cereus* (Yu et al., 2020 and Sornchuer & Tiengtip, 2021). Types of foods tested were cooked meat, vegetables, rice and noodles dishes, soups and curries, and miscellaneous foods. Although vegetable dishes are considered low-risk food, they are usually not subjected to heat treatment resulted in incomplete elimination of microorganisms (Yu et al.,

2020). Moreover, fruits and vegetables are usually in contact with soil that contains high levels of spore-forming bacteria such as *B. cereus* and other *Bacillus* spp. (Guinebretiere et al., 2003).

Antibiotic Resistance of *Bacillus cereus* in Food

There were five types of common antibiotics used to check the resistance of *B. cereus* extracted from food samples. The analysis was done on Mueller-Hinton agar using the antibiotics disc diffusion method (Figure 4). The antibiotics used were clindamycin (2 ug/ml), chloramphenicol (30 ug/ml), tetracycline (30 ug/ml), gentamycin (10 ug/ml), and vancomycin (30 ug/ml). The resistance level of strains towards antibiotics used was determined by comparing the diameter of the zone to the CSLI guideline. The percentage of antibiotic resistance were shown in Table 2.

Figure 4. (left) Antibiotics disc on Mueller-Hinton agar inoculated with B.cereus before 24 hours incubation time, (right) Resistance zone of B. cereus on Mueller-Hinton agar after 24 hours incubation time

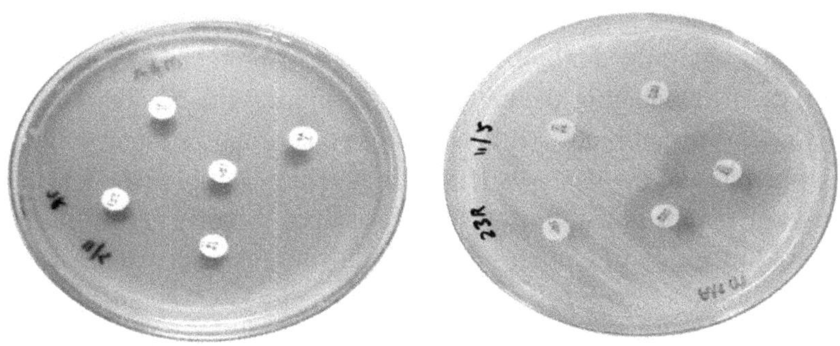

Table 2. Resistance level of Bacillus cereus strain towards different antibiotics

Antibiotics	Concentration (ug/ml)	No. of strains (%)		
		Sensitive	Intermediate	Resistant
Clindamycin	2	47	35	17
Chloramphenicol	30	78	17	4
Tetracycline	30	47	0	52
Gentamycin	10	83	17	0
Vancomycin	30	47	52	0

The results found that *B. cereus* isolates showed resistance to tetracycline and clindamycin, which is supported by a study done by Jawad et al., (2016) and Ahmed et al., (2020). There is no need for bacteria to be resistant to many antibiotics to be dangerous because resistance to even one antibiotic also can mean serious problems. This is because the bacteria can cause more harm to patients with antibiotic-resistant infections and caused longer care and recovery period due to the need for using second-and third-line treatments (US Department of Health & Human Services, 2019).

On the other hand, isolates were found to be highly sensitive to chloramphenicol, gentamycin, and vancomycin. These findings were supported by previous studies by Chon et al., (2012), Arsian et al., (2014), Gdoura-Ben et al., (2019), Fiedler et al. (2019), Zhao et al. (2020) and Sornchuer & Tiengtip (2021) who also reported the same results. Besides, there was a case report in 2016 found that vancomycin as the most suitable treatment for *B. cereus* bacteremia (Aygun et al., 2016). However, it was contrasted to a study done by Ali et al. (2017) which found isolates were resistant to vancomycin (87.5%) and clindamycin (91.6%). The difference in findings showed that isolates can develop new ways to resist antibiotics which caused relying on one type of antibiotic for certain bacteria may no longer effective.

The resistance mechanism of bacteria can work in several ways to survive from antibiotics by (i) developing new cell processes that avoid using the antibiotic's target, (ii) destroying the antibiotic with enzymes or protein that can break down the antibiotic, (iii) changing the entryways or limiting the number of entryways to restrict access of antibiotic, (iv) changing the antibiotic's target to make drug can no longer fit and (v) using its pump to get rid of antibiotic. At worst, some resistant germ can also pass directly their resistance to other germs that have not been exposed to antibiotics either by (i) transferring resistance genes via phages, (ii) transferring resistance genes when they connect by DNA conjugation, and (iii) picking up resistance genes released from nearby live or dead germs by another germ. (US Department of Health & Human Services, 2019). As the use of antibiotics also kills helpful bacteria that naturally protect humans, resistant germs have bigger advantage to multiply and emerge.

FUTURE RESEARCH DIRECTIONS

Antimicrobial resistance (AMR) microorganisms have been reported to be increasing worldwide. This happens due to the overuse of drugs especially in the animal farms such as poultry, cattle, fish, and others involving animals in large quantities. To top it all off, some of the bacterial pathogens develop multiresistant to more than one antibiotic which makes it harder to be eradicated. Infections of these superbugs among humans will cause huge problems because the treatment in health facilities might become useless and difficult. This situation can lead to severe health problems as might also cause death. Overuse of antibiotics can also release them to the environment through improper irrigation, which can lead to more AMR organisms. Continuous research on antimicrobial resistance testing to more types of antibiotics is crucial for finding the most suitable antibiotic for the treatment of food pathogens. Besides, future studies should include a larger randomized sample size across the country to permit better and broad statistical results. Overall, the One Health approach, which is the collaborative efforts of multiple disciplines working locally, nationally, and globally, to attain optimal health for people, animals, and the environment should be practiced in all sectors including the food industry. This initiative can ensure the food source is clean, safe, and healthy.

CONCLUSION

From this study, it was discovered that about 28% of the ready-to-eat foods sold are contaminated with *B. cereus* with the highest percentage detected was in the meat and rice-based sample. The bacterium was also isolated in poultry and gravy dishes at a low percentage. In addition, the samples were contaminated with a high percentage of aerobic bacteria. Even though there was no significant difference between the type of foods to microbial count, the food safety and hygiene practices of food handlers are questionable since 83% of the food tested were contaminated. The mean count of *B. cereus* was in the range of 10^2 to 10^3 which is still considered as safe based on the international standard of RTE food. However, the number of aerobic bacteria exceeded the allowable limit which is more than 10^4. Further study found that some isolates detected showed resistance to few antibiotics which were tetracycline, clindamycin, and chloramphenicol. However, no isolate was resistant to gentamycin and vancomycin. The finding of the antimicrobial resistance profile of *B. cereus* towards several antibiotics showed that relying on one type of antibiotic for certain bacteria may no longer be effective and is a cause for concern.

ACKNOWLEDGMENT

This work was supported by the Ministry of Education Malaysia through the Higher Institution Centre of Excellence (HICoE) grant scheme—project no. HICoE-ITAFoS/2017/FS6/6369114. The authors would like to acknowledge the head of department and staff of the Food Microbiology Unit, Ipoh Public Health Laboratory, Perak, and Faculty of Food Science and Technology, Universiti Putra Malaysia.

REFERENCES

Ali, A. E., Msarah, M. J., & Sahilah, A. M. (2017). Environment contaminant of *Bacillus cereus* isolated from ready to eat meat curry collected at various locations in Malaysia. *International Food Research Journal, 24*(6), 2640–2644.

Angelo, K. M., Nisler, A. L., Hall, A. J., Brown, L. G., & Gould, L. H. (2017). Epidemiology of restaurant-associated foodborne disease outbreaks, United States, 1998-2013. *Epidemiology and Infection, 145*(3), 523–534. doi:10.1017/S0950268816002314 PMID:27751201

Arslan, S., Eyi, A., & Küçüksari, R. (2014). Toxigenic genes, spoilage potential, and antimicrobial resistance of *Bacillus cereus* group strains from ice cream. *Anaerobe, 25*, 42–46. doi:10.1016/j.anaerobe.2013.11.006 PMID:24309214

Aygun-Fatma. (2016). D., Aygun-Fatih., & Cam, H. (2016). Successful treatment of *Bacillus cereus* bacteremia in a patient with propionic acidemia. *Case Reports in Pediatrics*, 1–2.

Bilung, L. M., Jaraee, J., & Vincent, M. (2018). Detection, genetic diversity and antibiotic resistance profiles of Bacillus cereus isolated from sago processing plants in Malaysia. *Malaysian Journal of Microbiology, 14*, 320–324.

Bilung, L. M., Tesfamariam, F., & Andriesse, R. (2018). Presence of *Bacillus cereus* from local unhusked (rough) rice samples in Sarawak, Malaysia. *Journal of Sustainability Science and Management, 13*(1), 181–187.

Biokar Diagnostic. (2019). *COMPASS Bacillus.: 1-2. Detection and enumeration of Bacillus cereus group in human and animal food products.* http://www.biokar-diagnostics.com

Bolton, E. (2009). Guidelines for Assessing the Microbiological Safety of Ready-to-Eat Foods Placed on the Market. Health Protection Agency.

Carlin, F., Brillard, J., Broussolle, V., Clavel, T., Duport, C., Jobin, M., Guinebretière, M. H., Auger, S., Sorokine, A., & Nguyen-Thé, C. (2010). Adaptation of Bacillus cereus, an ubiquitous worldwide-distributed foodborne pathogen, to a changing environment. *Food Research International, 43*(7), 1885–1894. doi:10.1016/j.foodres.2009.10.024

Choi, S., Kim, H., Kim, Y., Kim, B., Beuchat, L. R., & Ryu, J. H. (2014). Fate of bacillus cereus and naturally occurring microbiota on milled rice as affected by temperature and relative humidity. *Food Microbiology, 38*, 122–127. doi:10.1016/j.fm.2013.08.016 PMID:24290634

Christiansson, A. (2002). Bacillus Cereus. Encyclopedia of Dairy Sciences, 123–128. doi:10.1016/B0-12-227235-8/00032-8

Cockerill, F., Patel J., Alder J., Bradford P., Dudley M., Eliopoulos G., ... Zimmer, B. (2013). Performance Standards for Antimicrobial Susceptibility Testing; Twenty-Third Informational Supplement (M100-S23-2013). *Clinical and Laboratory Standards Institute, 49*(9).

Department of Standard Malayisia (DOSM). (2020). *Conpendium of Environment Statistics Malaysia 2020, released 27 November 2020*. https://www.dosm.gov.my

Fang, T. J., Wei, Q., Liao, C., Hung, M., & Wang, T. (2003). Microbiological quality of 18°C ready-to-eat food products sold in Taiwan. *International Journal of Food Microbiology, 80*(3), 241–250. doi:10.1016/S0168-1605(02)00172-1 PMID:12423926

Fiedler, G., Schneider, C., Igbinosa, E. O., Kabisch, J., Brinks, E., Becker, B., Stoll, D. A., Cho, G.-S., Huch, M., & Franz, C. (2019). Antibiotics resistance and toxin profiles of Bacillus cereus-group isolates from fresh vegetables from German retail markets. *BMC Microbiology, 19*(1), 1–13. doi:10.118612866-019-1632-2 PMID:31706266

Food Act 1983 and regulations 1985, Part VII Incidental Constituent: 39(2).

Food and Drug Administration. (2012). Bad Bug Book, Foodborne Pathogenic Microorganisms and Natural Toxins (2nd ed.). Author.

Food Standards Australia New Zealand (FSANZ). (2018). *Compendium of Microbiological Criteria for Food September 2018*. Author.

Gadaga, T. H., Samende, B. K., Musuna, C., & Chibanda, D. (2008). The microbiological quality of informally vended foods in Harare, Zimbabwe. *Food Control, 19*(8), 829–832. doi:10.1016/j.foodcont.2007.07.016

Gdoura-Ben Amor, M., Jan, S., Baron, F., Grosset, N., Culot, A., Gdoura, R., Gautier, M., & Techer, C. (2019). Toxigenic potential and antimicrobial susceptibility of Bacillus cereus group bacteria isolated from Tunisian foodstuffs. *BMC Microbiology, 19*(1), 1–12. doi:10.118612866-019-1571-y PMID:31445510

Gormley, F. J., Rawal, N., & Little, C. L. (2012). Choose your menu wisely: Cuisine-associated food-poisoning risks in restaurants in England and Wales. *Epidemiology and Infection, 140*(6), 997–1007. doi:10.1017/S0950268811001567 PMID:21854669

Guinebretiere, M. H., Girardin, H., Dargaignaratz, C., Carlin, F., & Nguyen-The, C. (2003). Contamination flows of Bacillus cereus and spore-forming aerobic bacteria in a cooked, pasteurized and chilled zucchini purée processing line. *International Journal of Food Microbiology*, *82*(3), 223–232. doi:10.1016/S0168-1605(02)00307-0 PMID:12593925

Hwang, J. Y., & Park, J. H. (2015). Characteristics of enterotoxin distribution, hemolysis, lecithinase, and starch hydrolysis of *Bacillus cereus* isolated from infant formulas and ready-to-eat foods. *Journal of Dairy Science*, *98*(3), 1652–1660. doi:10.3168/jds.2014-9042 PMID:25597976

Kotiranta, A., Lounatmaa, K., & Haapasalo, M. (2000). Epidemiology and pathogenesis of *Bacillus cereus* infections. *Microbes and Infection*, *2*(2), 189–198. doi:10.1016/S1286-4579(00)00269-0 PMID:10742691

Lee, H. K., Abdul Halim, H., Thong, K. L., & Chai, L. C. (2017). Assessment of food safety knowledge, attitude, self-reported practices, and microbiological hand hygiene of food handlers. *International Journal of Environmental Research and Public Health*, *14*(1), 55. doi:10.3390/ijerph14010055 PMID:28098788

Ministry of Health Malaysia. (2020). *Health facts 2020. Reference data for year 2019*. Ministry of Health Malaysia. Planning Division. Health Informatics Centre.

Okareh, O. T., & Erhahon, O. O. (2015). Microbiological Assessment of Food and Hand-Swabs Samples of School Food Vendors in Benin City, Nigeria. *Food and Public Heath*, *5*(1), 23–28.

Penido, A., Mendes, P., Campos, I., & Mendes, L. (2013). Enterotoxigenic Bacillus cereus from cooked chicken meat: A potential public health hazard. *Malaysian Journal of Microbiology*, *9*(2), 166–175.

Sandra, A., Afsah-Hejri, L., Tunung, R., Tuan Zainazor, T. T. C., Tang, J. Y. H., Ghazali, F. M., & Son, R. (2012). *Bacillus cereus* and *Bacillus thuringiensis* in ready-to-eat cooked rice in Malaysia. *International Food Research Journal*, *19*(3), 829–836.

Schneider, K. R., Schneider, R. G., Silverberg, R., Kurdmongkoltham, P., & Bertoldi, B. (2017). *Preventing Foodborne Illness: Bacillus cereus* and *Bacillus anthracis*. Institute of Food and Agricultural Sciences (IFAS), University of Florida. FSHN04-05:1–5.

Severi, E., Booth, L., Johnson, S., Cleary, P., Rimington, M., Saunders, D., Cockcroft, P., & Ihekweazu, C. (2012). Large outbreak of *Salmonella Enteritidis* PT8 in Portsmouth, UK, associated with a restaurant. *Epidemiology and Infection*, *140*(10), 1748–1756. doi:10.1017/S0950268811002615 PMID:22166322

Shilla, J. (2011). Survey beta lactamase production and resistance pattern into beta lactame antibiotics in *Bacillus cereus* strain isolated from staff hands and hospital environment in Iran. *African Journal of Microbiological Research*, *5*(19), 2980–2985. doi:10.5897/AJMR11.515

Sornchuer, P., & Tiengtip, R. (2021). Prevalence, virulence genes, and antimicrobial resistance of *Bacillus cereus* isolated from foodstuffs in Pathum Thani Province, Thailand. *Pharmaceutical Sciences Asia*, *48*(2), 194–203. doi:10.29090/psa.2021.02.19.119

Tirloni, E., Bernardi, C., Ghelardi, E., Celandroni, F., Cattaneo, P., & Stella, S. (2019). *Bacillus cereus* in fried rice meals: Natural occurrence, strain dependent growth and haemolysin (HBL) production. *Lebensmittel-Wissenschaft + Technologie*, *114*, 108393. doi:10.1016/j.lwt.2019.108393

Torkar, K. G., & Bedenić, B. (2018). Antimicrobial susceptibility and characterization of metallo-β-lactamases, extended-spectrum β-lactamases, and carbapenemases of *Bacillus cereus* isolates. *Microbial Pathogenesis*, *118*(March), 140–145. doi:10.1016/j.micpath.2018.03.026 PMID:29551437

Torso, L. M., Voorhees, R. E., Forest, S. A., Gordon, A. Z., Silvestri, S. A., Kissler, B., Schlackman, J., Sandt, C. H., Toma, P., Bachert, J., Mertz, K. J., & Harrison, L. H. (2015). *Escherichia coli* O157:H7 outbreak associated with restaurant beef grinding. *Journal of Food Protection*, *78*(7), 1272–1279. doi:10.4315/0362-028X.JFP-14-545 PMID:26197277

US Department of Health & Human Services. (2019). *Antibiotic resistance threats in the United States.* Centers for Disease Control and Prevention. https://www.cdc.gov/drugresistance/ biggest_threats.html

Yu, S., Yu, P., Wang, J., Li, C., Guo, H., Liu, C., Kong, L., Yu, L., Wu, S., Chen, M., Zeng, H., Rui, P., Zheng, Y., Wei, X., Zhang, J., Wu, Q., & Ding, Y. (2020). A study on prevalence and characterization of Bacillus cereus in Ready-to-Eat foods in China. *Frontiers in Microbiology*, *10*(January), 1–11. doi:10.3389/fmicb.2019.03043 PubMed

KEY TERMS AND DEFINITIONS

Antibiotic Resistance: This could happen when microorganisms like bacteria develop the ability to resist the drugs designed to destroy them. Antibiotic resistance is a major concern of overuse of antibiotics.

Food Poisoning or Foodborne Illness: An infection or irritation of the digestive tract that is caused by eating contaminated food. The illness is commonly caused by viruses, bacteria, and parasites.

Food Spoilage: This could occur when there are changes in the food through physical, chemical, enzyme deterioration, or microorganisms growth, that lead to the food becoming damage, inedible, or unsafe for human consumption.

Foodborne Illness Outbreak: This outbreak occurs when two or more people experience the same illness resulting from the consumption of the same contaminated food or drinks.

Ready-to-Eat (RTE) Foods: Food products that are pre-cleaned, precooked, mostly packaged, and ready for consumption without prior preparation or cooking. Examples of RTE foods are salads, cooked meats, smoked fish, desserts, sandwiches, cheese, and food that is cooked in advance to serve cold.

Chapter 3
Rapid and Non-Invasive Techniques:
Opportunities and Challenges to Authenticate Food Safety Issues in Restaurants

Muhammad Haseeb Ahmad
Government College University, Faisalabad, Pakistan

Rabia Shabir Ahmad
Government College University, Faisalabad, Pakistan

Amna Sahar
University of Agriculture, Faisalabad, Pakistan

Muhammad Bilal Hussain
Government College University, Faisalabad, Pakistan

Muhammad Imran
Government College University, Faisalabad, Pakistan

Muhammad Faizan Afzal
Government College University, Faisalabad, Pakistan

Muhammad Kamran Khan
Government College University, Faisalabad, Pakistan

Marwa Waheed
Government College University, Faisalabad, Pakistan

ABSTRACT

In this modern era of digitalization and consumer awareness regarding food safety issues, it has become important to build proper strategies that can ensure the quality and safety of the food items from farm to forks. People love to eat at restaurants not only during business meetings but also with their family for fun and entertainment. The choice and safety of the food is vital to attract the consumer in this competitive

DOI: 10.4018/978-1-7998-7415-7.ch003

environment. Previously, conventional methods have been employed for assurance of quality and safety parameters of the food. But in this modern era, there are many potential alternatives that can serve the purpose rapidly and non-destructively. Hence, this chapter describes the rapid and non-destructive methodologies such as fluorescence, NIRS, MIR, and Raman spectroscopy that can be used for the food safety evaluations.

INTRODUCTION

Due to advancements in modern technologies and consumer awareness about the foods and their ingredients, it has become important to ensure the quality and safety of the food items. Food quality and safety are the important criteria that help for the selection of the food that the consumer eats. If they eat low quality and substandard food products then they will suffer from different ailments and become a burden not only on their loved ones but also on society. Therefore, the government and the consumers are focusing on the quality and safety of food products. It is an important tool to ensure the quality of life and a way to compete with their market competitors.

Food safety and quality of any substance are distracting by the presence of any hazard or foreign particle. There are three types of hazards (physical, chemical and microbiological) that may be present in the food. These hazards may come in contact with food through contamination, mishandling unhygienic and poor sanitary practices during handling, processing, storage and serving of food. The ingestion of such hazards present in the food may lead to illness. There are intrinsic and extrinsic sources of these hazards (Edwards & Stringer, 2007). The intrinsic source of physical hazards includes fruit stalk in dried fruits as well as the presence of bone in boneless meat products. The extrinsic sources of physical hazards are glass, rubber or metal etc. The presence of a foreign body in food may be a risk factor that is dependent on the size, shape, type, hardness and sharpness of the object. The existence of external hazards in food leads to choking as well as in few cases surgery may be required to remove these foreign objects from the body (Bansal, et al., 2017; Hyman, et al., 1993; Olsen, 1998). In addition to this, chemical and biological hazards are more problematic as one cannot see them with a naked eye and the presence of such hazards in the food may seriously damage nutritional, organoleptic and textural characteristics. The ingestion of such hazards may lead to food poisoning and severe physiological damage, which ultimately reduce one´s immunity.

Food industries and foodservice sectors such as hotels and restaurants are very conscious about the product. The restaurants receive raw and processed materials. The raw materials include meat and fresh produce such as fruits and vegetables.

The handling and storage of such products need extensive care of temperature monitoring for maintaining the quality and safety of such raw materials. On the other hand, processed food items such as dairy products, frozen items and spices that also need proper attention for ensuring the safety of the served food items to the customers. Any mishandling during the processing and storage of such products may lead to severe damage. Food safety is the responsibility of growers, retailers, and the employees dealing with the production and serving of the food. Previously, various methods have been employed for assurance of the food safety issue that is time consuming, laborious and needs experts for their conduction. Therefore, there are some potential rapid and non-invasive methodologies, which are easy to use, environment friendly and need no sample preparation and experts for the analysis. The optical methods based on different spectroscopic techniques such as fluorescence, near-infrared, mid-infrared and hyperspectral imaging as well as the biomimetic sensing devices can have the potential to serve as rapid and non-destructive tools for combating the safety issues in restaurants and hotels. This chapter discusses in detail the principle and potential of all these rapid and non-invasive technologies for the assurance of food safety.

BACKGROUND

Food industries and food businesses promote a lot of techniques to ensure the safety of food products from any external objects. There are a lot of conventional methods to detect external objects like a magnet, electrical impedance, and surface penetrating radar as well as metal detector, etc. These are very simple and reliable techniques but these have many restricted tasks such as these are not able to detect the plastic and glass-like materials. Similarly, electrical impedance and surface penetrating radar enable to detect the metallic and foil packaging.

Recently, the assurance of safety and quality of food becomes a major problem related to public concerns. It is mentioned by the discipline of food safety that the availability of food should be safe, as well as it is not contaminated with toxins and chemicals that are hazardous to human health. Therefore, food quality, the nutritional value and the appearance of food are the attraction towards the consumers (Alander et al., 2013). All countries must assure the quality of imported food products to protect their consumers. It is also observed the exported foodstuff is being more reliable for the consumers globally. Thus, many countries imposed strict rules to assure the food standards that are associated with the safety and quality of food. The outbreaks of any disease via food becomes more dangerous for human health and also affect the economy (Saldaña et al., 2013). Foodborne illness and frauds are preventable and can be managed through employing good management practices.

These can also be assured using novel technologies in this modern era. Recently, various kinds of non-invasive techniques are used to investigate food safety and quality due to their specification and uniqueness in several tasks like analysis of different food parameters, reduce wastes and allow the repeated measurements on time (Nicolai et al., 2007). However, it is noticed that conventional techniques are not reliable due to some limitations like being more labor-intensive, requires material preparation and being time-consuming. Whereas the visual detecting techniques are best to achieve safety and quality of food.

It is reported that the foodstuff is evaluated by the most commonly used parameters which are safety and quality of that substance (Busby & Hyman., 2012). Fruits and vegetables are a reliable source of micronutrients and rich in antioxidants due to which enhance immunity and resulted in no chance of any foodborne illness. It is reported in many countries that the food because illness recognized in the fresh foodstuff products. The producers must ensure food safety and quality and prevent the production of these hazardous types of food products (Siddiqui et al., 2014).

Food safety and quality are constraining to evaluate the consumers' contentment, and these are affected by visual variables like surface, flavor, texture appearance and local government standards (Mahendran, 2012). Quality is a comprehensive term that may vary from one buyer to another, but the quality integrates characteristics that attract the customers. Many countries are willing to imposed legislation and acquired food certification to assure safety and quality. It is an actual standard to achieve the financial development of the nation is to nature of nourishment. Several experts believe that security is the key factor for this but the absence of this leads to real damage and demise of the consumer. An item may be harmful if it is not evaluated by the pathogenic organisms, chemical toxins and poisonous materials

The production of food is a very typical confusion worldwide, due to sufficient quality of producer which is the main problem for food sectors, food business and consumers. Currently, due to awareness about the food products the customer knows about the product which they purchase. Thus, to attract and satisfy the consumer quality is not compromised. To achieve these types of interest, modern technologies are being used that provide more security to the consumers associated with safety and quality as well as introduce the standard for nutrition (Harrison., 2003).

The European food industry is liable to produce the market demand due to the incredible technological progress and advancement of consumer's lifestyles. Generally, these advanced technologies are quickly adopted by consumers but their expectation about food products is frequently not reliable. It is noticed that they are willing food products should have some characteristics like increased functional and nutritional properties, sensory quality, wholesome image and safety ensured as well as inexpensive. But they are also interested to ensure the increased shelf life and convenient processing of food. These methods have some limitations because these are time-consuming, highly trained and educated staff is required and also limit to use potentially (Nychas et al., 2008; Papadopoulou et al., 2011). It is noticed that these molecular techniques are more costly, and expensive equipment is needed.

Food industries and food businesses promote a lot of techniques to ensure the safety of food products from any external objects. There are a lot of conventional methods to detect external objects like, magnet, electrical impedance, and surface penetrating radar as well as metal detector, etc. these are very simple and reliable techniques, but these have many restricted tasks such as these are not able to detect the plastic and glass-like materials. Similarly, electrical impedance and surface penetrating radar enable to detect the metallic and foil packaging.

These days, some noninvasive analytical techniques are used to evaluate food safety and quality. These approaches include spectroscopic techniques such as fluorescence, and vibrational spectroscopy (near-infrared, Fourier transform infrared and Raman). In addition to this, biomimetic sensors and hyper/multispectral imaging are being used for ensuring food safety and authenticity (Argyri, et al., 2014; Ellis & Goodacre, 2001; Nychas et al., 2008).

RAPID AND NON-INVASIVE TECHNOLOGIES

Fluorescence Spectroscopy

Fluorescence spectroscopy falls under the umbrella of rapid and non-invasive technologies. It is one of the techniques that is widely used for the authentication and characterization of food and food products. The main reason for its utilization in the food system as it requires very little and no sample preparation. It is simple and easy to apply on all types of the product either in solid and liquid forms. It is considered to be a cost-effective method, which can detect the contaminants in the food items (Horigome et al., 2016).

Principle of Fluorescence Spectroscopy

The basic principle of fluorescence spectroscopy is based on the Jablonski diagram. In this method, the light of a certain frequency is absorbed by the fluorophores (aromatic amino acids, NADH and vitamins) which may be present in the sample. Due to the absorbed excitation light, the electrons in the molecule become excited and goes to a higher energy state. This excited state is highly unstable and the electron comes back to the ground state by the emission of light. This emission of light is called fluorescence.

Application of Fluorescence Spectroscopy for Authentication of Food Safety and Quality of Food Products

Fluorescence spectroscopy is widely used in various food applications for ensuring the safety and quality of food products.

Cereals are the basic ingredients in various food products that may be served in restaurants. The most important cereals grains are wheat, rice, and corn. Among these cereals, wheat is widely used in various wheat-based products. The quality of wheat plays an important role in the quality of these wheat-based ingredients that may be served in restaurants.

Wheat from various varieties has different characteristics that affect the quality of the bakery and pasta product. Fluorescence spectroscopy has been used for analytical, rheological and nutritional parameters of the wheat flour. In rheological parameters of the wheat, dough development time and water absorption are the most important parameters that determine the accurate amount of water and proper mixing time which ultimately ensure the quality of the bakery product that has been prepared from these wheat flour (Ahmad et al., 2016a). The nutritional contents of wheat flour have been also determined using fluorescence spectroscopy that used the partial least square regression and locally weighted regression for obtaining the high value of the coefficient of determination (Ahmad et al., 2016b). In addition to this, sourdough fermentation is an important operation in which pH and lactate contents are important parameters which were determined with the help of fluorescence spectroscopy with acceptable root mean square error of prediction (Grote et al., 2014).

Dairy items such as milk, dried milk powder and cheese are the most important ingredients that can be utilized in various food products. So the quality and safety of these food ingredients are important for maintaining the organoleptic parameters. Fluorescence spectroscopy can be utilized to monitor the changes in milk due to the mild heat treatments (Kamal & Karoui, 2017). It can also be used for the monitoring of changes in protein isolate powders during storage (Babu & Amamcharla, 2018). Similarly, it has been employed for monitoring the changes in non-fat dry milk powder during storage. The changes in Millard browning, tryptophan and riboflavin spectral fingerprints have been used that illustrate the changes taking place during the storage at various temperatures (4, 22, 35, 50°C) for 8 weeks (Liu & Metzger, 2007). It has been also utilized in the determination of milk composition (Shaikh & Donnell, 2017). It is used as a rapid and no invasive tool for the determination of riboflavin in milk that replaces the chemical-based high liquid chromatographic method (Alvarado et al., 2020). It has also been used for monitoring the light-induced oxidation in cheese (Anderson et al., 2005). In addition to this, it can be employed to monitor the changes in the milk during processing and storage (Purna et al., 2005). As for as the authentication of the dairy products is concerned, it has been employed for the detection of plant oil in cheese (Dankowska et al., 2015). The aflatoxin in milk is one of the prevailing safety issues that can be tested with fluorescence spectroscopy (Manetta et al., 2005). Hence it can be inferred from the above discussion that fluorescence can be a value able tool for the characterization of various milk-based food items.

Meat and meat-based products are some of the most served food items that are being served in the hotel and restaurants. Therefore the quality of the meat and its products is very important for having excellent organoleptic characteristics. The meat is normally cooked to increase its taste and palatability. Therefore, fluorescence spectroscopy has been utilized that can monitor the changes which were subjected to various cooking temperatures (66, 90, and 237°C) for 1 min to 10 min (Amna et al., 2016).

Meat is a perishable food that undergoes microbial spoilage which can be monitored using fluorescence spectroscopy. The spectral signatures of meat stored at 15°C under aerobic conditions were taken at the tryptophan and NADH region. The high peaks in these regions give indications that due to microbial spoilage, NADH produce which is evident from these peaks and also creates the changes in the protein structure that generate peaks in the tryptophan region (Oto et al., 2013). The fluorescence spectroscopy was used to determine the microbial contamination on the surface meat that is stored at refrigerated temperature for 20 days. The fluorophores such as NADH and porphyrin show peaks related to the different microbiota. During the storage of the meat, porphyrin level increases that generate peaks which show that fluorescence spectroscopy can be used to investigate the microbial spoilage on the surface of the meat (Durek et al., 2016).

Minced meat is also used as an ingredient in various products in restaurants. It may spoil due to microbial activity. Hence fluorescence spectroscopy was used to monitor the microbial spoilage of minced meat using a portable spectrophotometer stored at refrigeration temperature and 15°C under aerobic and anaerobic conditions. The portable device can predict the microbial spoilage of the meat with a higher coefficient of determination (Ait Kaddour et al., 2011). Similarly, aerobic plate count was also investigated on the sliced beef and spectral signatures were taken in porphyrin, vitamin A, tryptophan and NADH regions which can be utilized to build the predictions models that generate a higher coefficient of determination showing the authenticity of the partial least square regression models for determination of microbial spoilage (Mala et al., 2016). The above discussion elaborates that fluorescence spectroscopy is a rapid and non-destructive tool for the investigation of microbial spoilage of meat during storage. These applications can be utilized for the authentication of meat products in restaurants regarding microbial spoilage.

All these methods have been applied at a laboratory scale and need to develop miniature devices so that such devices can be used at restaurants and hotels for ensuring the safety and quality of the food products.

Near-Infrared Spectroscopy (NIRS)

In recent years, near-infrared spectroscopy (NIRS) has been proved to be one of the most promising and efficient techniques not only for monitoring but also for controlling the quality and safety of the processing of agri-based foods. In comparison with conventional detection methods, it requires little sample preparation time, less time for detection and also exhibits good adaptability. It covers the electromagnetic spectra ranges between 780 and 2500 nm. Its spectra can be documented in transmission, reflection, or interactance modes presenting complex physical and chemical information with reference to the vibrational behavior of molecular bonds like OH, CH, and NH (Kademi et al., 2019). The prompt development in NIRS has led to innovation in the evaluation of food safety and its quality control. With the chemometrics development, NIRS has become even more famous and grabs the attention of food scientists and researchers.

Fundamentals of Near-Infrared Spectroscopy

The NIR spectrometers comprise of beam splitter system also known as a wavelength selector, a light source, optical detector, sample detector and data analyzing system (Wang & Paliwal, 2007). All these components possess different properties and these should be selected on the basis of their intended use, in order to offer a consistent and effective instrument. The most frequently used radiation sources of NIR, which are quite cheap in price but also give high-intensity radiation are QTH (quartz–tungsten–halogen) lamps. But these are not very productive at the industrial level due to their temperature sensitivity, low energy efficiency, long start-up times, vibration sensitivity and heat generation. Contrary to it, LED (light-emitting diodes) has resolved many of the issues presented by QTH light sources and gaining popularity in this regard, but their high cost is preventing their regular use by the relevant authorities (Nicolai et al., 2007; Wang & Paliwal, 2007). Beam splitter systems used in NIR spectrometers give light of single-color translating from multi-color input. Within the NIR spectrometer, sample detectors vary by speed of response, spectral response, and the minimum threshold for detectable radiant power. More often, NIR spectrometers use single-channel and multi-channel photon detectors. Various sample detectors are only for use in laboratories; though, portable devices which are recently developed are more simple in use, compact, have a more robust design and are cheap in price. These improvements have enabled the utilization of the technology of NIR spectroscopy in a variety of applications (Qu et al., 2015).

Applications of NIRS for Food Safety and Quality Control

The NIRS is being used successfully in various sectors of the food industry with a special focus on authentication, poison detection, adulteration, freshness determination and illegal treatments on different food materials, which are described as below:

Meat authentication is a reason of great concern for meat and its products, as its adulteration is a common phenomenon because of the great gap in its demand and supply, which is amplifying its adulteration by several means and causing major losses in this sector. Adulteration of meat is causing trust deficit issues among the public as it may cause great harm to human health. Apart from health issues, meat adulteration is in direct link with religious issues as consumption of certain meat (e.g. pork) is prohibited among Muslims. There is a dire need for reliable and rapid methods that could highlight peculiar adulteration would be of great prominence. Numerous methods were used including enzymatic, Immunological, electrophoretic and DNA-based techniques, but all the methods were quite demanding and laborious. NIRS being easy, rapid and non-destructive was involved in various studies, from which satisfactory results were attained (Ballin et al., 2009; Ballin, 2010).

Beef adulteration is of great concern for the consumers, as few retailers tend to provide customers with beef of inferior quality which is partially substituted with the meat of lower quality from other species (Cozzolino & Murray, 2004; Mamani-Linares et al., 2012) or beef has undergone some illegal treatment (use of harmful growth promoters) (Berzaghi et al., 2006). All these above-mentioned problems were resolved by NIRS by providing highly accurate meat classifications. Mamani-Linares et al. (2012) used VIS-NIRS (visible near-infrared spectroscopy) to differentiate among horse, beef and llama meat by examining reflectance spectra of their minced meat. In this experiment, more than 89% of accuracy was achieved. It was also used by Alomar et al., 2003, in which he tried to differentiate among various cattle meats including many breeds, ages and different muscle types. NIR was successful in discriminating and identifying beef samples based on their difference in water contents and intramuscular fat.

Fish and its products are the food items, which rank in the category of perishable food owing to which they need a lot of care for maintaining their quality and ensuring safety. A huge proportion of this food material is lost due to spoilage annually. Series of different postmortem chemical and physical changes take place including autolysis by endogenous enzymes and cellular structures breakdown along with an increase in microbial growth, which results in spoilage of fish and thus can't cope with consumer expectations (Qu et al., 2015).

Consequently, quality assessment of fish is vital for the safety of its products during processing transportation and storage. Contrary to traditional methods for the assessment of fish quality which are quite laborious and time-consuming, NIRS provides nondestructive and rapid measurements, which aid in the determination of real-time assessment of bacterial load on fish. Loss of freshness in these products is also aligned with high microbial load, which is followed by a combination of physical, chemical, and microbiologic spoilage processes, which could result in the death of an individual. That is why the government authorities and consumers are paying great attention to this critical issue (Cheng et al., 2013).

Many scientists (Uddin et al., 2007) are working on these lines and have successfully explored the use of VIS/NIRS for the assessment of the freshness of fish. Similarly, another study was held by Sivertsen et al. (2011) using the same equipment (VIS/NIRS), which was also equipped with an imaging spectrometer and handheld interactance probe. They were successful in differentiating between frozen-thawed and fresh cod fillets, keeping in view the freshness as criteria. They determined that oxidation of myoglobin and hemoglobin during cold storage and freezing-thawing proved to be the main reasons for change. Several researchers were also successful in differentiating between fresh and frozen-thawed (F-T) fish (inferior in quality) through NIRS using chemometric tools (Uddin et al., 2005).

Fruits and vegetables are the main sources of vitamins (fat and water-soluble) and minerals that play a key role in the growth, development and maintenance of the human body. During their growth phase, both fruits and vegetables have to go through attacks by the pests, which could reduce their quality and acceptability. Pesticides are chemicals that protect them against all the pests, which can cause damage to them. But their excessive use is resulting in pesticide residues on both natural food materials, which is a matter of food safety for end-users and this serious issue must be resolved. Many studies in recent times have used NIRS for detecting dichlorvos (Xue et al., 2012) and herbicide residues (Salguero-Chaparro et al., 2012) which is helping scientists in controlling this critical issue.

Another important parameter in the evaluation of freshness of fruits and vegetables, which can be measured shelf life. Numerous studies have been accomplished utilizing the NIRS technique on green asparagus (Sánchez et al., 2009), pineapples (Di Egidio et al., 2009) and nectarines. In a study using pineapples, it was observed that FT-NIR was used to study the shelf-life of fresh-cut pineapple in the region from 800 to 2564 nm. The changes in the product were owing to the modification in composition, water loss and microbiological and enzymatic processes (Di Egidio et al., 2009). So the NIRS has the ability to give a clue about the acceptability of the product after storage time.

NIRS has the ability to discriminate and classify edible oils of different breeds and grades, identifying harmful substances in them and detecting the presence of oil adulterants of inferior quality, also distinguishing between transgenic and non-transgenic oils. The most prominent issue among them is the authentication of virgin olive oils and separating them from the oils of poor quality. Sinelli et al. (2010) successfully used NIRS at wavelengths of 800 to 2222 nm and 750 to 2500 nm, respectively to classify and characterize virgin olive oils based on their fatty acids and other olfactory attributes. Moreover, 10 different edible oils were successfully differentiated after examining through FT-NIR combined with LDA, accomplishing a 93% accuracy rate (Yang et al., 2005). Furthermore, many experiments have been done to detect the adulteration in valuable edible oils have by NIRS, such as olive pomace, soybean, corn, in extra virgin olive oils (833–2500 nm), camellia oils (400–1700 nm) and in sesame seed oils (1111–2222 nm) and got exceptional results (Wang et al., 2006; Mignani et al., 2011; Luo et al., 2012).

Milk and affiliated dairy products are very important sources of proteins, vitamins and minerals, which are quite critical in maintaining the health status of the masses. But their high demand has paved the way for their adulteration with numerous life-threatening chemicals and adulterants to enhance its shelf life and also to maintain its organoleptic properties (Qu et al., 2015). Resultantly, many experiments have been performed by NIRS methods on raw milk, yogurt, powdered milk and more. Additionally, milk powder was analyzed for whey adulteration and reconstituted milk through NIRS and suitable chemometric methods like SIMCA, DPLS, LS-SVM and SVM. The addition of illegal mixtures in powdered milk was also explored by NIRS, especially melamine which poses a great threat to humans (Qu et al., 2015).

It is important to establish a speedy and effective method for quality and safety assessment as there are few illegal transactions of cereals and their products comprising illegal additives and microbial contaminants along with a few sham and shoddy commodities. Aflatoxin, a mycotoxin produced by *Aspergillus parasiticus* and *Aspergillus flavus*, ranks among one of the life-threatening toxins that frequently appear in cereal grains, under humid conditions. Sirisomboon et al. (2013) both explored the NIRS method to identify aflatoxin B1 (AFB1) and *Aspergillus* contamination (yellow-green) in raw grains like barley, maize, and rice which proved to be very useful for ensuring the safety of cereals and their products.

MIR Spectroscopy

Principle

MIR spectroscopy is a vibrational spectroscopic technique, uses a ray of MIR light that passes through the sample and helps in the measurement of transmitted and absorbed light. Fourier transform (FT) helps to increase the sensitivity, allows a throughput of higher energy and enhances the speed of spectral acquisition. Transmission, attenuated total reflectance and transflection are the three key methods for sampling this technique. It is a feasible substitute for label-free examination of liquid foods (Bakalis & Sun, 2018; Bell, 2012; Davis & Mauer, 2010).

Applications of MIRS for Food Safety and Quality Control

Edible oil obtained from plant, animal or synthetic fat can be utilized in cooking food or flavoring it. MIR technique has been applied successfully for investigating quality attributes of numerous oils including soybean, olive and other oil seeds. This technique coupled with PLSDA and PLSR was proved to be an effective method to determine the adulteration of extra virgin olive oil with other low-quality oils including sunflower oil and soybean oil (Hirri et al., 2016; Mendes et al., 2015; de Souza et al., 2015). It was also useful in detecting the quality level of four virgin olive oils of commercial grade including extra virgin, virgin, ordinary virgin, and lampante olive oils. This method could also be useful in determining the adulteration of extra virgin olive oil with old olive oil. Apart from this, adulteration in high-priced avocado oil with sesame oil, soybean oil or grapeseed oil was also accomplished by this technique (Su & Sun, 2019).

Honey, a natural food product produced by the honey by extracting the nectar either from the same plant source or from different plants to produce uni-floral or multi-floral honey respectively. The main ingredients of honey are water, glucose, fructose, sucrose, and numerous beneficial components minerals, amino acids, vitamins and polyphenols. It has also been used as an effective antimicrobial agent owing to the presence of Methylglyoxal, which helps to perform this function. This beneficial component can be determined by MIR spectroscopy, to detect the antimicrobial ability of that particular honey. Moreover, this technique also helps in the accurate quantification of adulteration of honey. Apart from this, MIR spectroscopy has also aided in the successful differentiation of botanical origin of variant types of honey (Bogdanov et al., 2008; Mandal & Mandal, 2011).

Juice is a type of beverage that is obtained from fresh fruits and vegetables. Numerous studies exhibit that MIRS is ideal for not only classifying but also characterizing variations in the chemical composition of fruit and vegetable juices. Igual et al. 2010 found that heat treatment by microwave can preserve vitamin C and ascorbic acid in fruit juices, while traditional treatments cause substantial reduction in these compounds. Spectroscopic techniques based on MIR and NIR regions have been used for the quantification of total anthocyanin and total phenolics in grape juice (Caramês et al., 2017). Other quality parameters comprising serum viscosity, soluble solids, Bostwick consistency, Ostwald value, fructose, glucose, reducing sugars and pH in tomato juice were quickly determined using MIR spectroscopy

(Ayvaz et al., 2016). Thus, MIR spectroscopy is a great device for the determination of the quality and authentications of fruit and vegetable juices.

Raman Spectroscopy

Raman spectroscopy is a type of vibrational spectroscopic technique that uses light which interacts with matter gives information regarding intra- and inter-molecular vibrations and can also provide understanding about the reaction (McMillan & Hofmeister, 2018). Hence, Raman spectroscopy provides a spectrum of specific vibrations of a molecule ("molecular fingerprint') and are appreciated for the identification of a substance (Vašková, 2011).

Principle of Raman Spectroscopy

Raman effect was devised by Sir C.V. Raman and got the Nobel Prize in Physics for his work in 1930. When a photon interacts with matter (solid, liquid, and gas), which results in dispersion and scattering at the same energy. This is called elastic scattering or Rayleigh scattering. A small number of photons, around 1 photon in 10 million may scatter at a different frequency as compared to the incident photon. This process is known as inelastic scattering or the Raman effect. Raman has been used for a vast array of applications in various fields of sciences including the discipline of food science and technology for the characterization and safety assessment of food products (Vandenabeele, 2013).

Applications of Raman Spectroscopy in Food Safety and Quality Control

In this modern era of digitalization, all the actors who are dealing with the handling and production of food need to develop some rapid and non-invasive technologies. The quality of the fresh produce is also important as these products are being consumed without any thermal treatments. In this regard, Raman spectroscopy has been employed for safety, structural analysis, quantification and classification of fruits and vegetables (Yang & Ying, 2011).

The foodborne pathogens may present on the surface of the fruits which need to quantify. FT-Raman method was used to detect the pathogenic microorganisms such as bacteria and viruses on the whole apple skin for the very first time. In addition to this, the discrimination of five strains of *Escherichia coli* was detected with 100% accuracy which shows the potential of FT-Raman spectroscopy for the analysis of microorganisms on fruit surface (Yang & Irudayaraj 2003). Studies by Bonora et al. (2009) confirmed the use of FT-Raman spectroscopy for the identification of microbial diseases in fruits. The spectral signatures taken for pulp obtained from elephantiasis fruits showed peaks at 1530 and 1661 cm^{-1} which is related to the infection and can be utilized as a marker of demonstrating diseased plants.

At present, the pesticides are being sprayed to get maximum yield but the pesticide residues may remain in the fruits and vegetables that may cause serious health hazards. Hence, the pesticidal residues can be detected with the help FT-Raman technique quite accurately (Zhang et al. 2006), On the other hand, Surface Enhanced Raman Spectroscopy (SERS) can also be utilized for the quantification of pesticidal residues both on the surface and of the fresh fruit as well as its juice (Shende et al. 2004). This technique can detect the pesticide residues in parts per billion.

FT-Raman spectroscopy was also employed to discriminate between pure and adulterated milk powders (Almeida et al., 2011). Melamine in milk can also be detected by using the same methodology that is normally added to increase the protein contents of powdered milk which causes severe health hazards (Ai et al., 2009). Previously chromatographic techniques were applied to determine melamine but now the Raman spectroscopy can detect such adulterations (Liu et al., 2009). The strong peaks were observed for melamine in milk powders at 676 cm^{-1} (Okazaki et al. 2009). Similarly, Qin et al. (2010) quantified melamine in milk powders with greater accuracy. In addition to this portable devices based on this technology were also devised to determine melamine adulteration in dried milk (Cheng et al. 2010). Such portable devices used two bands (673 and 982 cm^{-1}) that generate reproducible results. SERS has the potential to detect samples with a limit of detection (LOD) down to parts per billion (ppb) or even a single-molecule level.

Multispectral (MSI) and Hyperspectral Imaging (HSI)

Hyperspectral and Multispectral imaging (HSI - MSI) is a technique that provides spectral and spatial information from chemical targets. Chemical imaging technique is also a combination of computer vision and vibrational spectroscopy. The computer vision technique is an optical technique that describes the association between incident light and molecules in matter. On the other hand, spectroscopy is used to evaluate a small portion of the food sample but the techniques take the whole sample for detection they provide a comprehensive and more representative measurement. Computer vision technique also used for spatial information by imitates the human vision using three bands like red, green, blue to attain the features of samples. Some characteristics of a sample like color, size, shape and texture are detected by computer vision techniques in a visible range. There are a few bands involved like HSI shows more than 100 bands, as well as MSI, which has less than 20 bands.

MSI is generally reliable due to the HSI that provides very important wavelengths. In hyperspectral imaging, a variety of options are available to evaluate the task.

These choices involved some other techniques like fluorescence HIS, near-infrared HIS and Raman HIS that offer flexibility to determine the solutions of all kinds of measured complications. While the hyperspectral imaging technique has versatility due to its involvement in food inspection related to safety and quality. Several studies describe the role of hyperspectral imaging techniques in various aspects of food safety and quality (Gowen et al., 2015; Liu et al. 2015).

Hyperspectral imaging (HSI) is commonly used in restaurants due to its beneficial applications like used as the detection of grains, provide spatial and spectral information sensitive to minor components. As well as working on the principle of optical absorption and reflection. At the same time, it is not suitable because it takes a long time for pre-processing of the data and classification.

Biomimetic Sensors

It is noticed that in food quality assurance a highly trained sensory personal is required but this approach has few disadvantages such as inconsistency that may be due to personal stress or fatigue. According to food sectors, these employees (sensory panels) are more expensive and are not reliable for large-scale observations. This problem is resolved by introducing the biomimetic system in which sensors are designed to mimic the gustatory and olfactory systems of humans that are known as e-tongue and electronic nose respectively (Ghasemi-Varnamkhasti, et al., 2010).

A human can sense odor through a proper channel like when the odor of a chemical substance resembles the nasal cavity receptors then the signal is transferred from receptors to the brain via neurons. According to this principle, the e-nose has a chemical sensor that resembles the neurons related to show different sensitivity to unique odors. Thus, the interaction associated between odor substances and sensors leads to the generation of electrical signals that are recorded by the instrument.

Recently, in the food science discipline, e-nose is more used than e-tongue. There are distinct types of different sensors introduced but the commonly used types of sensors are conducting polymers composites, metal oxides and intrinsically conducting polymers. Thus, these sensors are more reliable due to their shortcoming which is useful in data analysis like the stability of sensor behavior, senor drift, sensitivity and selectivity. Every technique has advantages and drawbacks but ignoring the drawbacks of these sensors are more useful in some characteristics of foodstuff like shelf-life, ripeness, quality grading and freshness of the products (Baietto & Wilson, 2015; Baldwin, et al., 2011; Loutfi et al., 2015; Wilson, 2013).

To evaluate the shelf life, ripeness, quality grading and freshness of the products used in the restaurants this technique is suitable. It is a noninvasive and unexpansive technique but trained personal is required for it.

FUTURE RESEARCH DIRECTIONS

All these rapid and non-invasive technologies described above are gaining a lot of interest in this digital era. These methods are simple, easy to use, environment friendly which requires no chemicals and no complex preparation of the sample. The above-mentioned application has been experimented with at a laboratory scale. Sensors based on spectroscopic techniques can be used to characterize and authenticate food products. But very few applications have been adopted for industrial applications. The need of the time is to develop some handheld portable devices that can scan the food product and the desired characteristics. Hence, researchers and scientists focus on the development of such devices so that food safety can be assured at all levels including the foodservice sector. They should focus on the different optimization techniques such as genetic algorithms, ant colony optimization, etc. for the desirable wavelength selection from the spectral information that can ultimately result in the development of handheld devices for food safety and quality evaluation.

CONCLUSION

In this chapter, the use of noninvasive techniques to ensure food safety and quality in restaurants is discussed. These noninvasive techniques are more reliable than the traditional instrumental analysis and chemical analysis methods to evaluate food safety and quality. Traditionally used analysis methods have some drawbacks like time-consuming, laborious and inflated cost. Thus, in this modern era, noninvasive techniques show a better result than previously used methods. These noninvasive techniques are more reliable for accessing the food quality and also improve the internal quality of foodstuff. In restaurants, noninvasive techniques are increasing the interest of food businesses to produce quality foodstuff. The instruments used in these techniques are very costly these days.

REFERENCES

Ahmad, M. H., Nache, M., Hinrichs, J., & Hitzmann, B. (2016b). Estimation of the nutritional parameters of various types of wheat flours using fluorescence spectroscopy and chemometrics. *International Journal of Food Science & Technology*, *51*(5), 1186–1194. doi:10.1111/ijfs.13080

Ahmad, M. H., Nache, M., Waffenschmidt, S., & Hitzmann, B. (2016a). A fluorescence spectroscopic approach to predict analytical, rheological and baking parameters of wheat flours using chemometrics. *Journal of Food Engineering*, *182*, 65–71. doi:10.1016/j.jfoodeng.2016.03.006

Ai, K., Liu, Y., & Lu, L. (2009). Hydrogen-bonding recognition-induced color change of gold nanoparticles for visual detection of melamine in raw milk and infant formula. *Journal of the American Chemical Society*, *131*(27), 9496–9497. doi:10.1021/ja9037017 PMID:19537721

Alamprese, C., Casale, M., Sinelli, N., Lanteri, S., & Casiraghi, E. (2013). Detection of minced beef adulteration with turkey meat by UV–vis, NIR and MIR spectroscopy. *Lebensmittel-Wissenschaft + Technologie*, *53*(1), 225–232. doi:10.1016/j.lwt.2013.01.027

Alander, J. T., Bochko, V., Martinkauppi, B., Saranwong, S., & Mantere, T. (2013). A review of optical nondestructive visual and near-infrared methods for food quality and safety. *International Journal of Spectroscopy*, *2013*, 341402. doi:10.1155/2013/341402

Almeida, M. R., Oliveira, K. D. S., Stephani, R., & de Oliveira, L. F. C. (2011). Fourier-transform Raman analysis of milk powder: A potential method for rapid quality screening. *Journal of Raman Spectroscopy: JRS*, *42*(7), 1548–1552. doi:10.1002/jrs.2893

Alomar, D., Gallo, C., Castaneda, M., & Fuchslocher, R. (2003). Chemical and discriminant analysis of bovine meat by near infrared reflectance spectroscopy (NIRS). *Meat Science*, *63*(4), 441–450. doi:10.1016/S0309-1740(02)00101-8 PMID:22062513

Alvarado, U., Zamora, A., Liu, J., Saldo, J., & Castillo, M. (2020). Rapid Quantification of Riboflavin in Milk by Front-Face Fluorescence Spectroscopy: A Preliminary Study. *Foods*, *9*(1), 6. doi:10.3390/foods9010006 PMID:31861817

Andersen, C. M., Vishart, M., & Holm, V. K. (2005). Application of fluorescence spectroscopy in the evaluation of light-induced oxidation in cheese. *Journal of Agricultural and Food Chemistry*, *53*(26), 9985–9992. doi:10.1021/jf051143d PMID:16366684

Ankita, J. L. (2018). *Development of rapid detection methods for acidic food spoilage and calcium carbide based artificial ripening of fruits* (Doctoral dissertation). Department of Food Science and Technology.

Argyri, A. A., Panagou, E. Z., & Nychas, G. J. (2014). Monitoring microbial spoilage of foods by vibrational spectroscopy (FT-IR & Raman). *Novel Food Preservation and Microbial Assessment Techniques, 386*.

Ayvaz, H., Sierra-Cadavid, A., Aykas, D. P., Mulqueeney, B., Sullivan, S., & Rodriguez-Saona, L. E. (2016). Monitoring multicomponent quality traits in tomato juice using portable mid-infrared (MIR) spectroscopy and multivariate analysis. *Food Control*, *66*, 79–86. doi:10.1016/j.foodcont.2016.01.031

Babu, K. S., & Amamcharla, J. K. (2018). Application of front-face fluorescence spectroscopy as a tool for monitoring changes in milk protein concentrate powders during storage. *Journal of Dairy Science*, *101*(12), 10844–10859. doi:10.3168/jds.2018-14885 PMID:30316594

Baietto, M., & Wilson, A. D. (2015). Electronic-nose applications for fruit identification, ripeness and quality grading. *Sensors (Basel)*, *15*(1), 899–931. doi:10.3390150100899 PMID:25569761

Baldwin, E. A., Bai, J., Plotto, A., & Dea, S. (2011). Electronic noses and tongues: Applications for the food and pharmaceutical industries. *Sensors (Basel)*, *11*(5), 4744–4766. doi:10.3390110504744 PMID:22163873

Ballin, N. Z. (2010). Authentication of meat and meat products. *Meat Science*, *86*(3), 577–587. doi:10.1016/j.meatsci.2010.06.001 PMID:20685045

Ballin, N. Z., Vogensen, F. K., & Karlsson, A. H. (2009). Species determination–Can we detect and quantify meat adulteration? *Meat Science*, *83*(2), 165–174. doi:10.1016/j.meatsci.2009.06.003 PMID:20416768

Bansal, S., Singh, A., Mangal, M., Mangal, A. K., & Kumar, S. (2017). Food adulteration: Sources, health risks, and detection methods. *Critical Reviews in Food Science and Nutrition*, *57*(6), 1174–1189. doi:10.1080/10408398.2014.967834 PMID:26054861

Bell, R. (2012). *Introductory Fourier transform spectroscopy*. Elsevier. Academic Press.

Berzaghi, P., Segato, S., Cozzi, G., & Andrighetto, I. (2006). Mid and near infrared spectroscopy to identify illegal treatments in beef cattle. *Veterinary Research Communications*, *30*(1), 109–112. doi:10.100711259-006-0022-z

Bogdanov, S., Jurendic, T., Sieber, R., & Gallmann, P. (2008). Honey for nutrition and health: A review. *Journal of the American College of Nutrition*, *27*(6), 677–689. doi:10.1080/07315724.2008.10719745 PMID:19155427

Bonora, S., Francioso, O., Tugnoli, V., Prodi, A., Di Foggia, M., Righi, V., Nipoti, P., Filippini, G., & Pisi, A. (2009). Structural characteristics of 'Hayward'kiwifruits from elephantiasis-affected plants studied by DRIFT, FT-Raman, NMR, and SEM techniques. *Journal of Agricultural and Food Chemistry*, *57*(11), 4827–4832. doi:10.1021/jf9002957 PMID:19413311

Buzby, J. C., & Hyman, J. (2012). Total and per capita value of food loss in the United States. *Food Policy*, *37*(5), 561–570. doi:10.1016/j.foodpol.2012.06.002

Caramês, E. T., Alamar, P. D., Poppi, R. J., & Pallone, J. A. L. (2017). Rapid assessment of total phenolic and anthocyanin contents in grape juice using infrared spectroscopy and multivariate calibration. *Food Analytical Methods*, *10*(5), 1609–1615. doi:10.100712161-016-0721-1

Cheng, J. H., Dai, Q., Sun, D. W., Zeng, X. A., Liu, D., & Pu, H. B. (2013). Applications of non-destructive spectroscopic techniques for fish quality and safety evaluation and inspection. *Trends in Food Science & Technology*, *34*(1), 18–31. doi:10.1016/j.tifs.2013.08.005

Cheng, Y., Dong, Y., Wu, J., Yang, X., Bai, H., Zheng, H., Ren, D., Zou, Y., & Li, M. (2010). Screening melamine adulterant in milk powder with laser Raman spectrometry. *Journal of Food Composition and Analysis*, *23*(2), 199–202. doi:10.1016/j.jfca.2009.08.006

Cozzolino, D., & Murray, I. (2004). Identification of animal meat muscles by visible and near infrared reflectance spectroscopy. *Lebensmittel-Wissenschaft + Technologie*, *37*(4), 447–452. doi:10.1016/j.lwt.2003.10.013

Dankowska, A., Małecka, M., & Kowalewski, W. (2015). Detection of plant oil addition to cheese by synchronous fluorescence spectroscopy. *Dairy Science & Technology*, *95*(4), 413–424. doi:10.100713594-015-0218-5 PMID:26097644

Davis, R., & Mauer, L. J. (2010). Fourier transform infrared (FT-IR) spectroscopy: A rapid tool for detection and analysis of foodborne pathogenic bacteria. Current *Research. Technology and Education Topics in Applied Microbiology and Microbial Biotechnology*, *2*, 1582–1594.

de Souza, L. M., de Santana, F. B., Gontijo, L. C., Mazivila, S. J., & Neto, W. B. (2015). Quantification of adulterations in extra virgin flaxseed oil using MIR and PLS. *Food Chemistry*, *182*, 35–40. doi:10.1016/j.foodchem.2015.02.081 PMID:25842305

Di Egidio, V., Sinelli, N., Limbo, S., Torri, L., Franzetti, L., & Casiraghi, E. (2009). Evaluation of shelf-life of fresh-cut pineapple using FT-NIR and FT-IR spectroscopy. *Postharvest Biology and Technology*, *54*(2), 87–92. doi:10.1016/j.postharvbio.2009.06.006

Durek, J., Fröhling, A., Bolling, J., Thomasius, R., Durek, P., & Schlüter, O. K. (2016). Non-destructive mobile monitoring of microbial contaminations on meat surfaces using porphyrin fluorescence intensities. *Meat Science*, *115*, 1–8. doi:10.1016/j.meatsci.2015.12.022 PMID:26773794

Edwards, M. C., & Stringer, M. F. (2007). Observations on patterns in foreign material investigations. *Food Control*, *18*(7), 773–782. doi:10.1016/j.foodcont.2006.01.007

Ellis, D. I., & Goodacre, R. (2001). Rapid and quantitative detection of the microbial spoilage of muscle foods: Current status and future trends. *Trends in Food Science & Technology*, *12*(11), 414–424. doi:10.1016/S0924-2244(02)00019-5

Ghasemi-Varnamkhasti, M., Mohtasebi, S. S., & Siadat, M. (2010). Biomimetic-based odor and taste sensing systems to food quality and safety characterization: An overview on basic principles and recent achievements. *Journal of Food Engineering*, *100*(3), 377–387. doi:10.1016/j.jfoodeng.2010.04.032

Gowen, A. A., Feng, Y., Gaston, E., & Valdramidis, V. (2015). Recent applications of hyperspectral imaging in microbiology. *Talanta*, *137*, 43–54. doi:10.1016/j.talanta.2015.01.012 PMID:25770605

Grote, B., Zense, T., & Hitzmann, B. (2014). 2D-fluorescence and multivariate data analysis for monitoring of sourdough fermentation process. *Food Control*, *38*, 8–18. doi:10.1016/j.foodcont.2013.09.039

Harrison, I. (2003). Non-destructive testing for fruit quality assurance. *Innovations in Food Technology*, *19*, 86–87.

Hirri, A., Bassbasi, M., Platikanov, S., Tauler, R., & Oussama, A. (2016). FTIR spectroscopy and PLS-DA classification and prediction of four commercial grade virgin olive oils from Morocco. *Food Analytical Methods*, *9*(4), 974–981. doi:10.100712161-015-0255-y

Horigome, J., Kozuma, M., & Shirasaki, T. (2016). Fluorescence pattern analysis to assist food safety. *Hitachi Review*, *65*(7), 249.

Hyman, F. N., Klontz, K. C., & Tollefson, L. (1993). Food and Drug Administration surveillance of the role of foreign objects in foodborne injuries. *Public Health Reports*, *108*(1), 54. PMID:8434098

Igual, M. G. M. E., García-Martínez, E., Camacho, M. M., & Martínez-Navarrete, N. (2010). Effect of thermal treatment and storage on the stability of organic acids and the functional value of grapefruit juice. *Food Chemistry*, *118*(2), 291–299. doi:10.1016/j.foodchem.2009.04.118

Isoni Auad, L., Cortez Ginani, V., dos Santos Leandro, E., Stedefeldt, E., Costa Santos Nunes, A., Yoshio Nakano, E., & Puppin Zandonadi, R. (2019). Brazilian food truck consumers' profile, choices, preferences, and food safety importance perception. *Nutrients*, *11*(5), 1175. doi:10.3390/nu11051175 PMID:31130664

Kademi, H. I., Ulusoy, B. H., & Hecer, C. (2019). Applications of miniaturized and portable near infrared spectroscopy (NIRS) for inspection and control of meat and meat products. *Food Reviews International*, *35*(3), 201–220. doi:10.1080/87559129.2018.1514624

Kamal, M., & Karoui, R. (2017). Monitoring of mild heat treatment of camel milk by front-face fluorescence spectroscopy. *Lebensmittel-Wissenschaft + Technologie*, *79*, 586–593. doi:10.1016/j.lwt.2016.11.013

Liu, D., Zeng, X. A., & Sun, D. W. (2015). Recent developments and applications of hyperspectral imaging for quality evaluation of agricultural products: A review. *Critical Reviews in Food Science and Nutrition*, *55*(12), 1744–1757. doi:10.1080/10408398.2013.777020 PMID:24915395

Liu, X., & Metzger, L. E. (2007). Application of fluorescence spectroscopy for monitoring changes in nonfat dry milk during storage. *Journal of Dairy Science*, *90*(1), 24–37. doi:10.3168/jds.S0022-0302(07)72605-X PMID:17183072

Liu, Y., Chao, K., Kim, M. S., Tuschel, D., Olkhovyk, O., & Priore, R. J. (2009). Potential of Raman spectroscopy and imaging methods for rapid and routine screening of the presence of melamine in animal feed and foods. *Applied Spectroscopy*, *63*(4), 477–480. doi:10.1366/000370209787944398 PMID:19366516

Loutfi, A., Coradeschi, S., Mani, G. K., Shankar, P., & Rayappan, J. B. B. (2015). Electronic noses for food quality: A review. *Journal of Food Engineering*, *144*, 103–111. doi:10.1016/j.jfoodeng.2014.07.019

Luo, J., Liu, T., & Liu, Y. (2011, October). FT-NIR and confocal microscope raman spectroscopic studies of sesame oil adulteration. In *International Conference on Computer and Computing Technologies in Agriculture* (pp. 24-31). Springer.

Mahendran, R., Jayashree, G. C., & Alagusundaram, K. (2012). Application of computer vision technique on sorting and grading of fruits and vegetables. *Journal of Food Processing & Technology*, *10*, 2157–7110.

Mala, D. M., Yoshimura, M., Kawasaki, S., Tsuta, M., Kokawa, M., Trivittayasil, V., ... Kitamura, Y. (2016). Fiber optics fluorescence fingerprint measurement for aerobic plate count prediction on sliced beef surface. *Lebensmittel-Wissenschaft + Technologie*, *68*, 14–20.

Mamani-Linares, L. W., Gallo, C., & Alomar, D. (2012). Identification of cattle, llama and horse meat by near infrared reflectance or transflectance spectroscopy. *Meat Science*, *90*(2), 378–385. doi:10.1016/j.meatsci.2011.08.002 PMID:21889854

Mandal, M. D., & Mandal, S. (2011). Honey: Its medicinal property and antibacterial activity. *Asian Pacific Journal of Tropical Biomedicine*, *1*(2), 154–160. doi:10.1016/S2221-1691(11)60016-6 PMID:23569748

Manetta, A. C., Di Giuseppe, L., Giammarco, M., Fusaro, I., Simonella, A., Gramenzi, A., & Formigoni, A. (2005). High-performance liquid chromatography with post-column derivatisation and fluorescence detection for sensitive determination of aflatoxin M1 in milk and cheese. *Journal of Chromatography. A*, *1083*(1-2), 219–222. doi:10.1016/j.chroma.2005.06.039 PMID:16078711

McMillan, P. F., & Hofmeister, A. M. (2018). Infrared and Raman spectroscopy. In F. C. Hawthorne (Ed.), *Spectroscopic methods in mineralogy and geology* (pp. 99–160). De Gruyter.

Mendes, T. O., da Rocha, R. A., Porto, B. L., de Oliveira, M. A., dos Anjos, V. D. C., & Bell, M. J. (2015). Quantification of extra-virgin olive oil adulteration with soybean oil: A comparative study of NIR, MIR, and Raman spectroscopy associated with chemometric approaches. *Food Analytical Methods*, *8*(9), 2339–2346. doi:10.100712161-015-0121-y

Mignani, A. G., Ciaccheri, L., Ottevaere, H., Thienpont, H., Conte, L., Marega, M., Cichelli, A., Attilio, C., & Cimato, A. (2011). Visible and near-infrared absorption spectroscopy by an integrating sphere and optical fibers for quantifying and discriminating the adulteration of extra virgin olive oil from Tuscany. *Analytical and Bioanalytical Chemistry*, *399*(3), 1315–1324. doi:10.100700216-010-4408-y PMID:21107823

Nicolai, B. M., Beullens, K., Bobelyn, E., Peirs, A., Saeys, W., Theron, K. I., & Lammertyn, J. (2007). Nondestructive measurement of fruit and vegetable quality by means of NIR spectroscopy: A review. *Postharvest Biology and Technology*, *46*(2), 99–118. doi:10.1016/j.postharvbio.2007.06.024

Nychas, G. J. E., Skandamis, P. N., Tassou, C. C., & Koutsoumanis, K. P. (2008). Meat spoilage during distribution. *Meat Science*, *78*(1-2), 77–89. doi:10.1016/j.meatsci.2007.06.020 PMID:22062098

Okazaki, S., Hiramatsu, M., Gonmori, K., Suzuki, O., & Tu, A. T. (2009). Rapid nondestructive screening for melamine in dried milk by Raman spectroscopy. *Forensic Toxicology*, *27*(2), 94–97. doi:10.100711419-009-0072-3

Olsen, A. R. (1998). Regulatory action criteria for filth and other extraneous materials: I. Review of hard or sharp foreign objects as physical hazards in food. *Regulatory Toxicology and Pharmacology*, *28*(3), 181–189. doi:10.1006/rtph.1998.1249 PMID:10049789

Oto, N., Oshita, S., Makino, Y., Kawagoe, Y., Sugiyama, J., & Yoshimura, M. (2013). Non-destructive evaluation of ATP content and plate count on pork meat surface by fluorescence spectroscopy. *Meat Science*, *93*(3), 579–585. doi:10.1016/j.meatsci.2012.11.010 PMID:23273467

Papadopoulou, O., Panagou, E. Z., Tassou, C. C., & Nychas, G. J. (2011). Contribution of Fourier transform infrared (FTIR) spectroscopy data on the quantitative determination of minced pork meat spoilage. *Food Research International*, *44*(10), 3264–3271. doi:10.1016/j.foodres.2011.09.012

Prieto, N., Andrés, S., Giráldez, F. J., Mantecón, A. R., & Lavín, P. (2008). Discrimination of adult steers (oxen) and young cattle ground meat samples by near infrared reflectance spectroscopy (NIRS). *Meat Science*, *79*(1), 198–201. doi:10.1016/j.meatsci.2007.08.001 PMID:22062613

Purna, S. G., Prow, L. A., & Metzger, L. E. (2005). Utilization of front-face fluorescence spectroscopy for analysis of process cheese functionality. *Journal of Dairy Science*, *88*(2), 470–477. doi:10.3168/jds.S0022-0302(05)72708-9 PMID:15653511

Qin, J., Chao, K., & Kim, M. S. (2010, June). *Development of a Raman chemical imaging system for food safety inspection.* American Society of Agricultural and Biological Engineers.

Qu, J. H., Liu, D., Cheng, J. H., Sun, D. W., Ma, J., Pu, H., & Zeng, X. A. (2015). Applications of near-infrared spectroscopy in food safety evaluation and control: A review of recent research advances. *Critical Reviews in Food Science and Nutrition*, *55*(13), 1939–1954. doi:10.1080/10408398.2013.871693 PMID:24689758

Sahar, A., Rahman, U., Kondjoyan, A., Portanguen, S., & Dufour, E. (2016). Monitoring of thermal changes in meat by synchronous fluorescence spectroscopy. *Journal of Food Engineering*, *168*, 160–165. doi:10.1016/j.jfoodeng.2015.07.038

Saldaña, E., Siche, R., Luján, M., & Quevedo, R. (2013). Computer vision applied to the inspection and quality control of fruits and vegetables. *Brazilian Journal of Food Technology*, *16*(4), 254–272. doi:10.1590/S1981-67232013005000031

Salguero-Chaparro, L., Gaitán-Jurado, A. J., Ortiz-Somovilla, V., & Peña-Rodríguez, F. (2013). Feasibility of using NIR spectroscopy to detect herbicide residues in intact olives. *Food Control*, *30*(2), 504–509. doi:10.1016/j.foodcont.2012.07.045

Sánchez, M. T., Pérez-Marín, D., Flores-Rojas, K., Guerrero, J. E., & Garrido-Varo, A. (2009). Use of near-infrared reflectance spectroscopy for shelf-life discrimination of green asparagus stored in a cool room under controlled atmosphere. *Talanta*, *78*(2), 530–536. doi:10.1016/j.talanta.2008.12.004 PMID:19203619

Shaikh, S., & O'Donnell, C. (2017). Applications of fluorescence spectroscopy in dairy processing: A review. *Current Opinion in Food Science*, *17*, 16–24. doi:10.1016/j.cofs.2017.08.004

Shende, C., Gift, A., Inscore, F., Maksymiuk, P., & Farquharson, S. (2004). Inspection of pesticide residues on food by surface-enhanced Raman spectroscopy. In *Monitoring food safety, agriculture, and plant health* (Vol. 5271, pp. 28–34). International Society for Optics and Photonics. doi:10.1117/12.511941

Siddiqui, M. W., Longkumer, M., Ahmad, M. S., Barman, K., Thakur, P. K., & Kabir, J. (2014). Postharvest biology and technology of sapota: A concise review. *Acta Physiologiae Plantarum*, *36*(12), 3115–3122. doi:10.100711738-014-1696-4

Sinelli, N., Cerretani, L., Di Egidio, V., Bendini, A., & Casiraghi, E. (2010). Application of near (NIR) infrared and mid (MIR) infrared spectroscopy as a rapid tool to classify extra virgin olive oil on the basis of fruity attribute intensity. *Food Research International*, *43*(1), 369–375. doi:10.1016/j.foodres.2009.10.008

Sirisomboon, C. D., Putthang, R., & Sirisomboon, P. (2013). Application of near infrared spectroscopy to detect aflatoxigenic fungal contamination in rice. *Food Control*, *33*(1), 207–214. doi:10.1016/j.foodcont.2013.02.034

Sivertsen, A. H., Kimiya, T., & Heia, K. (2011). Automatic freshness assessment of cod (Gadus morhua) fillets by Vis/Nir spectroscopy. *Journal of Food Engineering*, *103*(3), 317–323. doi:10.1016/j.jfoodeng.2010.10.030

Su, W. H., Bakalis, S., & Sun, D. W. (2018). Fourier transform mid-infrared-attenuated total reflectance (FTMIR-ATR) microspectroscopy for determining textural property of microwave baked tuber. *Journal of Food Engineering*, *218*, 1–13. doi:10.1016/j.jfoodeng.2017.08.016

Su, W. H., & Sun, D. W. (2019). Mid-infrared (MIR) spectroscopy for quality analysis of liquid foods. *Food Engineering Reviews*, *11*(3), 142–158. doi:10.100712393-019-09191-2

Uddin, M., Okazaki, E., Turza, S., Yumiko, Y., Tanaka, M., & Fukuda, Y. (2005). Non-destructive visible/NIR spectroscopy for differentiation of fresh and frozen-thawed fish. *Journal of Food Science*, *70*(8), c506–c510. doi:10.1111/j.1365-2621.2005.tb11509.x

Uddin, M., Turza, S., & Okazaki, E. (2007). Rapid determination of intact sardine fat by NIRS using surface interactance fibre probe. *International Journal of Food Engineering*, *3*(6). Advance online publication. doi:10.2202/1556-3758.1248

Vandenabeele, P. (2013). *Practical Raman spectroscopy: An introduction*. Wiley. doi:10.1002/9781119961284

Vašková, H. (2011). A powerful tool for material identification: Raman spectroscopy. *International Journal of Mathematical Models and Methods in Applied Sciences*, *5*, 1205–1212.

Wang, L., Lee, F. S., Wang, X., & He, Y. (2006). Feasibility study of quantifying and discriminating soybean oil adulteration in camellia oils by attenuated total reflectance MIR and fiber optic diffuse reflectance NIR. *Food Chemistry*, *95*(3), 529–536. doi:10.1016/j.foodchem.2005.04.015

Wang, W., & Paliwal, J. (2007). Near-infrared spectroscopy and imaging in food quality and safety. *Sensing and Instrumentation for Food Quality and Safety*, *1*(4), 193–207. doi:10.100711694-007-9022-0

Wilson, A. D. (2013). Diverse applications of electronic-nose technologies in agriculture and forestry. *Sensors (Basel)*, *13*(2), 2295–2348. doi:10.3390130202295 PMID:23396191

Xue, L., Cai, J., Li, J., & Liu, M. (2012). Application of particle swarm optimization (PSO) algorithm to determine dichlorvos residue on the surface of navel orange with Vis-NIR spectroscopy. *Procedia Engineering*, *29*, 4124–4128. doi:10.1016/j.proeng.2012.01.631

Yang, D., & Ying, Y. (2011). Applications of Raman spectroscopy in agricultural products and food analysis: A review. *Applied Spectroscopy Reviews*, *46*(7), 539–560. doi:10.1080/05704928.2011.593216

Yang, H., & Irudayaraj, J. (2003). Rapid detection of foodborne microorganisms on food surface using Fourier transform Raman spectroscopy. *Journal of Molecular Structure*, *646*(1-3), 35–43. doi:10.1016/S0022-2860(02)00575-6

Yang, H., Irudayaraj, J., & Paradkar, M. M. (2005). Discriminant analysis of edible oils and fats by FTIR, FT-NIR and FT-Raman spectroscopy. *Food Chemistry*, *93*(1), 25–32. doi:10.1016/j.foodchem.2004.08.039

Zhang, P. X., Zhou, X., Cheng, A. Y., & Fang, Y. (2006). Raman spectra from pesticides on the surface of fruits. Journal of Physics: Conference Series, 28(1), 2. doi:10.1088/1742-6596/28/1/002

Zhang, X., Wu, D., Zhou, X., Yu, Y., Liu, J., Hu, N., Wang, H., Li, G., & Wu, Y. (2019). Recent progress in the construction of nanozyme-based biosensors and their applications to food safety assay. *Trends in Analytical Chemistry*, *121*, 115668. doi:10.1016/j.trac.2019.115668

KEY TERMS AND DEFINITIONS

Food Safety: A discipline that deals with food handling, preparation, as well as storage to prevent harmful infectious diseases and to ensure that food has adequate nutrients for a healthy diet.

Foodborne Disease: Due to the consumption of food or drink that are contaminated through pathogens like bacteria, viruses and parasites during the food processing to consumption.

Fluorescence: The emission of light by a substance that has absorbed light or other electromagnetic radiation. It is a form of luminescence. In most cases, the emitted light has a longer wavelength, and therefore lower energy, than the absorbed radiation.

Near-Infrared Spectroscopy: (NIRS): A spectroscopic method that uses the near infrared region of the electromagnetic spectra (from 780 nm to 2500 nm).

Raman Spectroscopy: A a non-destructive chemical analysis technique which provides detailed information about chemical structure, phase and polymorphy, crystallinity, and molecular interactions.

Chapter 4
Food Microbial Hazards, Safety, and Quality Control:
A Strategic Approach

Aysha Sameen
University of Agriculture, Faisalabad, Pakistan

Amna Sahar
https://orcid.org/0000-0002-9299-4000
University of Agriculture, Faisalabad, Pakistan

Farwa Tariq
University of Agriculture, Faisalabad, Pakistan

Usman Mir Khan
University of Agriculture, Faisalabad, Pakistan

Tayyaba Tariq
University of Agriculture, Faisalabad, Pakistan

Bushra Ishfaq
University of Agriculture, Faisalabad, Pakistan

ABSTRACT

Food is any material or substance eaten or drunk to provide energy and nutrients for the body's growth, development, and maintenance. Food can be considered safe if it is free from all hazardous substances that can affect consumer health. Food safety issues can place a high burden of responsibility on traders, government bodies, and international organizations. This chapter covers the hazards, their types, foodborne diseases, and strategies to ensure food safety and quality. Different food quality and safety assurance programs are discussed as well like quality management systems, HACCP certification, ISO 9000 family, good manufacturing practices (GMP)/good hygiene practices (GHP), total quality management (TQM), good working practices (GWP), good lab practices (GLP), etc. Moreover, the role of some novel processing technologies is also focused on in this regard.

DOI: 10.4018/978-1-7998-7415-7.ch004

INTRODUCTION

Food is any material or substance eaten or drunk to provide energy and nutrients for the body's growth, development, and maintenance. It is usually of an animal or plant origin consisting of nutrients like fats, proteins, carbohydrates, vitamins, and minerals (Abdulmumeen et al., 2012). Food is the basic precondition for survival but food quality and safety issues embrace us all time (Jeukendrup & Gleeson, 2018). According to the Food and Agriculture Organization of the United Nations (FAO), food quality can be defined as "a complex characteristic of food that determines its value or acceptability to consumers" (McLoughlin et al., 2017; Nielsen, 2010). Basic quality control and quality assurance programs are used by food industries like other industries for achieving food quality e.g. ISO 9000 Quality Management System Standard, HACCP system, GMPs, etc. (Hubbard, 2012; Nguz, 2007).

Food can be considered safe if it is free from all hazardous substances that can affect consumers' health. In pre-historic times, the food safety chain was very short; it compromised hunter-gatherers and their families. When societies grew complex and larger, this chain became more diffused due to an expanded international trade network and long shipping distances (Gorris, 2005). This chain can be a reason for introducing unfamiliar food safety hazards during food transportation. This situation places a high burden of responsibility on traders, government bodies, and international organizations. There is a need to improve food safety regulations and systems to ensure the safety of consumers by taking multidimensional approaches at all levels of the food chain, from farm to fork (Gorris, 2005; Jeukendrup & Gleeson, 2018; Unnevehr & Huirne, 2002).

FOOD HAZARDS

Food hazard can be characterized by any physical, chemical, or biological agent that has the potential to exert harmful effects on human health (Peter et al., 2013).

Physical Hazards

Physical hazards are any poisonous or deleterious foreign objects incorporated into food that have clinical evidence of injury or traumatic injury including perforation of tissues present in the oral cavity and gastrointestinal tract if ingested. This includes hard or sharp foreign objects and choking hazards. Contaminants like dirt, hair, insects, wood splinters, glass, and metal are included in the category of physical hazards (Das et al., 2019). Potential sources of these hazards include:

- Foreign objects that are present in raw materials
- Objects that are broken from containers, machinery, and equipment present at the processing plant
- Objects linked with maintenance operations such as a piece of glass from a bulb

Analytical considerations for the detection of physical hazards in food can be the visual examination of raw material and other objects involved in processing, separation of food components from heavier foreign objects by sedimentation, filtration, and other such processes. The state is responsible for providing a legislative framework that defines certain conditions involved in the processing and provision of food (Alston & Quinn, 2017).

Chemical Hazards

These hazards include agriculture chemicals, heavy metals, sanitizing chemicals, and some naturally occurring chemicals. Agriculture chemicals are used in the production of crops like fertilizers, growth regulators, miticides, fungicides, and insecticides. They can also be added to food during post-harvest handling and processing at industries. Heavy metals can be incorporated into the water from geogenic sources, waste materials of industries, organic fertilizers, domestic effluents, and pharmaceutical sources. Chemicals for sanitization and cleaning can be hazardous, if not used correctly (Pretty, 2012). Naturally occurring hazards include allergens like weeds, alkaloids, phytohaemagglutinin, mycotoxins like aflatoxins, and mushroom toxins. Mycotoxins are secondary metabolites from fungi contaminating fruits and vegetables like spices, dried fruits, soybeans, cereals, corn, tree nuts, and peanuts.

Biological Hazards

Biological hazards the microbes like viruses, bacteria, and parasites responsible for foodborne diseases. Moreover, some toxins produced by fungi are biological hazards (Merna & Al-Thani, 2011; EFSA Panel on Biological Hazards [BIOHAZ], 2013). Some microorganisms associated with foods are as given below:

- Bacteria: *Staphylococcus aureus*, *Bacillus cereus*, *Clostridium* species, *Listeria monocytogenes*, *Escherichia coli*, *Shigella,* and *Salmonella*
- Viruses: Coronaviruses, Adenoviruses, Parvoviruses, Coxsackie viruses, Enteroviruses, Astroviruses, Rotaviruses, Norwalk-like virus, and Hepatitis A

- Parasites: Isospora, Sarcocystis, Toxoplasma, Entamoeba, Giardia, and Cryptosporidium (Sun et al., 2019).

Microorganisms can be shifted to food due to incidental contamination from dust, soil, and surroundings, poor production, inappropriate handling practices, use of contaminated water, unsanitary handling of food, sewage fluids, infected animals, and humans (Bultman et al., 2013). Freshly harvested food has few to millions of bacteria/gram and from none to hundreds of mold spores (Yang et al., 2017). When grains, flour, and other food commodities are stored in a high moisture environment, it can lead to the growth of microorganisms leading to alteration of product properties (Okoye & Oni, 2017).

FOODBORNE DISEASES

Foodborne illness is any illness resulting due to consumption of food or water contaminated with pathogenic bacteria, parasites and viruses, mycotoxins, poisonous mushrooms, and some species of beans. Individuals of all life stages are at risk of such diseases especially children and older adults (WHO, 2007). Major foodborne diseases are brucellosis, cholera, traveler's disease, leptospirosis, salmonellosis, shigellosis, tuberculosis, botulism, etc. Poor hand hygiene practices are one of the major causes of the spread of the disease. Children who are malnourished and have weak immune systems are more at risk of foodborne diseases and diarrhea leading to increased child mortality (WHO, 2007). Foodborne diseases can lead to malnutrition; this is a vicious cycle of mortality and disease (Prüss-Ustün et al., 2011). Foodborne diseases are one of the major reasons for the increased mortality rate in children under 5 years. This rate is higher in developing countries (72/1000 live births) than developed ones (6/1000 live births) (UNICEF, 2010). Unsafe food consumption leads to more than 200 diseases from diarrhea to cancers. Foodborne diseases affect almost 600 million people worldwide and 420,000 die from these illnesses every year. Hepatitis A that spreads mostly through uncooked or raw seafood can lead to severe liver damage. Mycotoxins can be present in staple foods like cereals or corn and long-term exposure to them can lead to compromised immunity and abnormal development or even certain cancers (WHO, 2019).

STRATEGIES TO ENSURE GLOBAL FOOD QUALITY AND SAFETY

Food Preparation Practices at Home

Unhygienic practices during domestic food preparations are also a key factor in promoting foodborne diseases. If a person deals with food with contaminated hands then he is surely putting all his family at risk of infection. The use of unsafe water for cleaning, food processing, and cooking increases the chances of exposure to hazards (Prüss-Ustün et al., 2011). Consume cold foods cold and hot foods hot; be familiar with the signs of unsafe food to immediately identify it and discard it. These signs include bad smell, unusual taste, mold, and rancidity (King & Burgess, 1993). Wilna & Abdulkadir (2016) outline measures that can ensure food safety and quality at the domestic level.

Restaurants and Street Vendors

As the world population is increasing like a snowball, new technologies are introduced in every aspect of life to make it easy and comfortable. In this advanced era, people are mostly fond of eating junk foods; they are not committed to cooking food in their houses and tend to consume food away from home in restaurants, hotels, motels, and particularly from street vendors (Troedsson, 2009). This trend is mostly present among teenagers, young, and baby boomers whose taste buds' desires rich, spicy, crunchy, junk, food especially meat and meat-based products. Street-vended foods (SVFs) that encompass a big range of ready to eat beverages and fruits prepared and sold in streets are growing global concern. Their final preparation is done when the customer arrives, orders then purchases, and takes away. Fruits and vegetables sold outside the markets and streets come under this umbrella. Although these foods are affordable and accessible but they can contain hazards as food safety requirements are often neglected. This can lead to unprecedented incidences of foodborne illnesses making food safety a global concern and developmental issue leading to financial consequenccs (Troedsson, 2009).

Lack of safety skills and hygiene knowledge and lack of cooking and handling skills such as temperature, time, personal hygiene, reheating and thawing of frozen products, etc. can lead to serious problems (Osaili et al., 2013). Food business operators shall ensure hygiene maintenance, training of food handlers, appropriate handling of food and equipment, environmental hygiene, and personal hygiene care during all stages of production, processing, and distribution of food (Azanza et al., 2000). Local authorities and food inspectors must visit SVFs and restaurants regularly to improve food quality.

FAO Strategies to Promote Global Food Safety

FAO focuses on adequate food control programs worldwide for ensuring that all individuals on this planet are food secure in terms of quantity, quality, affordability, and safety. This is the key factor for promoting public health. In this regard, Codex Alimentarius Commission implemented the Codex Alimentarius General Principles of Food Hygiene in 1995 for providing a strong foundation for food safety and hygiene and highlighting the main control points at every step in the food chain (Codex Alimentarius, 2012). It further suggested that Hazard Analysis Critical Control Point (HACCP) system must be implemented for improving food suitability and safety for human consumption. Critical control points and food hazards are identified by HACCP. Critical control points are the points on which measures should be taken to eliminate or prevent hazards or to decrease them to an acceptable limit. So by applying HACCP throughout the food chain, effective use of resources, food safety, and time management of food safety concerns can be obtained (FAO, 1997).

WHO Strategies to Promote Global Food Safety

The world has become a global village and there are more opportunities for international trade than ever before. Food and food products trade between continents and countries puts the world at a challenge to maintain food quality and safety across the borders. A single or few countries cannot maintain this safe chain, because this is not a national concern and there is a need to address it through closer linkages at the international level among food safety authorities (WHO / FAO, 2011). In 2000, WHO was urged by the 53rd World Health Assembly for implementing a strategy to improve food safety practices and reduces the burden of foodborne diseases worldwide. In 2007, fifty countries adopted the Beijing Declaration on Food Safety, a prestigious international forum. In 2006, an international meeting was held by the collaboration of multiple international partners and the WHO Department of Food Safety and Zoonoses to launch an initiative for providing a strategic framework for estimation of the worldwide burden of foodborne illnesses (Prüss-Ustün et al., 2011). International Food Safety Authorities Network (INFOSAN) was launched by WHO with the collaboration of FAO for facilitating the collaboration of food safety authorities at the national as well as the international level (WHO/FAO, 2011). In 2006, a food hygiene message was developed by WHO to ensure the safety of customers and industry. These five ways of safer food are given below:

- Keep clean (this covers the personal hygiene as well as an area for food preparation)
- Raw and cooked food should be separated

- Cook thoroughly (this can reduce the risk of infection from food hazards)
- Store food at a suitable temperature
- Use safe raw materials and water

CHALLENGES FOR DEVELOPING COUNTRIES

Developing countries have more burdens of foodborne diseases due to poor food safety conditions. They face various challenges while ensuring food safety and quality such as rapid population growth, urbanization, poor hygiene, and sanitary practices, pre and post-harvest food losses due to lack of resources, insufficient scientific expertise and adequate facilities, lack of coordination between government departments dealing with food, and environment hygiene issues (Henson, 2003; WHO, 2007). Government should take on a regulatory and advisory role in this matter and provide supporting infrastructure to industries and food handlers at any stage from farm to fork. This responsibility of maintaining food safety is shared between the government, industry, and consumers. Government is responsible for food legislation, research and acquiring information, guidance to industry, consumer education, and provision of health care services. Industry and trade stakeholders must ensure good handling of food during processing and distributing, training of staff especially food handlers, and appropriate labeling. Consumers must try to educate themselves about the products they are eating through safe domestic food handling practices and community participation (Henson, 2003; Wilna & Abdulkadir, 2016). World Trade Organization (WTO) agreement emphasizes on recognition of international food standards in international markets. Phytosanitary, sanitary, certificate of origin is an essential part of import documents. The purpose of these regulations is to ensure harmony in world food markets and providing equal chances of participation in international trade to developing countries (FAO, 2020; Henson, 2003). In 2004, WHO and FAO developed the Codex Trust Fund for assisting developing countries to improve their food quality and safety in terms of facility and training support systems (WHO, 2004). This will increase the trade between developing and developed countries that will lead to the better economic condition of developing countries and fewer outbreaks of foodborne diseases (WHO, 2007; Wilna & Abdulkadir 2016).

AN ANALYTICAL MODEL OF FOOD QUALITY

Food will be considered of good quality if it satisfies the needs and expectations of consumers. Analytical model of food quality comprises of consumer requirements (Peri, 2006), that include:

- *Safety requirements* normally mean the absence of harmful substances in food. In case of risk factors, consumer health can be affected and industry can be punished by law.
- *Commodity requirements* account for the genuineness and authenticity of the product. Voluntary regulations, laws, and customary practices set these requirements.
- *Nutritional requirements* are of key importance because food is taken to fulfill the nutritional needs of the body. Functional foods have additional health benefits beyond energy provision. Strong reactions are there in case of any fraud in this regard leading to product rejection.
- *Sensory requirements* are the vital mean of interaction between consumers and products. Sensations are transformed to perceptions by the brain in a space closely connected to other brain functions like memory, values, emotions, culture, etc. These complex interactions are responsible for our memory, ideas, emotions, and sensory reactions associated with food.
- *Production context requirements* are related to tradition, origin, and culture of food/area where it is produced or use of organic agriculture. These requirements have a strong emotive and psychological effect on consumers. Their appeal to consumers depends on ''how'', ''where'' and ''when'' of a food product rather than on "what".
- *Ethical requirements* are related to the system of values conditioning behaviors of a consumer. They include organic agriculture, the defense of biodiversity, defense of the environment, the well-being of animals, etc. However, unfortunately, there are fewer ways of verifying for a consumer that ethical requirements are met or not. There is no way to assess, either biological agriculture was respected or not during the production of an apple. There are more chances of deceit and fraud. That's why guarantee requirements are becoming more popular.
- *Product/packaging system requirements* are essential for product recognition, advertisement, and use. They include aesthetic requirements as well, relating to its presentation and information provided on the label. The products that are easy to handle are preferred by customers and their desire for ease is the most fertile ground for advertising and marketing experts.

- *Product/market system requirements* include product availability at the right place, right time, in the desired amount, and affordable price.

FOOD STANDARDS

Food standards have also been established along with the food laws and regulations to ensure the quality and safety of food. Food standards are not legally required until or unless they are part of the food regulations. One of the best examples of food standards is Codex Standards. For the implementation of the joint Food Standards Program of the World Health Organization (WHO) / Food and Agricultural Organization (FAO), the Codex Alimentarius Commission has been given the authority (Havinga, 2006). Codex Alimentarius can be defined as a set of standards for the safety, quality, and suitability of food. They consist of codes of good hygiene practices such as general food hygiene principles, standards for the usage of animal drugs in foods, standards for the determination of maximum residue levels (MRL) when a pesticide is used on food products, and standards for the regulation of food additives. Food standards promote the international trade of food and are approved by countries all over the world (Newslow, 2013).

Quality Management Systems (QMS)

Quality management systems (QMS) are essential in all departments of the food-related industry to make sure the safety and quality of food (Orris & Whitehead, 2000). Major quality management systems in the food sector are International Food Standard (IFS), Safe Quality Food (SQF), Global Food Safety Initiative (GFSI), International Organization for Standardization - ISO 22000:2005, and British Retail Consortium (BRC) (Baert et al., 2005). Management related aspects in food industries are focused on by Total Quality Management (TQM) and International Organization for Standardization (ISO). On the other hand, HACCP and GMP cover technical aspects (Hoogland et al., 1998). Guidelines are covered by Good Manufacturing Practices whereas TQM makes use of rewards and assessments. The only quality management system comprised of a complete 14 stage plan is HACCP (Van der Spiegel et al., 2003).

Global Food Safety Initiative (GFSI)

The Global Food Safety Initiative (GFSI) is coordinated by CIES also known as Food Business Forum, which is an organization of the world's biggest retailers and was initiated in 2000. The primary mission of GFSI is the continuous advancement of the systems required for the management of food safety to restore the faith of consumers in the safest food delivery. It is primarily aimed at consumer protection and to gain and sustain consumers' trust. The fundamental principle of GFSI is the fact that insurance of food safety should not be a competitor in nature because the internationalization of the market may result in a single problem affecting the entire food supply chain (CIES, 2008). Four major standards of GFSI are IFS and BRC, SQF, and HACCP. The criteria for all of these food safety standards are defined by experts in food safety. It is primarily aimed at the safe production of food products consequently reducing cost and audits in the supply chain (CIES, 2008).

IFS AND BRC Standards

British Retail Consortium (BRC) standard and International Food Standard (IFS) is based on the HACCP system, GHP / GMP principles, and ISO 9001 (Drabas & Wojciechowski 2006). These standards include the requirements as given below;

- The requirement to do threat analysis allergies associated hazards
- Work efficiency monitoring
- Cover facial hair with masks during working
- Carry out products' application tests
- Documentation of stocks of raw material and final products
- Use of FIFO principle "first-in, first-out"
- Establishment of a system allowing the industry to get information about GMO
- Elaboration of all procedures used in case of unusual circumstances
- Identification and elaboration of the area of the industry where hazardous materials like glass occur
- Applying of metal detectors
- Devising a procedure for qualification, verification, and approval of suppliers.

Safe Quality Food (SQF)

A system was developed in 1995 by the Australian government and farmers' associations 1995 for controlling whole and safe quality food system agro-food chain. It later remained as SQF 2000. The basis of this system is requirements established by ISO 9000 and Codex Alimentarius. Since farmers' associations were involved in the development of these systems, therefore; they can be applied in primary production as well. SQF is managed by Food Marketing Institute (FMI) in Washington since 2003. Due to variation in processes, size, products, and circumstances using a single standard are impractical; several standards of SQF have been made (Safe Quality Food Institute, 2008).

- SQF 1000 for smaller businesses and primary production
- SQF 2000 for big industries
- SQF 3000 for restaurants and retail.

HACCP Certification

Dutch food-processing industry launched a standard of certification called Hazard Analysis and Critical Control Point system (HACCP) in 1996 with the name of "Criteria for the assessment of an operation HACCP-system". It was renamed "Requirements for a HACCP-based food safety system" after six years (SCV, 2008). HACCP is a proactive, systematic, and preventative approach to food safety that relies on identifying and controlling hazards in the food chain from harvesting to consumption. This system consists of 7 steps and 12 principles that are given below (1-12 are principles and 1-7 are steps) (Burgess et al., 2019; Das et al., 2019).

1. Assembling HACCP Team
2. Product description (describe food and its distribution)
3. Identification of intended use and target customers
4. Construct a flow diagram
5. Verification of flow diagram
6. Conduction of hazard analysis
7. Determination of the Critical Control Points (CCPs)
8. Establishment of critical limit(s)
9. Establishment of a system to monitoring the control of the CCP
10. Identify where CCPs are lacking by monitoring
11. Establishment of procedures for verifying the efficiency of the HACCP system
12. Documentation of all the procedures and records.

STANDARDS FROM THE ISO 9000 FAMILY

ISO 9000 is a family of standards of quality management that is implemented in various organizational systems. Their application was allowed in a diverse range of enterprises regardless of their operating branches. Standards comprised by the ISO 9000 family are as follow:

- ISO 9000, this standard supports the basis as well as terminology of quality management systems (ISO 9000:2000)
- ISO 9001, this standard is specifically implemented in enterprises as well as it proposes requirements concerned with the quality management system (ISO 9001:2000)
- ISO 9004, is associated with the specification of the guidelines to improve the already implemented system of the company or organization (ISO 9004:2000)
- ISO 19011, provides the recommendations regarding audit (ISO 19011:2002)

Although standards mentioned are optional still customers, contractors, as well as recipients, urge their supplier to adopt systems fulfilling with standards of ISO 9000 because they can then be assured that their suppliers work in a specific, distinguished manner by keeping in view the following eight principles:

- Customer-centric
- Leadership (leaders establish the unity of the aim and operation of the organization)
- Engrossment of the workers
- Process approach
- System approach to management
- Continuous improvement
- Decision making based on facts
- Mutually beneficial cooperation with suppliers

Apart from these, it streamlines the qualification procedure for customers as well as suppliers along with a reduction of cost associated with these operations. Management of organization and resources, product realization and measurement processes, analysis, and improvements are the main areas that are being covered by a quality management system based on standards of ISO 9000 (Knaflewska & Pospiech, 2007).

ISO 22000:2005

This standard was published in September 2005. The goal of it is to merge the codes of the quality management system being used in the food industries (Fabisz-Kijowska & Kijowski, 2006). However, it is optional as it surpasses the agenda of GHP/GMP and HACCP. It encompasses;

- The Prerequisite Programme (PRP), i.e. the above-mentioned GHP/GMP principles and GAP (Good Agricultural Practice), GVP (Good Veterinarian

Practice), GPP (Good Production Practice), GDP (Good Distribution Practice), GTP (Good Trading Practice)

- The HACCP system
- The identification system (traceability system)
- The quality management system ISO 9001:2000

ISO 22000:2005 incorporates both the HACCP and quality management system (ISO 9001:2000). There are also cross-references between ISO 22000, ISO 9004, and terms and definitions from ISO 9000. The implementation of the most effective system of food quality and safety that was designed into the current framework of the management system may be beneficial for both organizations as well as interested parties. Additionally, its implementation into the enterprise may be independent of the other existing management system (Fabisz-Kijowska & Kijowski, 2006).

Good Manufacturing Practices (GMP) And Good Hygiene Practices (GHP)

Good Hygiene Practice (GHP) denotes all food processing procedures must be carried out in hygienic conditions to guarantee food safety. Good Manufacturing Practice (GMP) refers to the procedures to be undertaken and conditions to be fulfilled to ensure that food production, packaging, raw materials, and other related materials having contact with food are conducted properly to ensure safe food production at the end of the process (Turlejska, 2003). Manufacturing practices include well-defined and controlled protocols. It includes validated processes that show their compliance with specifications as well as consistency. These practices are well controlled and are carefully evaluated for any change made to them. Necessary validations are performed for all those changes, having an impact on food quality. Procedures and all related guidelines are written in a language that is clear-cut and easy to comprehend. Operators are well trained to perform as well as document all the procedures (Good Documentation Practices) (Abdellah et al., 2015).

During manufacturing, records are well maintained either manually or by using different instruments. These records show that all necessary steps were taken during a procedure and all instructions were followed and food was of required quantity and quality. Any deviances in the procedure are also documented (De Oliveira et al., 2016). The area covered by GMP requirements (Turlejska, 2003) comprises of following mentioned below:

- The working site, environment, and infrastructure
- Equipment and machines
- Disinfecting and washing procedures

- Water supplies
- Waste management
- Pest control and protection
- Training of staff
- Personnel training and hygiene
- Documentation and record keeping.

Six Sigma

In the middle of the 1980s, the concept of six sigma emerged from the work of Motorola. The Method utilizes the application of statistical tools in business progressions, with the motive to eliminate risks. A key goal of Six Sigma is quality improvement, enhancing the pace of production, and making the product more economical. Six Sigma focuses on quality improvement by waste removal; waste is identified and removed which is of no value to the customer. The technique helps the company to carry out operations more efficiently at lower costs, improve quality, enhance customer satisfaction, and upsurge profitability (Paiva, 2013; Tjahjono & Ball 2010; Werkema, 2004). The Greek letter "Sigma (σ)" is used by statisticians to quantify the change in a process.

A cycle known as DMAIC (Define, Measure, Analyze, Improve, Control) is utilized as a guide for professionals, to apply on the most innovative projects and highly ambitious targets set by the company (Figure 1). The cycle can be described as follows:

Define: definition of the problem and the situation that needs improvement keeping in view the company's targets and strategic ambitions

Measure: application of established and consistent methods for information and data handling and processing

Analyze: analysis of the obtained information from the measurement tools to identify the loopholes and gaps in the current performance of the system and developing a method to reduce the flaw to achieve the desired goal. The analysis must be done with the aid of statistical methods and an approach

Increment: install processes, utilize the project management tools, and planning techniques for newly established processes

Control: implement effective controls to improve the cycle and create a continuously improving cycle (Kwak & Anbari, 2006; Paiva, 2013; Tjahjono & Ball 2010; Werkema, 2004).

Figure 1. DMAIC Cycle (Six Sigma Approach)

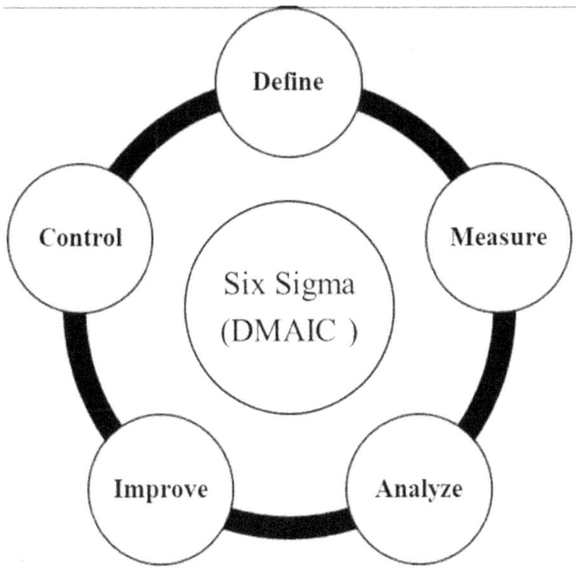

Allied Quality Assurance Publication (AQAP)

There should be cooperation between user, ordered, and supplier in the field of quality assurance of the product. There should be cooperation in the quality systems as described in ISO 9000 series. Adjustments must be made in quality issues to reduce the risk level associated. The representative must supervise the quality assurance system of the deliverer especially in the areas with higher risks. Assessment and analysis of risk that is associated with the deliverer and product to ensure that the necessities of the contract will be fulfilled (Wierzbicki, 2003).

Total Quality Management (TQM)

The term total quality management (TQM) was first introduced in North America during the 1980s. This term refers to the management approach to improve quality that was utilized in Japan for acquiring long-term success. TQM possesses both quality concepts and management principles including empowerment of people, strategic planning, leadership, customer focus process management, and improvement (Dark et al., 2015; Schilling & Neubauer, 2017; Tricker, 2019).

These concepts and principles first evolved during the twentieth century with the considerable involvement of many experts related to the field of quality management. During the 1980s and 1990s, many food businesses in North America approved the

TQM approach and proposed the framework for using it in their system of Quality Management, for acquiring maximum advantage in the global market (Child, 2015; Wallace et al. 2018; Tricker, 2019).

Implementing QMS

Before implementation of QMS, the two important requirements need to be ensured:

- *Customers' requirements* – an organization's ability to provide desired product and service to meet their expectations and needs
- *Organization's requirements* – optimum cost with effective utilization of available resources like materials, technology, workforce, and information.

A quality management system has four distinct components: quality control, quality planning, quality assurance, and improvement. The major constraints in implementing advanced food safety and quality management programs are lack of human resources and finance due to high cost, less personnel, lack of knowledge, experience, and time restrictions. The analysis of key factors influencing the application of these integrated or advanced systems can be helpful to understand the association between certain elements like internal (communication, staff involvement, leadership), external (relationships with authorities, suppliers, and customers), structural (ownership structure and size). In Figure 2, the principles of TQM are presented. The accountability for implementing and regulating food laws is allocated to the agencies regulating the government. These enforcement actions are divided into two classes:

- Inspection and auditing of establishments that are involved in processing, handling, and food storage to make sure that necessary sanitary and well-controlled conditions are applied; audits are utilized by a few regulatory agencies enforcing HACCP-based regulations (Marriott et al., 2018)
- Food analysis and inspection for detrimental substances to make sure conformity to acknowledged limits along with tolerances.

Despite efforts made by government agencies for enforcing food laws as well as regulations, the misbranded foods that are harmful for human use sometimes penetrate food distribution or supply chain (Kotsanopoulos & Arvanitoyannis, 2017). Therefore, regulating government agencies and food companies need to take mandatory action to protect customers against these violations (Hubbard, 2012; Kotsanopoulos & Arvanitoyannis, 2017).

Figure 2. Principles of TQM

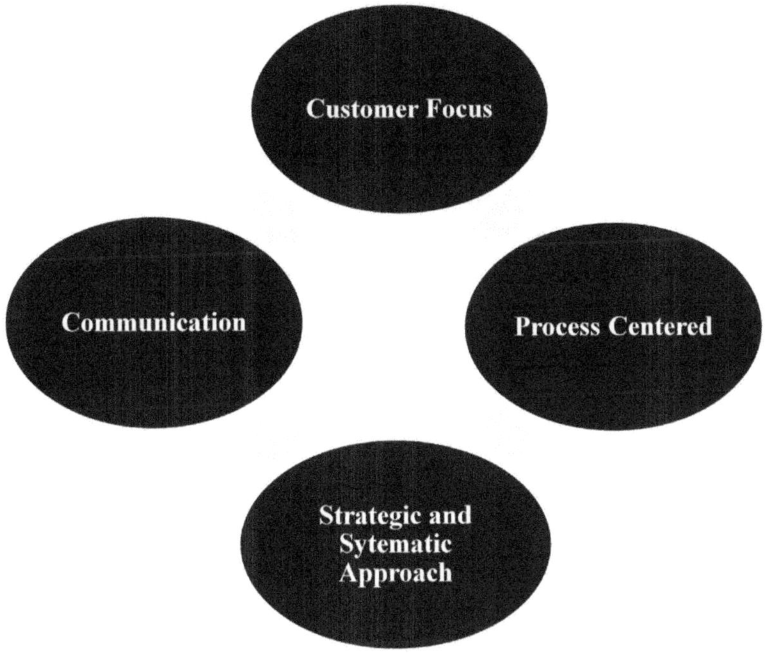

Good Working Practices (GWP)

Working practices may be regulated by compliance between employers and workers' representatives, or the practices, which are considered without any formal agreement. The processes that are used to produce goods following precise procedures of manufacturing and marketing authorizations, to obtain the indispensable standard of the products (Bidwell et al., 2013). Protocols must be followed and if required must be recorded while taking care of products and handling the raw materials including processes like quarantine, sampling, cleaning storage, dispensing labeling, packaging, processing, and distribution. Deviation from the authorized protocol must be avoided to the fullest extent. Approved methods must bc followed to carry out any discreet process in case of divergence from normality (Karmacharya, 2014). The batch number should be used to label and identify the equipment and materials; it is also encouraged to mention the materials and capacity of the equipment. Mention the stage of the production process where it seems reasonable and applicable. In many cases recording of previously processed products is also useful (Sonnemann et al., 2018). Only authorized and relevant personnel should be allowed inside the production facility and intra-process controls are applied and carried out inside the production area (Press, 2003). Contamination of both the raw material at the start of a process and the finished product must be avoided. Strict measures must be in place to avoid cross contaminations on both technical and organizational levels like keeping a check on gaseous, vaporized, and liquid effluents which can cause contamination of the product. Standard operating procedures must include periodical checks of the production units to check for the sources of contamination and production areas should be regularly scrutinized from time to time for environmental sustainability (Patel & Chotai, 2013). Before starting packing operations ensure that documents used, workplace, and packaging lines are clean. No previous product or unnecessary documents, products, or materials are lying around that are not required for the operation at hand (Cohen, 2014; Moerman & Wouters, 2016).

Good Lab Practices (GLP)

Any quality system related to organizational conditions and processes under which environmental safety and non-clinical health studies and research are planned, practically done, examined, documented, archived, and finally reported must be according to recommendations. A chemical must be bought in minimum quantity, even if it seems cheap. Store chemicals within correctly and well-labeled containers having appropriately identified hazard warning (either utilizing circled letter written with help of permanent marker or any of the labels with special hazard code having yellow color; also store chemicals according to the instructions by the supplier). All the chemicals containers should be properly sealed when transporting and store hazardous chemicals like volatile compounds within the fire-proof cabinet. On the other hand, if a chemical is non-flammable or inflammable in little amounts, store in the fume cupboard. Flammable solvents having flashpoints less than 32°C (e.g. ethanol, acetone, propanols, diethyl ether, petroleum ether, isoamyl alcohol, butanols) their storage must be done in bottles with not greater than 500 ml unless within the fire-proof cabinet (De Oliveira et al., 2016).

Good Documentation Practice (GDocP)

Good documentation practice is normally abbreviated as GDP but it is recommended to abbreviate it as GDocP so that it might not be confused with "good distribution practice", which is also being abbreviated as GDP. These are the practices that are used in the food, medical, pharmaceutical industries to create and maintain necessary documents. GMP prospects in addition to the legal requirements are inspected by authorities but not the Law, and if departures are seen, then necessary comments are made. In the last few years, it has been seen that these GDocP is also expanding to the cosmetic industries and ingredient manufacturers (Trafialek & Kolo zyn-Krajewska, 2011).

Good Distribution Practices (GDP)

The distributed food products must be authorized according to the defined set of rules. Storage conditions are well checked and maintained throughout transportation. Chances of any contamination are minimized and adequate turnover of stored food is ensured. To minimize the quality risk of the product, safe and secure areas are chosen for storage. It is ensured that a good quality product is delivered to the right addressee within the time limit provided. An effective tracing system should be used to identify any faulty product (Blanchfield, 2005).

PROCESSING TECHNOLOGIES FOR IMPROVING FOOD QUALITY AND SAFETY

Consumer demands for healthy, fresh, and nutritious food have triggered the food industries to adopt new technologies for high quality products. These technologies can help in enhancing functional and nutritional characteristics, reducing the processing cost and reduction of the foodborne pathogen. Some of these technologies are discussed below (Considine et al., 2008).

High Temperature Short Time (HTST) Pasteurization and Ultra High Temperature (UHT)

Traditional pasteurization was used for milk and milk products for 30 minutes in the large tank. Now the modified form of pasteurization is used called High Temperature Short Time (HTST) pasteurization. In this type of pasteurization, metal plates are used for attaining the temperature of 161° F for 15 seconds. This increases the shelf life but the product still needs refrigeration. Ultra High Temperature (UHT) is an aseptic method done in aseptic containers above 135 °C (275 °F) for 2-3 seconds; packaging is also done by following food safety standards. No refrigeration is required after this type of treatment (IDFA, 2020).

High Pressure Processing (HPP)

This processing can inactivate enzymes and microorganisms and modify structures exerting no effect on nutritional and sensory attributes like texture, the color of products. This helps in controlling spoilage of food, ensuring food safety, and increasing the shelf life of the product (Cheftel, 1995; Hogan et al., 2005; Patterson, 1999). Color changes in meat/poultry are there due to change in heme displacement/

release, changes in myoglobin, and ferrous atom oxidation. HPP can help in coping up with these color change issues.

Ultrasonication

It is a technique that can pasteurize and preserve food products by inactivation of enzymes and microorganisms at low temperatures improving food quality. The baking industry is dependent on control of dough properties to achieve consistency and food quality. Ultrasonic measurements like attenuation and velocity at w100 kHz help in assessing flour quality (Alava et al., 2007). They can be used to determine the rheological properties of doughs made from different types of flour. It can help in discriminating different types of flours (Chandrapala et al., 2012; Gomez et al., 2008).

Vacuum Frying

Traditional frying of potatoes and other foods can lead to acrylamide due to the high content of reducing sugars and amino acid asparagine. Acrylamide formation can be reduced by using the vacuum frying method. This method does not compromise the taste and nutritional properties of foods (Granda et al., 2004).

Nanotechnology

Nano-particles can be used in the food industry to improve nutritional and flow properties, color, flavor, increase shelf life, and extending shelf life. Healthy foods with reduced fat, sugar, and salt content can be produced by using this technology to prevent diseases. Nutraceuticals can be incorporated in carriers including phytosterols, beta-carotene, and lycopene are used in foods for preventing the accumulation of cholesterol (Mozafari et al., 2006). The anti-cancer activity of curcumin was increased by hydrophobically modified starch encapsulation (Yu & Huang, 2010). TiO2 nanoparticles can be used against foodborne pathogens like L. monocytogenes, Vibrio parahaemolyticus, and Salmonella choleraesuis (Kim et al., 2003).

Biomimetics

Food taste and perception are based on the five senses we are gifted by nature. Biomimetics as the 'abstraction of good design from nature' is to tap and analyze nature's potential for innovative solutions. They are human sense inspired technologies to given the decision about the quality of the final product (Ghasemi-Varnamkhasti et al., 2010).

Microwave Processing

A significant reduction in cooking consumption and cooking time can be achieved by using microwaves. Microwave technologies include sterilizing, heating, and drying leading to better food quality and safety. It can retain higher levels of bioactive compounds, the activity of antioxidants, and vegetable color than conventional methods. It can inactivate enzymes and bioactive active components (Guo et al., 2017).

Radio Frequency Electric Field (RFEF) Processing

Commercial heating methods generally focus on convention and conduction methods transferring heat from outer surfaces of products to inner ones leading to more time consumption overheating and dehydration. On the other hand, interaction between foods and electromagnetic energy for controlling bacteria, appearance, improving sensory, and nutritional attributes (Di Rosa et al., 2019).

Ohmic Heating

Ohmic heating also called electrical resistance heating, Joule heating, direct electrical resistance heating is a procedure of heating food products by passing current dissipating energy directly to food. Potential applications for ohmic heating include heating, pasteurization, sterilization, fermentation, dehydration, evaporation, and blanching. It is a faster heating method along with the sustainability of nutritional value and color properties. However, great start-up cost, lack of awareness, and lack of its application in foods having oil and fat limit its use (Kaur & Singh, 2016).

Pulsed Electric Field (PEF)

This commercialized nonthermal intervention technology was introduced in 2005. This method has excellent microbial inactivation for liquid products. It can kill molds, yeasts, and vegetative bacteria. Pumpable and fluid can be pasteurized using PEF. It assures the retention of product freshness and quality compared to non-thermal procedures (Barbosa-Canovas & Swanson, 2019).

Ultraviolet (UV) Light Irradiation

Radiations from the ultraviolet region of the electromagnetic spectrum are used in ultraviolet processing. These radiations can reduce the growth of the microbial population by damaging their DNA. UV irradiation serves as a universal ligand to active cell surface receptors and to induce the assembly of signaling complexes at the plasma membrane. These complexes activate cytosolic protein kinase cascades that relay signals to the cell nucleus, thereby regulating the activity of a variety of transcription factors. Key cellular responses are up-regulation of matrix-degrading metalloproteinase and down-regulation of extracellular matrix biosynthesis. The resultant net deficit of structural integrity of skin connective tissue is one of the major factors that contribute to the phenotype of photo aging (Xu & Fisher, 2005).

RELATIONSHIPS BETWEEN QUALITY SYSTEMS IN THE FOOD INDUSTRY

Specific approaches are highly required in agri-food production industries to satisfy consumers' expected quality parameters (Van der Speigel et al., 2003). It is essential to integrate above mentioned various quality systems to improve the overall performance of the industry. HACCP is considered an important part of quality systems because it not only assures the safety of food products but also ensures the better implementation of the whole quality system. There is a clear distinction between assurance and management. The word assurance describes the product itself and includes various safety assurance systems (GHP, GMP, and HACCP) and also the Quality Assurance Control Points. Thus, the term assurance not only refers to the safety but also the quality of the end product.

An integrated model regarding essential requirements for food organization is indicated in Table 1. The assurance of safer food production is a mandatory requirement for the agri-production industry and can be achieved by following an organizational and systematic structure, controlling processes, activities, and resources mentioned by the quality and hygienic systems like ISO 9001, HACCP, and ISO14000 series.

FOOD QUALITIES AND CONSUMERS' PERCEPTION

The perceptions of consumers vary with time (Fife-Shaw & Rowe, 1996). Commonly interlinked risks occur frequently and are less harmful. Moreover, these risk factors are known to be in control of individuals and are often underestimated (Alvensleben, 2002). Nowadays people are more curious about their health, specifically in terms of nutrition. It is assumed that this drastic change in consumer perception is mainly because of safety measures taken by manufacturers and the awareness of consumers. Food manufacturers in the industry are working ambitiously to ensure food safety in products. Reliability and trust are the main factors on which effective communication regarding food risk and safety between people and manufacturers is based. There are various dimensions of trust such as honesty, competence, and concern with public welfare (Frewer & Miles, 2001). The best way to develop trust is to actively provide information to consumers regarding food safety in peaceful times and proper information should be provided on labels (Röhr et al., 2005).

Table 1. Essentials of food quality and safety assurance in the industry

Quality management	The commitment of top management Sharing of responsibilities, plan, material and human resources Effective quality system assurance
Production areas and equipment	Guarantee the presence of all equipment and facilities, installation and maintenance of equipment, sanitization, cleaning
Staff	Development of a proper organizational framework Description of executive positions Training of staff Development of attitude and behavioural competencies Supervision of personal health and hygiene
Documentation	Develop, approval, updating, distribution, maintenance of document
Regulations	Follow contractual requirements Apply mandatory requirements
Quality control	Apply GLP and good sampling techniques Inspect the analytical methods Validate the process Maintenance, checking, calibration of monitoring and measuring devices
Processing	Process validation Avoid cross-contamination Procure quality raw materials Fulfil label requirements Ensure good quality of end product Identification and tracking of batches
Consumer complaints	Handle the complaints from customers Documentation of recalls and withdrawals Analyze the decisions
Self-inspection	Conduct internal audit Check corrective and compliance
Supplier relationships	Identify key suppliers Communicate open and clear Share future plans and information Start joint improvement and development activities

Source: Rotaru et al., 2015

FUTURE RESEARCH DIRECTIONS

Food safety rules and standards have been the primary drivers of new food business changes in recent decades. However, this is changing; consumers increasingly have emerged as recent trends in many businesses, including the food industry.

Plasma-activated water (PAW), which is water or solutions that have been treated with atmospheric cold plasma, is a procedure that causes minimal changes in food, making it a viable alternative to existing disinfection methods. PAW has gotten a lot of attention lately because of its possible microbicidal qualities in the culinary, agricultural, and biomedical areas. As a result, urgent research efforts should be directed toward the development and scaling up of lower-cost PAW production methods for commercial applications. One of the most important future study paths is the generation mechanism of reactive species in PAW, which might potentially assist to improve the antibacterial capacity of PAW by improving the plasma process parameters for PAW production. Furthermore, multi-omics technologies such as transcriptomics, proteomics, and metabolomics should be used to better unravel PAW's antimicrobial activity mechanisms. Overall, PAW offers prospective solutions to improve the safety and quality of food products as a potentially environmentally friendly and effective antibacterial agent.

Due to its transparency, decentralization, tamper-proof nature, and encryption security, blockchain technology is projected to confront some obstacles and hurdles in a variety of industries. Blockchain is a potential technology for food safety regulation, but there are still many food-related concerns, impediments, and challenges to overcome. Nonetheless, it is believed to be a practical option for reducing food safety issues.

Current and projected future developments in food technology, consumption, and commerce that may have an impact on foodborne disease are examined, and significant driving variables are highlighted, with a focus on the European Union and, to a lesser extent, the United States and worldwide challenges. System-based methodologies are used to establish an understanding of factors, and their impact is examined in relation to current events and future trends projections.

CONCLUSION

Ensuring food quality and food safety is a necessity today especially for developing countries to take part in international trade. Both negative and positive experiences from all countries can be used for the betterment of systems globally. All types of hazards can lead to deleterious effects on the body; strategies should be devised to avoid them. Foodborne diseases exert a significant effect on the health, morbidity, mortality and economic development of a country. Food safety must be ensured to avoid the risk of any hazards in food. The effectiveness of a quality and safety assurance system is based on the relationship of external, internal and structural factors. Food quality assurance and certification has an important role in building the trust of the customer. The industry needs to select a suitable management system for evaluating and measuring the performance of the company.

REFERENCES

Abdellah, A., Noordin, M. I., & Ismail, W. A. W. (2015). Importance and globalization status of good manufacturing practice (GMP) requirements for pharmaceutical excipients. *Saudi Pharmaceutical Journal*, *23*(1), 9–13. doi:10.1016/j.jsps.2013.06.003 PMID:25685037

Abdulmumeen, H. A., Risikat, A. N., & Sururah, A. R. (2012). Food: Its preservatives, additives and applications. *International Journal of Chemical and Biochemical Science*, *1*, 36–47.

Alava, J. M., Sahi, S. S., Garcia-Alvarez, J., Tur, A. O., Chavez, J. A., & Garcıa, M. J. (2007). Use of ultrasound for the determination of flour quality. *Ultrasonics*, *46*(3), 270–276. doi:10.1016/j.ultras.2007.03.002 PMID:17462688

Alston, P., & Quinn, G. (2017). The nature and scope of states parties' obligations under the international covenant on economic, social and cultural rights. In *Economic, Social and Cultural Rights* (pp. 3–76). Routledge. doi:10.4324/9781315257044-2

Alvensleben, R. V. (2002). Verbraucherbild—Verbraucherverantwortung—Verbrauchererziehung. Ziele und Fakten. *ZLR-Zeitschrift fu¨r das Gesamte Lebensmittelrecht*, *2*, 139–150.

Amenta, V., Aschberger, K., Arena, M., Bouwmeester, H., Moniz, F. B., Brandhoff, P., Gottardo, S., Marvin, J. P. H., Mech, A., Pesudo, L. Q., Rauscher, H., Schoojans, R., Vettori, M. V., Weigel, S., & Peters, R. J. (2015). Regulatory aspects of nanotechnology in the agri/feed/food sector in EU and non-EU countries. *Regulatory Toxicology and Pharmacology*, *73*(1), 463–476. doi:10.1016/j.yrtph.2015.06.016 PMID:26169479

Azanza, M. P., Gatchalian, C. F., & Ortega, M. P. (2000). Food safety knowledge and practices of street food vendors in a Philippines university campus. *International Journal of Food Sciences and Nutrition*, *51*, 235–246. doi:10.1080/09637480050077121 PMID:11027035

Baert, K., Devlieghere, F., Jacxsens, L., & Debevere, J. (2005). Quality management systems. In *The food industry in safety in the agri-food chain* (pp. 877–879). Wageningen Academic Publishers.

Barbosa-Canovas, G.V., & Swanson, B. (2019). Nonthermal inactivation of endoproteases by pulsed electric field technology. *Pulsed Electric Fields in Food Processing: Fundamental Aspects and Applications*, 135.

Bidwell, M., Briscoe, F., Fernandez-Mateo, I., & Sterling, A. (2013). The employment relationship and inequality: How and why changes in employment practices are reshaping rewards in organizations. *The Academy of Management Annals*, *7*(1), 61–121. doi:10.5465/19416520.2013.761403

Blanchfield, J. R. (2005). Good manufacturing practice (GMP) in the food industry. In Handbook of Hygiene Control in the Food Industry (pp. 324-347). Woodhead Publishing.

Bultman, M. W., Fisher, F. S., & Pappagianis, D. (2013). The ecology of soil-borne human pathogens. In *Essentials of Medical Geology* (pp. 477–504). Springer. doi:10.1007/978-94-007-4375-5_20

Burgess, C. M., Arroyo, C., Bolton, D. J., Danaher, M., O'Connor, L., O'Mahony, P. J., & Tlustos, C. (2019). Food safety: A public health issue of growing importance. *Introduction to Human Nutrition*, 388.

Chandrapala, J., Oliver, C., Kentish, S., & Ashokkumar, M. (2012). Ultrasonics in food processing – Food quality assurance and food safety. *Trends in Food Science & Technology*, *26*(2), 88–98. doi:10.1016/j.tifs.2012.01.010

Cheftel, J. C. (1995). Review: High pressure, microbial inactivation and food preservation. *Food Science & Technology International*, *1*(2-3), 75–90. doi:10.1177/108201329500100203

Child, J. (2015). *Organization: Contemporary principles and practice*. John Wiley and Sons. doi:10.1002/9781119176862

CIES. (2008). *The food business forum*. http://www.ciesnet.com

Codex Alimentarius. (2012). *About Codex*. http://www.codexalimentarius.org/about-codex/en/

Cohen, M. X. (2014). *Analyzing neural time series data: Theory and practice*. MIT Press. doi:10.7551/mitpress/9609.001.0001

Considine, K. M., Kelly, A. L., Fitzgerald, G. F., Hill, C., & Sleator, R. D. (2008). High-pressure processing–effects on microbial food safety and food quality. *FEMS Microbiology Letters*, *281*(1), 1–9. doi:10.1111/j.1574-6968.2008.01084.x PMID:18279335

Dark, G., McLean, D., & Weatherhead, S. (2015). *Kitchen operations*. Pearson Higher Education AU.

Das, A. K., Nanda, P. K., Das, A., & Biswas, S. (2019). Hazards and safety issues of meat and meat products. In *Food safety and human health* (pp. 145–168). Academic Press. doi:10.1016/B978-0-12-816333-7.00006-0

De Oliveira, C. A. F., Da Cruz, A. G., Tavolaro, P., & Corassin, C. H. (2016). Food safety: Good Manufacturing Practices (GMP), Sanitation Standard Operating Procedures (SSOP), Hazard Analysis and Critical Control Point (HACCP). In Antimicrobial food packaging (pp. 129-139). Academic Press.

Di Rosa, A. R., Bressan, F., Leone, F., Falqui, L., & Chiofalo, V. (2019). Radio frequency heating on food of animal origin: A review. *European Food Research and Technology*, *245*(9), 1787–1797. doi:10.100700217-019-03319-8

Drabas, E., & Wojciechowski, A. (2006). IFS and BRC – International standards of food products quality assurance. *Kalejd. Miesn*, *2*, 86–94.

EFSA Panel on Biological Hazards (BIOHAZ). (2013). Scientific Opinion on the maintenance of the list of QPS biological agents intentionally added to food and feed (2013 update). *EFSA Journal*, *11*(11), 3449.

Fabisz-Kijowska, A., & Kijowski, J. (2006). Food safety requirements by new international standard. *Mieso Wedl*, *5*, 8–12.

FAO, Food and Agricultural Organization. (1997). *Hazard analysis and critical control point (HACCP) System and guidelines for its application*. https://www.fao.org/docrep/005/y1579e/y1579e03.htm

FAO, Food and Agricultural Organization. (2020). *The importance of food quality and safety for developing countries*. https://www.fao.org/docrep/meeting/x1845e.htm

Fife-Schaw, C., & Rowe, G. (1996). Public perceptions of everyday food hazards: A psychometric study. *Risk Analysis*, *16*(4), 487–500. doi:10.1111/j.1539-6924.1996.tb01095.x PMID:8819341

Frewer, L. J., & Miles, S. (2001). Risk perception, communication and trust. How might consumer confidence in the food supply be maintained? In *Food, people and society* (pp. 401–413). Springer. doi:10.1007/978-3-662-04601-2_24

Ghasemi-Varnamkhasti, M., Mohtasebi, S. S., & Siadat, M. (2010). Biomimetic-based odor and taste sensing systems to food quality and safety characterization: An overview on basic principles and recent achievements. *Journal of Food Engineering*, *100*(3), 377–387. doi:10.1016/j.jfoodeng.2010.04.032

Gomez, M., Oliete, B., Garcia-Alvarez, J., Ronda, F., & Salazar, J. (2008). Characterization of cake batters by ultrasound measurements. *Journal of Food Engineering*, *89*(4), 408–413. doi:10.1016/j.jfoodeng.2008.05.024

Gorris, L. G. (2005). Food safety objective: An integral part of food chain management. *Food Control*, *16*(9), 801–809. doi:10.1016/j.foodcont.2004.10.020

Granda, C., Moreira, R. G., & Tichy, S. E. (2004). Reduction of acrylamide formation in potato chips by low-temperature vacuum frying. *Journal of Food Science*, *69*(8), E405–E411. doi:10.1111/j.1365-2621.2004.tb09903.x

Guo, Q., Sun, D. W., Cheng, J. H., & Han, Z. (2017). Microwave processing techniques and their recent applications in the food industry. *Trends in Food Science & Technology*, *67*, 236–247. doi:10.1016/j.tifs.2017.07.007

Havinga, T. (2006). Private regulation of food safety by supermarkets. *Law & Policy*, *28*(4), 515–533. doi:10.1111/j.1467-9930.2006.00237.x

Henson, S. (2003). *The economics of food safety in developing countries.* ESA Working Paper No. 03-19. Agricultural Development Economics Division, FAO. https://ftp.fao.org/docrep/fao/007/ae052e/ae052e00.pdf

Hogan, E., Kelly, A. L., & Sun, D. W. (2005). *High pressure processing of foods: An overview. In Emerging Technologies for Food Processing (pp. 3–31).* Academic Press.

Hoogland, J. P., Jellema, A., & Jorgen, M. T. G. (1998). Quality assurance systems. In W. M. F. Jongen & M. T. G. Meulenberg (Eds.), *Innovative of Food Productions Systems: Product Quality And Consumer Acceptance*. Wageningen Pers.

Hubbard, M. R. (2012). *Statistical quality control for the food industry*. Springer Science and Business Media.

IDFA, International Dairy Foundation. (2020). *Pasteurization.* https://www.idfa.org/pasteurization

ISO 19011:2002. *Guidelines for Quality and/or Environmental Management Systems Auditing*. https://www.academia.edu/36204649/Guidelines_for_quality_and_or_environmental_management_systems_auditing

ISO 9000:2000. *Quality Management Systems – Fundamentals and Vocabulary.* https://www.iso.org/files/live/sites/isoorg/files/archive/pdf/en/watermarksample.pdf

ISO 9004:2000. *Quality Management Systems – Guidelines for Performance Improvements*. https://www.iso.org/standard/28692.html

Jeukendrup, A., & Gleeson, M. (2018). Sport nutrition. Human Kinetics, 18-30.

Karmacharya, J. B. (2014). Good manufacturing practices (GMP) for medicinal products. *Promising Pharmaceuticals*, 101-148.

Kaur, N., & Singh, A. K. (2016). Ohmic heating: Concept and applications: A review. *Critical Reviews in Food Science and Nutrition*, *56*(14), 2338–2351. doi:10.1080/10408398.2013.835303 PMID:25830778

Kim, B., Kim, D., Cho, D., & Cho, S. (2003). Bactericidal effect of TiO2 photocatalyst on selected food-borne pathogenic bacteria. *Chemosphere*, *52*(1), 277–281. doi:10.1016/S0045-6535(03)00051-1 PMID:12729712

King, F. S., Burgess, A., Quinn, V. J., & Osei, A. K. (Eds.). (2015). *Nutrition for developing countries*. Oxford University Press. doi:10.1093/med/9780199685226.001.0001

Knaflewska, J., & Pospiech, E. (2007). Quality assurance systems in food industry and health security of food. *Acta Scientiarum Polonorum. Technologia Alimentaria*, *6*(2), 75–85.

Kotsanopoulos, K. V., & Arvanitoyannis, I. S. (2017). The role of auditing, food safety, and food quality standards in the food industry: A review. *Comprehensive Reviews in Food Science and Food Safety*, *16*(5), 760–775. doi:10.1111/1541-4337.12293 PMID:33371608

Kwak, Y. H., & Anbari, F. T. (2006). Benefits, obstacles, and future of six sigma approach. *Technovation*, *26*(5-6), 708–715. doi:10.1016/j.technovation.2004.10.003

Marriott, N. G., Gravani, R. B., & Schilling, M. W. (2006). *Principles of food sanitation* (Vol. 22). Springer.

McLoughlin, C., Miura, T., Junpei, N., Mendez, A., & Homel, W. (2017). *True Kaizen: Management's role in improving work climate and culture*. Productivity Press. doi:10.1201/9781315180373

Merna, T., & Al-Thani, F. F. (2008). *Corporate risk management*. John Wiley & Sons.

Moerman, F., & Wouters, P. C. (2016). Emerging Trends and Methods in Food Factory Design. In *Innovation and future trends in food manufacturing and supply chain technologies* (pp. 41–79). Woodhead Publishing. doi:10.1016/B978-1-78242-447-5.00003-4

Mozafari, M. R., Flanagan, J., Matia-Merino, L., Awati, A., Omri, A., Suntres, Z. E., & Singh, H. (2006). Recent trends in the lipid-based nanoencapsulation of antioxidants and their role in foods. *Journal of the Science of Food and Agriculture*, *86*(13), 2038–2045. doi:10.1002/jsfa.2576

Newslow, D. (2013). *Food safety management programs: Applications, best practices, and compliance*. CRC Press. doi:10.1201/b16231

Nguz, K. (2007). Assessing food safety system in sub-Saharan countries: An overview of key issues. *Food Control*, *18*(2), 131–134. doi:10.1016/j.foodcont.2005.09.003

Okoye, J., & Oni, K. (2017). Promotion of indigenous food preservation and processing knowledge and the challenge of food security in Africa. *Journal of Food Security*, *5*(3), 75–87.

Osaili, T. M., Jamous, D. O. A., Obeidat, B. A., Bawadi, H. A., Tayyem, R. F., & Subih, H. S. (2013). Food safety knowledge among food workers in restaurants in Jordan. *Food Control*, *31*(1), 145–150. doi:10.1016/j.foodcont.2012.09.037

Paiva, C. L. (2013). Quality management: Important aspects for the food industry. *Food Industries*, 191–218.

Patel, K. T., & Chotai, N. P. (2013). GMP requirements for "buildings and facilities" for api-comparison of schedule m, india and ich guideline and approach for compliance to different regulatory expectations. *Pharma Science Monitor*, *4*(1).

Patterson, M. (1999). High-pressure treatment of foods. In The encyclopaedia of food microbiology (pp. 1059–1065). New York: Academic Press.

Peri, C. (2006). The universe of food quality. *Food Quality and Preference*, *17*(1-2), 3–8. doi:10.1016/j.foodqual.2005.03.002

Peter, R., Mateja, A., & Mojca, J. (2013). Food chain safety management systems: The impact of good practices. In *Advances in Food Process Engineering Research and Applications* (pp. 607–625). Springer. doi:10.1007/978-1-4614-7906-2_30

Pretty, J. (2012). *The pesticide detox: towards a more sustainable agriculture*. Earthscan. doi:10.4324/9781849773188

Prüss-Ustün, A., Vickers, C., Haefliger, P., & Bertollini, R. (2011). Knowns and unknowns on burden of disease due to chemicals: A systematic review. *Environmental Health*, *10*(1), 1–15. doi:10.1186/1476-069X-10-9 PMID:21255392

Röhr, A., Lüddecke, K., Drusch, S., Müller, M. J., & Alvensleben, R. V. (2005). Food quality and safety: Consumer perception and public health concern. *Food Control*, *16*(8), 649–655. doi:10.1016/j.foodcont.2004.06.001

Safe Quality Food Institute. (2008). https://www.sqfi.com/

Schilling, E. G., & Neubauer, D. V. (2017). *Acceptance sampling in quality control*. CRC Press. doi:10.1201/9781315120744

SCV, The Foundation for the Certification of Food Safety Systems. (2008). *HACCP requirements*. http://www.foodsafetymanagement. info

Sonnemann, G., Tsang, M., & Schuhmacher, M. (2018). *Integrated lifecycle and risk assessment for industrial processes and products*. CRC Press. doi:10.1201/9780429436949

Sun, Y., Wu, S., & Gong, G. (2019). Trends of research on polycyclic aromatic hydrocarbons in food: A 20-year perspective from 1997 to 2017. *Trends in Food Science & Technology*, *83*, 86–98. doi:10.1016/j.tifs.2018.11.015

Tjahjono, B., Ball, P., Vitanov, V. I., Scorzafave, C., Nogueira, J., Calleja, J., Minguet, M., Narasimha, L., Rivas, A., Srivastava, A., Srivastava, S., & Yadav, A. (2010). Six Sigma: A literature review. *International Journal of Lean Six Sigma*, *1*(3), 216–233. doi:10.1108/20401461011075017

Trafiąlek, J., & Kołożyn-Krajewska, D. (2011). Implementation of safety assurance system in food production in Poland. *Polish Journal of Food and Nutrition Sciences*, *61*(2), 115–124. doi:10.2478/v10222-011-0012-x

Tricker, R. (2019). *Quality management systems: A practical guide to standards implementation*. Routledge. doi:10.4324/9780429274473

Troedsson, H. (2009). *Forging shared international standards for food safety*. WHO speech. http://www.wpro.who.int/china/media_centre/speeches/speech_20090621.htm

Turlejska, H. (2003). Zasady GHP/GMP oraz systemu HACCP jako narzedzia zapewnienia bezpieczenstwa zdrowotnego _ywnosci [GHP/GMP principles and HACCP system as a tolls for food safety assurance]. *Poradnik Dla przedsiebiorcy*.

UNICEF (United Nation's Children's Fund). (2010). *Progress for children: Achieving the MDGs with equity*. UNICEF.

Unnevehr, L. J., & Huirne, R. B. M. (2002). *New approaches to food safety economics: Overview and new research directions*. Summary of papers presented at a Frontis workshop on New Approaches to Food Safety Economics, Wageningen, The Netherlands.

Van der Spiegel, M., Luning, P. A., Ziggers, G. W., & Jongen, W. M. F. (2003). Towards a conceptual model to measure effectiveness of food quality systems. *Trends in Food Science & Technology*, *14*(10), 424–431. doi:10.1016/S0924-2244(03)00058-X

Wallace, C. A., Sperber, W. H., & Mortimore, S. E. (2018). *Food safety for the 21st century: Managing HACCP and food safety throughout the global supply chain.* John Wiley and Sons. doi:10.1002/9781119053569

Werkema, C. (2004). *Criando a cultura seis sigma.* Nova Lima: Werkema Editora.

WHO. (2004). Food safety in developing countries: Building capacity. *Weekly Epidemiological Record, 79*(18), 173–180. PMID:15168565

WHO. (2007). *Countries urged to be more vigilant about food safety.* www.who.int/mediacentre/news/releases/2007/pr39/en/

WHO. (2019). *Food Safety.* https://www.who.int/news-room/fact-sheets/detail/food-safety

WHO/FAO. (2011). *The international food safety authorities network (INFOSAN) progress report 2004-2010.* http://whqlibdoc.who.int/publications/2011/9789241501286_eng.pdf

Wierzbicki, S. (2003). AQAP w Sokołów S.A [AQAP at Sokołów S.A.]. *Bezp. Hig. żywn., 6*(1), 3.

Wilna, H. O., & Abdulkadir, A. E. (2016). Food quality and food safety. In N. J. Temple & N. Stey (Eds.), *Community Nutrition for Developing Countries*. Athabasca University Press and UNISA Press.

Xu, Y., & Fisher, G. J. (2005). Ultraviolet (UV) light irradiation induced signal transduction in skin photoaging. *Journal of Dermatological Science. Supplement, 1*(2), S1–S8. doi:10.1016/j.descs.2005.06.002

Yang, W., Li, D., & Mugambi, A. (2017). *Spoilage Microorganisms in Cereal Products. In Food Spoilage Microorganisms: Ecology and Control.* CRC Press.

Yu, H., Huang, Y., & Huang, Q. (2009). Synthesis and characterization of novel antimicrobial emulsifiers from polylysine. *Journal of Agricultural and Food Chemistry, 58*(2), 1290–1295. doi:10.1021/jf903300m PMID:20020765

Section 2

Food Safety: Cognitive and Behavioral Perspectives

This section presents four chapters highlighting issues relating to knowledge, attitude, and practices of the use of polystyrene food packaging among food operators, standard food safety practices in a restaurant, a literature review on the level of food safety knowledge among food handlers, and a study of consumers awareness and opinion toward food safety practices and policy.

Chapter 5
Knowledge, Attitude, and Practice (KAP) of Polystyrene Food Packaging Usage Among Food Operators

Ungku Fatimah Ungku Zainal Abidin
Universiti Putra Malaysia, Malaysia

Maimunah Sanny
Universiti Putra Malaysia, Malaysia

Nur Hanani Zainal Abedin
Universiti Putra Malaysia, Malaysia

ABSTRACT

Polystyrene is commonly used on a daily basis for the packaging of takeaways as if people assume that it is safe. Although studies show the migration of styrene into food could cause adverse health impacts such as cancer, neurotoxicity, and hormone-related problems, less is known about public awareness of its risk. This chapter presents a study conducted to assess the level of knowledge, attitude, and practice (KAP) of polystyrene food packaging usage among food operators (i.e., hawkers, restaurants, and night market). Data were collected using self-administered to survey 115 food operators. Results show that the food operators have a moderate knowledge and attitude in polystyrene usage. Their practice in polystyrene usage is still poor. Food operators from restaurants had the highest level of KAP compared to hawker and night markets. The finding could provide some guides in developing interventions to educate food operators about food safety issues related to polystyrene.

DOI: 10.4018/978-1-7998-7415-7.ch005

INTRODUCTION

A packed schedule and busy lifestyle had caused us to have lesser time spend to cook especially during weekdays; classes, work or when so much activity is to be completed in a day. Therefore, take-away is always the best option. Consumers could eat their food anywhere and everywhere that they want according to their preference. Food packaging material such as plastic is a medium mostly used to take away food in developing nations including Malaysia (Jayaraman et al., 2011). Plastic is still widely used among foodservice operators especially if they provide takeaway and delivery service. Food packaging had several functions where it protects food from being contaminated and spoiled, ease of transportation and provides a uniform measurement of contents. There are many types of favorite food packaging in Malaysia; for example, polystyrene, plastics, oil paper, paper box, etc.

Polystyrene food packaging is one of the products derived from monomer styrene and had side effects if being exposed.-This food packaging had been very popular, not only among food operators but also customers and the usage is prevalent nationally before it was banned in certain areas (Haqim, 2015; Soon, 2016). The authorities of federal territories such as Putrajaya, Kuala Lumpur and Labuan, as well as the state of Selangor, Pulau Pinang and Malacca, had recently forbidden the usage of polystyrene in their states (Ahmad, 2016). In almost all food operations, polystyrene could be found as a medium of packing food before it is banned in some states in Malaysia (Haqim, 2015). The purpose of this study is to assess food operators' level of knowledge, attitude and practice (KAP) about polystyrene food packaging. The specific objectives are as follow:

i. To evaluate the level of KAP among food operators
ii. To determine the relationship between KAP
iii. To compare the level of KAP between different types of food operations (i.e., hawker, restaurant and night market).

BACKGROUND

Polystyrene

Polystyrene is a common type of food packaging that food service operators regularly employ to package their take-out meals. This packaging is a transparent, rigid, and brittle addition polymer of styrene with a relatively low melting temperature. Styrene is a monomer that is utilised as a fundamental building block in the plastics industry and has been deemed safe for use in food packaging. Foaming produced from styrene generates an opaque, hard, lightweight material that is resistant to impact and provides thermal insulation. Protective packaging such as egg cartons, containers, disposable plastic utensils, lids, cups, plates, bottles, and food trays are common examples available in the market (Marsh & Bugusu, 2007). The drinking cup, bowl, and foam food are all examples of polystyrene employed in this industry.

Polystyrene is a polymer composed of many different styrene monomers that is commonly utilised by manufacturers and food service operators. They feature a number of characteristics that are ideal for food service, including lightweight, portability, and heat-resistant material. Additionally, polystyrene is opaque, hard, and impact-resistant, making it an ideal material for protective packaging (Marsh & Bugusu, 2007).

Despite the benefits of polystyrene for both operators and customers, there are significant risks associated with its use, necessitating the use of alternate packaging. The problem is to develop containers that maintain the temperature of the food while remaining economically packaged (Boyce, Charles & Binkley, 2008). The currently utilised white polystyrene containers failed to preserve food's temperature, spilled frequently, and were easily broken by sharp cutlery (Sheridan, 2003).

Additionally, environmental-conscious individuals demanded green packaging in order to avoid food waste and to have packaging that degrades quickly. While manufacturers of biodegradable items frequently claim to be environmentally friendly, there is currently no solution that can replace polystyrene while remaining environmentally benign. Even biodegradable paper boxes for food packing have a higher life cycle assessment (LCA) and do not totally decompose when contaminated (Bavani, 2016). To date, no alternative to polystyrene that is environmentally friendly and fully degrade when polluted.

Chemical Migration of Polystyrene into Food

Chemical migration occurs due to the movement of chemicals from the substance used inside the polystyrene container to the food. Only if the packaging is handled incorrectly and comes into close touch with food will the chemical migrate. When metals, glass, ceramics, rubber, plastic, and paper come into contact with certain types of foods, small amounts of their chemical contents can be released. Chemical migration was facilitated by packaging that came into direct touch with food, which is referred to as primary packaging. Food and beverage products can be quite aggressive and can interact violently with other materials, depending on the components with which they come into touch (Ardic & Kahve, 2015; Castle, 2006).

The chemical styrene may be soluble in food because it undergoes partial breakdown under specific conditions, such as when fat is present, when alcohol is present, or when Vitamin A is present. Besides temperature and food acidity, other factors that may contribute to the solubility of styrene in food or beverages include long-term exposure of polystyrene to food or beverages. Styrene migrates rapidly, especially when food or beverages are hot, as this causes partial decomposition of styrene.

Due to its solubility in oil and fat, styrene migrates into food. According to Ardic and Kahve (2015), the more the fat content of food, the higher the level of styrene migration into the food. Foods with high-fat content, such as entrees, soups, and beverages, styrene will migrate at a faster rate than other foods. Aside from that, the ethanol that is typically included in alcoholic beverages has the ability to dissolve styrene. Vitamin A, also known as beta carotene, which is widely believed to be beneficial when eaten, can be harmful to the human body if not consume properly. Because vitamin A degraded by the addition of heat, it produces toluene, which is highly reactive and aggressively dissolves the styrene. Intentional and unintentional food contamination may come from acidic food such as coffee and spices, oil fried dishes, extreme hot conditions and food that contains Bisphenol A (BPA) (Ardic & Kahve, 2015; Castle, 2006; Khaksar & Khansari, 2009).

It is widely known that high-temperature food or drinks that contact directly with plastic surfaces could cause chemical migration (Ardic & Kahve, 2015; Castle, 2006). Styrene monomer will migrate into hot beverages based on the amount of fat present, the temperature, and the duration of the beverage. Temperature is the most important component to consider because the level of migration increases as the temperature rises. Similarly, it was demonstrated that food packaged in polystyrene trays and wrapped in supermarket packaging absorbs styrene from the container (Ardic & Kahve, 2015; Castle, 2006; Khaksar & Khansari, 2009;). The rates of chemical migration are also affected by a number of factors: a) direct or indirect contact with food by packaging material, b) characteristics of the materials in contact with food, c) migrant chemical property, d) migrant initial concentration in packaging material, e) contact time and temperature components in contact with the packaging material, and f) acid content in food. High acidity food can increase the rate of styrene migration in the packaging material (Ardic & Kahve, 2015; Kamidi, 2015).

Styrene can poison a person over an extended period of time because it has been classified as a human carcinogen by the Food and Drug Administration (FDA), and the International Agency for Research on Cancer (IARC). Investigations of the harmful effect of styrene on the haematological, central neurological and peripheral nerve systems, gastrointestinal, reproductive organs, and lymphatic systems have raised widespread concern (Emma et al. 2001). Only a little amount of evidence suggests that styrene can be harmful to the developing child, the female reproductive system, or the endocrine system. Other research have suggested that styrene may have an impact on neurobehavioral development and may have neurotoxic properties (Brown et al. 2000).

Consumer Perception of Food Packaging

In the food packaging industry, there are two levels of consumers: the primary consumer is the food operator who uses the packaging to pack the food purchased by the consumer, and the secondary consumer is the customer who consumes the food from the packaging. Despite the fact that sales continue to grow, customer interest in dining does not increase. Take-out and curb-side services are fueling the market (Prewitt, 2002). The increasing number of take-out sales at food establishments that continue to expand demonstrates that the use of food packaging is also increasing. On the contrary, there are certain worries about take-out meals, one of which being the sort of container that is being used to deliver the food (Boyce et al, 2008).

According to Boyce et al. (2008), consumers have inadequate awareness of food safety and food security. Furthermore, the end customer believes that they do not need to be concerned about food safety issues and has delegated this responsibility to food operators and health inspectors. Because of their lack of information about food safety issues, they have a low level of knowledge, and they have denied themselves the right to eat food of the highest possible quality that is also safe to consume. On the other hand, operators may not care because their major focus is their clientele. They will just continue to use them as long as there is a demand and no objection from the customer.

Despite the fact that Malaysia is a developing country, Malaysians have the habit of purchasing hot food items in plastic food packaging (Jayaraman et al., 2011). In Malaysia, this practice had been observed practically everywhere, and people may have concluded that there was nothing wrong with it. The actual issue is whether or not the average consumer is aware of the safety of plastic bag that is used for packing hot things (Jayaraman et al., 2011). Furthermore, food should not be packaged at extremely high temperatures in any sort of food packagings, such as polystyrene, plastic bags, or paper boxes, under any circumstances. Consumers should avoid using polystyrene boxes to pack food that would be exposed to high heat because this may result in the chemical migration of food packaging (Jayaraman et al., 2011).

Perceived Mechanism of Food Packaging

Food packaging serves as a conveyance for the food while it is being transported. Given the increasing popularity of delivery services, business owners are searching for more effective packing (Boyce et al. 2008). The attributes that consumers have been seeking in food packaging include the ability of the packaging to act as an insulator for the food, the ability to transport product without spillage, and the ability to keep food independently within the same package (Boyce et al, 2008). Additionally, customers have expressed an interest in food packaging that can moderately preserve convenience foods while maintaining their fresh-like appearance and taste (Vermeiren et al., 1999). Apart from that, it should be convenient, low-cost, and simply accessible for food service companies.

Food packaging protects food against three types of external impacts: chemical, biological, and physical. Chemical protection is used in situations where compositional changes caused by external factors, such as gases, moisture, or light, can be minimised or eliminated. Biological protection, on the other hand, acts as a barrier against bacteria and other pathogens that could cause disease and spoilage. Aside from that, it also serves as physical protection, preventing mechanical damage, stress, and vibration that may occur throughout the process of distribution (Bugusu & Marsh, 2007). Packaging is advantageous in the sense that it protects food from spoiling caused by external forces; yet, chemicals from packaging may have a negative impact on the quality and safety of the food they are protecting (Ardic & Kahve, 2015). Physical contamination could be easily observed and avoided if precautions were taken, but chemical and biological contamination could not be observed with the human eye and could not be controlled. These are the variables that should be the most concerned about, and food packaging should be effective in preventing this type of contamination from occurring.

Problem Statement

Polystyrene is commonly used as a medium of packaging for food in almost all food operations before it is banned in some states in Malaysia. There is a need to study the food operator's usage of this type of packaging especially in holding a hot edible item. While studies indicate that styrene migration into food may have negative health consequences such as cancer, neurotoxicity, and hormone-related disorders, little is known regarding public knowledge of the risk. The risk of migration is higher as the food is held inside polystyrene packaging for some time before it is consumed by the consumer. Hence, there is a need to know if food operators had knowledge on the usage of plastic food packaging to pack takeaway food specifically polystyrene in this study.

To date, a limited number of research on knowledge, attitude and practice (KAP) of plastic food packaging usage has been conducted among food operators such as restaurants, hawkers and night markets in Malaysia. Most studies on the usage of plastic food packaging, particularly on KAP, were conducted among consumers. Although food operators are also considered as customers because they bought and used plastic food packaging, they are classified as the primary user who is different from the end-user. In short, there is a lack of data on food operators KAP to understand the risk associated with the use of polystyrene.

METHODOLOGY

Research Design

A cross-sectional survey study was conducted to gather empirical data on the level of knowledge, attitude, and practice (KAP) in connection to the use of polystyrene food packaging among food operators. The survey approach was chosen because it is cost-effective, can be sent to a large number of respondents, and provides more accurate responses (Dillman, 2013). It is one of the most widely used methodologies in order to evaluate and determine the specific associations between variables in a corresponding study (Neuman, 2003).

Study Sample and Sampling

The population of this research is food operators in four selected states in Malaysia that have not banned the usage of polystyrene for food packaging, namely Negeri Sembilan, Perak, Kedah, and Kelantan. The selection of these locations was made with the goal of ensuring that the KAP of the food operators at the time of the study's execution could be examined. Hawkers, restaurants, and night markets were among the food establishments that have been targeted. The usage of polystyrene food packaging is high among these types of food operators. The sampling strategies utilized in this study were convenience sampling techniques. There are no resources available to build a sampling frame for polystyrene users in the food service industry, even if the population had been more narrowly defined. It is therefore difficult to use a random sampling technique. Respondents were chosen among food operators from small operation enterprises in which polystyrene is likely to be used for food packaging. The targeted number of samples was 150 respondents, but the useable valid data obtained was 115 respondents.

Data Collection

A paper survey questionnaire was employed as a tool for data collection. The questionnaire contains four sections and was prepared in two language options (i.e., English and Bahasa Malaysia). The first section contains questions on the respondent's demographic and operation profile. Demographic questions included were age, gender, work position, and years of experience with the current operation.

Three questions on operation profile pertaining to the duration of operation and estimated sales monthly were included using a multiple-choice answer format. The second section contains ten questions on the polystyrene usage level. Respondents were asked to estimate the amount of polystyrene used according to the types of food packed for the customer. The type of food they used to pack and the frequency of usage were also captured. Besides, a question on reasons for choosing polystyrene for their operation was also included in the questionnaire. Respondents were allowed to select more than one answer choice for this question.

In the third section, questions measuring the level of knowledge which contains 12 items on knowledge about styrene, polystyrene and chemical migration of styrene were included. Most of the knowledge questions were adapted from a previous study (Haqim, 2016), but some additional questions were developed by the researcher. Respondents must indicate whether each statement is true or false. The fourth section constitutes a series of questions about the attitude of food operators towards polystyrene by using five points Likert scale, where 1 = strongly disagree and 5 = strongly agree. The attitude of food operators towards polystyrene was measured based on how safe they think of polystyrene and the health hazard that might be caused by styrene compounds. Lastly, eight questions on practices were included in the final section of the questionnaire. Items in this section measure how food operators handled polystyrene food packaging and their decision-making in the usage of polystyrene for the operation. Respondents are given five points of Likert scale where 1 = never, 2 = occasionally, 3 = sometimes, 4 = usually, and 5 = frequently. The survey questionnaire was pre-tested with food operators (n=10) prior to data collection to check for face validity.

Data Analysis

The data were analyzed by using the Statistics Package for Social Sciences (Version 23). Descriptive analysis was done to provide frequency, percentage and mean scores from the data. Descriptive analysis was used to report the level of knowledge score, the mean of attitude and practice and the level of food packaging usage. This study applied the ANOVA test to compare the level of KAP among the types of food service operators. Next, correlation and regression analysis were used to determine the relationship between knowledge, attitude and practice.

RESULTS AND DISCUSSIONS

Profile of the Respondents

Table 1 shows the profile of respondents in this study. Slightly more than of the respondents are female (56.5%) and within the age range 19 and 29 years old (26.1%). Most of the respondents surveyed are the owner or supervisor of the food operations (67%), followed by a cook or kitchen helper (19.1%). The majority of them who participated in this study have one to ten years of work experience in food business operations (77.4%).

Table 1. Summary of respondents' profile (n=115)

Characteristics	Frequency (n)	Percentage (%)
Gender		
Male	50	43.5
Female	65	56.5
Age		
19 – 29 years old	30	26.1
30 – 39 years old	29	25.2
40 – 49 years old	27	23.5
More than 50 years old	29	25.2
Work Position		
Owner/supervisor	77	70.0
Cook/kitchen helper	22	19.1
Others	12	10.4
Work Experience in Food Business Operation		
1 – 10 years	89	77.4
11 – 20 years	15	13.0
21 – 30 years	7	6.1
More than 30 years	4	3.5

Note. Some percentages do not add up to 100% due to missing value

Respondents' Food Operation Information

Respondents' food operation information is presented in Table 2. Almost half of the food operations are restaurants (47.2%), followed by hawker stalls (30.3%) and night market (22.5%). Most of the respondents surveyed have operated their business within less than 10 years (85%). Only 14.8% of the food premises operated between 11 and 20 years while the remaining 10% operated for more than 10 years. The majority of food operations achieved less than RM10K sales per month (82.6%) and has less than five total numbers of workers (81.7%). Less than 20% of the operations obtained more than RM10K sales and have more than 5 workers.

Table 2. Summary of respondents' food operation profile (n=115)

Characteristic	Frequency (n)	Percentage (%)
Type of operation		
Hawker	39	30.3
Restaurant	42	47.2
Night Market	34	22.5
Period of operation		
Less than 10 years	85	73.9
11 to 20 years	17	14.8
21 to 30 years	6	5.2
More than 30 years	6	5.2
Sales (per month)		
Less than RM 10K	95	82.6
RM 11K to RM 20K	14	12.2
More than RM 20K	6	5.2
Number of workers		
Less than 5	94	81.7
6 to 10	19	16.5
More than 10	2	1.7

Note. Some percentages do not add up to 100% due to missing value

The Level of Knowledge Relating to Polystyrene Food Packaging Usage

Respondents' level of knowledge related to polystyrene food packaging usage was evaluated by asking them to choose true or false statements. Each response was then graded based on the correct answer was chosen by the respondents. The percentage of respondents providing a correct and incorrect response is tabulated in Table 3. The results are arranged in descending order from highest to lowest percentage of the correct answer. The highest percentage of respondents (82.6%) answered correctly was on the statement that "Styrene will not leach out from cold food". This indicates that the majority of the respondents knew styrene monomer does not undergo breakdown to migrate into food at room temperature or cold food and that they knew temperature does affect the migration.

The second highest statement answered correctly is the statement of "Polystyrene is made up of styrene that can cause adverse health impact", where 55.7% of the respondents answered correctly. The statement tested respondents' general knowledge of components of polystyrene. From this statement, the study found that most of the respondents had the idea of styrene could give negative impacts on our health.

Table 3. Food operators' knowledge on polystyrene food packaging (n=115)

Statement	Percentage of respondent % (n)	
	Correct answer	Incorrect answer
Styrene can leach out when cold food is packed in polystyrene food packaging	82.6 (95)	14.8 (17)
Polystyrene food packaging is made up of styrene that can cause adverse health impact	55.7 (64)	41.7 (48)
Styrene can leach out when oily hot food is packed in polystyrene food packaging	53.9 (62)	43.5 (50)
Human can be exposed to styrene from polystyrene food packaging	51.3 (59)	46.1 (53)
Polystyrene is a type of synthetic polymer that is used in making polystyrene food packaging	49.6 (57)	47.8 (55)
Polystyrene food packaging contains compound called styrene that is toxic	46.1 (53)	51.3 (59)
Polystyrene is safe to be used as food or drink packaging	44.3 (51)	53.0 (51)
The safety of polystyrene is no different from other types of food packaging such as paper or plastics	40.0 (46)	57.7 (66)
Styrene can migrate into food upon prolong exposure at high temperature	38.3 (44)	59.1 (68)
There are some food safety issues related to the use of polystyrene food packaging	37.4 (43)	60.0 (69)
Polystyrene used in food/drink packaging are made from natural gases	29.6 (34)	67.8 (78)
Some states in our country banned food provider from using polystyrene food packaging because of health impact	26.1 (30)	71.3 (82)

Note: Total percentage of respondent does not add up to 100% due to missing data

The next highest statement that was answered correctly is "Styrene can leach out when oily hot food is packed in polystyrene" with 53.9% of them answer correctly. This statement tested respondents' awareness about the factors that contribute to the migration of styrene into food. Next, a total of 51.3% of respondents correctly answered that "human is exposed to styrene through polystyrene food packaging". Only less than 50% of respondents answered correctly for the remaining statements given. The lowest correct response was on the statement "Some states banned food providers from using polystyrene because of health impact" with a percentage of 26.1% only.

The Level of Attitudes Relating to Polystyrene Food Packaging Usage

Respondents' level of attitudes towards polystyrene food packaging usage was evaluated on eight items using a Likert scale (1 = Strongly disagree; 5 = Strongly agree). Table 4 shows the level of attitude arranged in descending order of highest to lowest mean. The highest rating is at a mean score of 3.76 where around 60.9% of respondents believe that "it is safe to pack cold food in polystyrene food packaging". The following highest rating means is 3.24 in which respondents agree (42.6%) that "Polystyrene food packaging is safer compared to any other packaging (e.g., oil paper, plastic bags, polypropylene container)". In other words, respondents moderately believed that polystyrene food packaging is safer compared to any other packaging. Respondent agrees that the risk of all food packaging might be different. Results in Table 4 implied that respondents have a low attitude towards the safety of polystyrene food packaging usage.

The third highest mean score rating is 3.17 where 46.9% of respondents believe that "the use of polystyrene as food packaging is safe". In general, respondents believe that the usage of polystyrene as plastic packaging is safe. Most of the food operators had no idea that the usage of polystyrene could cause adverse health impacts to humans.

On the other hand, the results show that majority of respondents disagree (20.9%) or are neutral (51.3%) that "Continuous usage of polystyrene could cause adverse health impact such as neurotoxicity" with a mean score rating of 2.96. Most of the respondents also disagree (38.3%) that "it is safe to pack oily hot food in polystyrene food packaging". Finally, most of the food operators disagree (29.6%) and neutral (43.5%) that "Continuous usage of polystyrene could cause health impact such as neurotoxicity" with a mean score rating of 2.85. Based on this result, it could be concluded that respondents do not believe that polystyrene could cause neurotoxicity towards the human body as the majority of the respondents, disagree with this statement.

Table 4. Respondents' attitude towards polystyrene food packaging usage

Statement	Respondent % (n)					Mean	SD
	1	2	3	4	5		
It is safe to pack cold food in polystyrene food packaging	0 (0)	7.0 (8)	27.0 (31)	42.6 (49)	18.3 (21)	3.76	0.85
Polystyrene food packaging is safer compared to any other packaging (e.g., oil paper, plastic bags, polypropylene containers)	1.7 (2)	21.7 (25)	28.7 (33)	37.4 (43)	5.2 (6)	3.24	0.93
The use of polystyrene as food packaging is safe	9.6 (11)	26.1 (30)	12.2 (14)	33.0 (38)	13.9 (16)	3.17	1.26
Continuous usage of polystyrene could cause adverse health impacts such as cancer*	2.6 (3)	21.7 (25)	38.3 (44)	28.7 (33)	3.5 (4)	3.09	0.89
The use of polystyrene food packaging does not cause adverse health impacts on human	3.5 (4)	27.0 (31)	29.6 (34)	29.6 (34)	4.3 (5)	3.05	0.97
Continuous usage of polystyrene could cause adverse health impacts such as gastrointestinal effects*	2.6 (3)	20.9 (24)	51.3 (59)	18.3 (21)	1.7 (2)	2.95	0.77
It is safe to pack oily hot food in polystyrene food packaging	6.1 (7)	38.3 (44)	14.8 (17)	27.8 (32)	7.8 (9)	2.93	1.14
Continuous usage of polystyrene could cause adverse health impacts such as neurotoxicity*	2.6 (3)	29.6 (34)	43.5 (50)	17.4 (20)	1.7 (2)	2.85	0.82

Note: Scale used: 1- Strongly disagree, 2- Disagree, 3- Neutral, 4- Agree, 5- Strongly agree
*Reverse code items
Scale Reliability (Cronbach-alpha) = 0.62

The Level of Practice Relating to Polystyrene Food Packaging Usage

The result for respondents' practices toward polystyrene food packaging usage is presented in Table 5 with mean scores arranged in descending order. The level of food operator general practice relating to the use of polystyrene was evaluated using 5 points Likert scale (1= never; 5= frequently). Overall, the respondents have a moderately poor practice as the majority of them reported sometimes or usually used polystyrene for their business operation.

The highest mean score rating of 3.41 with 32.2% respondents rate they frequently use polystyrene when packing food for take-away in business. Only 14.8% of the respondents reported they never use polystyrene for their operation while around 14% occasionally used them. This result implied that respondents have moderately good practice towards polystyrene usage. Apart from that, the statement about using polystyrene to pack oily hot food was also rated high with a mean score rating of 3.28. Respondents claimed they usually (25.2%) and frequently (21.2%) use polystyrene to pack oily hot food, which indicates poor practice of polystyrene usage while only 10.4% of the respondents rated never use polystyrene frequently in packing of oily hot food.

Table 5. Respondents' practice towards polystyrene food packaging usage

Statement	Respondent % (n)					Mean	SD
	1	2	3	4	5		
I use polystyrene when packing food for take-away for my business.	14.8 (17)	13.9 (16)	15.7 (18)	18.3 (21)	32.2 (37)	3.41	1.47
I use polystyrene to pack oily hot food for my business.	15.7 (18)	10.4 (12)	21.7 (25)	25.2 (29)	21.7 (25)	3.28	1.38
If I have options for food packaging for my business (polystyrene, oil paper, plastic, etc.), I choose polystyrene.	10.4 (12)	16.5 (19)	27.8 (32)	27.0 (31)	13.0 (15)	3.17	1.20
When purchasing food packaging for my business, I do consider purchasing biodegradable packaging*	16.5 (19)	20.0 (23)	24.3 (28)	24.3 (28)	9.6 (11)	2.90	1.25
I tell the customer to bring their own food container if they wanted to take-away their food*	62.6 (72)	8.7 (10)	10.4 (12)	9.6 (11)	3.5 (4)	1.76	1.21
I avoid using polystyrene as much as I could for my business*	14.8 (17)	19.1 (22)	28.7 (33)	14.8 (17)	14.4 (20)	3.01	1.32

Note: 1: Never 2: Occasionally 3: Sometimes 4: Usually 5: Frequently
*Reverse code items
Scale Reliability (Cronbach-alpha) = 0.55

The next highest mean score rating is the statement "If I have options for food packaging for my business (polystyrene, oil paper, plastic, etc.), I choose polystyrene" with a mean score of 3.17. The majority of respondents choose polystyrene even if they were given other options possible due to the cost-effective use of polystyrene which shows good practice among the operators. The statement with the lowest mean score rating is "I tell the customers to bring their own food container if they wanted to take-away their food" ($M = 3.01$). Around 62% of the food operators never tell their customers to bring their own container, which indicates a poor practice to reduce the usage of polystyrene while only 3.5% of respondents encourage their customers to bring their own container to take-away food.

In short, the practice could be improved if food operators are aware and have related knowledge about particular problems especially those related to health. Necessary education will increase the knowledge and positive attitudes, therefore good handling practice (Azaman et al., 2016). In this study, food operators had a moderate level of knowledge and attitude with a lack of good practice in the usage of polystyrene for food packaging.

The Relationships between KAP relating to Polystyrene Food Packaging Usage

Regression analysis was performed to test the relationship between knowledge, attitude and practice. Prior to testing the relationship, correlation analysis between knowledge, attitude and practice was conducted as presented in Table 6. There is a moderate correlation between all variables ($p<0.000$) indicating knowledge, attitude and practice measured in this study are distinguished constructs. As summarized in Table 7, the influence of knowledge on attitude is significant ($F=52.435$, $p=0.000$). The beta value -0.517 shows that when the knowledge is increased by 1%, the respondent's attitude will increase by 51.7%, assuming that other variables are constant. Similarly, the effect of attitude on practice is significant ($F=52.435$, $p=0.000$). As shown in Table 7, beta value (0.364) indicates that, when attitude increases by 1%, respondents' practice will rise by 36.4%, assuming that other variables are constant. Hence, the results indicate there is a significant influence of knowledge on attitude and subsequently attitude on practice.

Table 6. Correlation between KAP variables

	Knowledge	Attitude	Practice
Knowledge	1.000	0.573** (p<0.000)	0.352** (p<0.000)
Attitude		1.000	0.364** (p<0.000)
Practice			1.000

Note. **Correlation is significant at the 0.01 level

Table 7. Relationship between KAP variables

Variable	Predictor	Beta	t	Sig.	F (Sig.)
Attitude	Knowledge	0.517	-7.241	0.000	52.435 (0.000)
Practice	Attitude	0.364	4.038	0.000	16.307 (0.000)

Note. Regression is significant at the 0.001 level

Comparisons of KAP Level Between Types of Food Operation

The types of operation in this study are divided into three categories (i.e., hawker, restaurant and night market). Table 8 shows the differences in the level of KAP among the three types of food operations (see Table 8). Restaurant had the highest mean score of knowledge (m= 6.22) compared to hawker (m=5.973) and night market (m= 5.696) at F-value = 5.957 (p=0.016). However, the difference in the level of attitude among all types of operation was insignificant (p>0.05). Finally, the difference in practice toward polystyrene food packaging usage was significant (F-value= 7.084, p<0.01) with night market had the highest mean score (m=3.50) indicating they have poor practices of polystyrene usage (1=never; 5=frequent) compared to other types of food operation.

Table 8. Comparisons of KAP among food operators

Variable	Type of operation			F-value	*p*-value
	Hawker (n=37)	Restaurant (n=41)	Night Market (n=34)		
Knowledge	5.973	6.220	5.696	5.957	0.016*
Attitude	3.153	3.055	3.276	2.116	0.149
Practice	3.278	3.329	3.500	7.084	0.009*

Note. *F-value is significant at the 0.05 level. Post-hoc analyses were conducted using Duncan test.

Figure 1 shows the percentage of polystyrene usage level among food operators namely hawker, restaurant and night market (see Figure 1). The usage level of polystyrene is divided into three categories (i.e., low, moderate and heavy). Specifically, low level of usage is less than 50 packages per day, moderate level of usage is between 51 and 100 polystyrene packs daily while heavy users used more than 100 polystyrene packs per day. The highest percentage of low user is restaurant (65.8%). Hawker had the lowest percentage of low user (40.7%). For moderate user, the highest percentage is the hawker (51.9%) followed by restaurant (29.3%) and night market (23.5%). Night market had the highest percentage of heavy user of polystyrene food packaging as they represent 23.5% of total respondents. On the other hand, only a few of hawkers and restaurants are heavy user of polystyrene food packaging with the percentage of 7.4% and 4.8%, respectively.

Figure 1. Percentage of usage level according to types of operation
Note. Low user (less than 50 packs per day), Moderate user (51-100 per day), Heavy user (More than 100 per day)

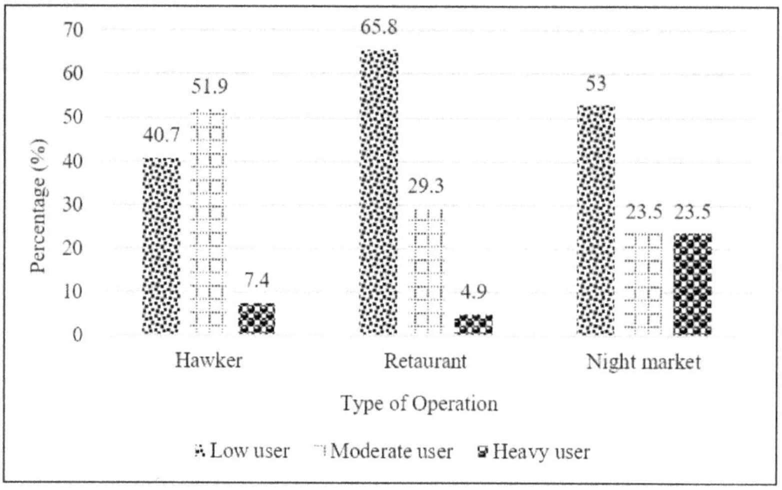

Restaurants are mostly classified as low user is possibly because they provide seating and usually, the customers who came are more likely to dine-in instead of take-away. On the other hand, night market operator is said to be the heavy user as night market only provide take-away services as they operated only by opening table to display their product on-site without any seating for customers. Hawker is a moderate user of polystyrene. Usually, hawkers operate along the street or at a certain site and offer take-away service most of the time. However, they also provide limited seats to dine-in for their customers.

The types of food packed in polystyrene packaging among the different types of operation were further analyzed. Respondents were asked to list the types of food commonly packed in polystyrene and state the frequency level for each (i.e., less than 30, 31-60, 61-100, and more than 100). Four different types of foods were commonly recorded from the survey (i.e., hot food, mixed rice, chicken rice and non-hot food). Hot food mostly represents cooked to order food (e.g., *mamak* fried noodles, fried rice, *kuey teow*, etc.) while non-hot food refers to food with a long holding time before it is sold to customers. Figure 2 shows the frequency of food packed with polystyrene packaging among hawkers (see Figure 2). The type of foods packed in polystyrene more than 100 packs per day are chicken rice and mixed rice, but only a small number of hawkers reported the usage. The largest number of hawkers packed only less than 30 per day is hot food followed by non-hot food packed between 30 and 60 per day.

Figure 2. Frequency of food packed with polystyrene among hawkers surveyed (n = 37)

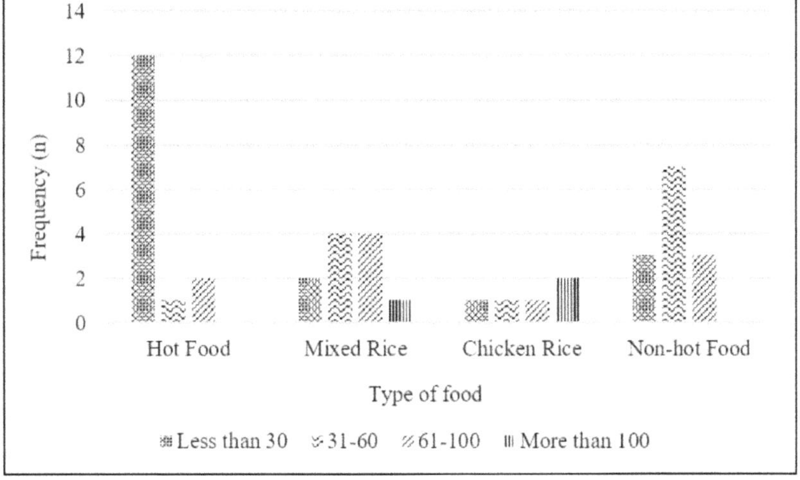

Figure 3 shows the frequency of food packed in polystyrene among the restaurants. Overall, hot food was packed using polystyrene by a large number of respondents from restaurants (n=22) followed by mixed rice (n=14) and non-hot food (n=12). About half of those who packed hot food would use less than 30 packs per day. Based on a total of 41 respondents surveyed, very few restaurants use more than 100 packs of polystyrene per day for various types of food.

Figure 3. Frequency of food packed with polystyrene among restaurants surveyed (n = 41)

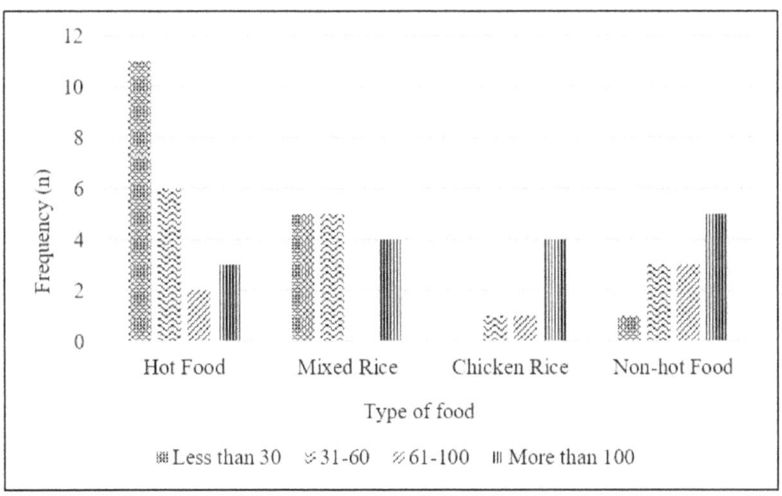

Finally, the frequency of food packed with polystyrene among night markets is shown in Figure 4. Similarly, the largest number of respondents from the night market use polystyrene to pack hot food (n=17), while a very limited number of respondents packed chicken rice (n=2). Both hot food and non-hot food are packed more than 100 per day by a few respondents from the night market. Only a small number of respondents who packed chicken rice with polystyrene possibly because this type of food was sold in limited stalls at the night market.

Figure 4. Frequency of food packed with polystyrene among night market surveyed (n = 34)

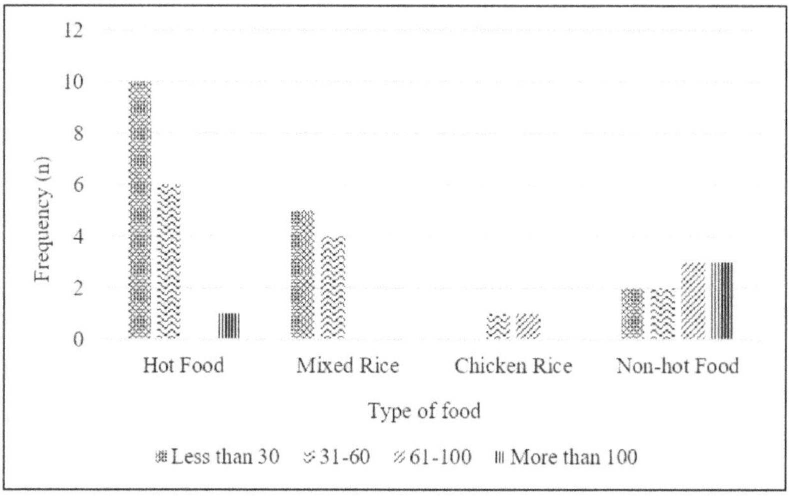

Figure 5 presents the factors that influence polystyrene food packaging usage among different types of food operations (see Figure 5). For hawkers, the highest frequently marked is the price factor. A possible reason for explaining the importance of price is that because polystyrene is one of the cheapest packaging compared to other food packagings. In business operation, often the price is the deciding factor as businesses prioritize profit, hence they need to lower the cost, especially among small food operations. The next highest-rated factor is the packaging feature of the polystyrene itself. A total of 20 respondents among hawkers reported that the features also influence their usage of polystyrene. Polystyrene features that are favorable including lightweight, convenient and able to maintain temperature for some times are the significant reason for it used.

Figure 5. Factors that influence polystyrene food packaging usage

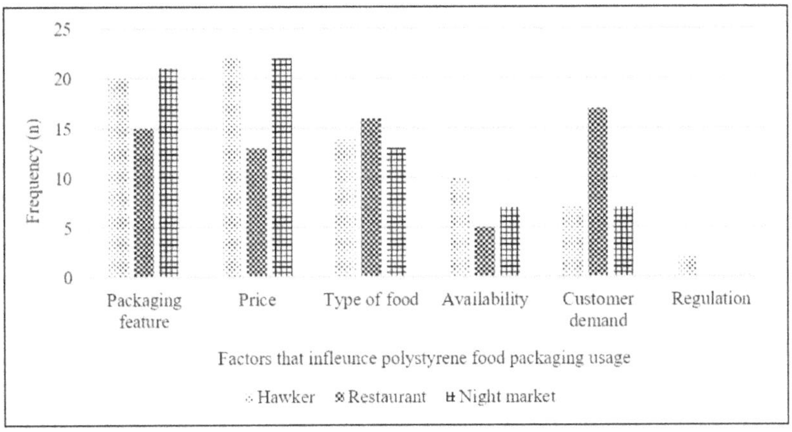

The type of food is the third important reason to use polystyrene among the hawkers. Respondents agree that their decision in choosing polystyrene as their food packaging is because of the type of food that they serve. Respondents from the night market appear to rate the same factors as the hawkers. Price, packaging features as well as the type of food packed are the three most frequently marked factors that influence night market use of polystyrene.

Customer demand is the most important factor that influences polystyrene usage particularly among restaurants, where 17 of them use polystyrene as requested by their customer. It is common that foodservice operators follow customer demand in many aspects of their operation particularly product and service. In return, it gives benefits to operations and brings in sales. The next important factor is the type of food followed by packaging features, which is consistent with other types of food operations.

The least important factor is the regulation as rated by all types of food operators. The regulation does not influence food operation because they had to choose polystyrene probably as the packaging widely available. Respondent also specified other factors described as the food operators would follow the trend and competitors whereby, they just follow what other food operators are doing.

CONCLUSION

In general, the findings of this study show the level of knowledge is at a moderate level among all of the food operators involved in this study. Food operators do not have sufficient general knowledge especially the proper way to handle polystyrene as food packaging and its safety effect. This is indicated by the fact that less than half of food operators provide the correct answer for most knowledge questions tested. Similarly, the attitude of the food operators towards polystyrene is also at a moderate level. Food operators have a moderate attitude particularly on the aspect of polystyrene health effects. From the results obtained, it is also found that level of practice among food operators is at a poor level as many of them would use polystyrene as food packaging materials for their business despite its safety and health effects. The finding is partly consistent with the research conducted by Haqim et al. (2016) that evaluated university students, a different group of population.

Overall, this study shows the level of KAP on the usage of polystyrene food packaging among food operators is still moderate, thus further research is needed to help the industry improve their knowledge and attitude, hence applying proper practice when handling polystyrene food packaging. In this study, it was also found that there is a significant relationship between knowledge and attitude as well as the relationship between attitude and practice. This finding supports the application of the KAP model in assessing the current status of polystyrene food packaging usage among food operators in Malaysia.

This study also compares the level of KAP among different types of food operators (i.e., hawker, restaurant and night market). The restaurant had the highest level of knowledge and the hawker had the best level of practice. The difference in the level of attitude was insignificant across all types of food operators. Besides the level KAP, other factors that influence the use of polystyrene food packaging were also investigated. The findings show price, packaging features and the suitable type of food packed are the common factors that influence the hawker and night market to choose polystyrene. Factors related to customer demand, price and packaging features play important role in choosing polystyrene as packaging among restaurants.

This study demonstrates that KAP on polystyrene food packaging usage among food service operators is still at a moderate level. It is, therefore, necessary to train the industry in order to increase their knowledge and attitude, resulting in the application of suitable practices while managing polystyrene food packaging. Providing more awareness among food operators who handle polystyrene could be advantageous to the business, as they are the ones who have the ability to reduce the risk of styrene leaking into food through chemical migration. If food operators have a better understanding, their attitude will improve as well, resulting in better practices when using polystyrene as food packaging in their business. Besides, this study could provide the policy makers with an understanding on the status of public health and safety risk of chemical migration of polystyrene. Additionally, the findings offer some data and information to aid in the creation of successful interventions or initiatives, such as public awareness campaigns and food handler training.

ACKNOWLEDGMENT

The authors acknowledge the Food Safety and Quality Division (FSQD), Ministry of Health Malaysia for technical support and funding [6300872]. Thank you to all participating food operators in the state of Negeri Sembilan, Perak, Kedah and Kelantan.

REFERENCES

Ahmad, S. (2016). Penggunaan polisterin diharamkan di WP mulai tahun depan. *Berita Harian Online*. http://www.Bharian.Com.My/Node/201159

Ardic, M., & Kahve, H. H. (2015). Chemical migration in food technology. *Academic Journal of Science*, *4*, 163–168.

Azaman, N. N. M., Kamarulzaman, N. H., Mad Nasir Shamsudin, M. N., & Selamat, J. (2006). Stakeholders' knowledge, attitude and practice (KAP) towards aflatoxins contamination in peanut-based products. *Food Control*, *70*, 249–256. doi:10.1016/j.foodcont.2016.05.058

Bavani, M. (2016). *Polystyrene not the real culprit*. http://www.thestar.com.my/metro/community /2016/07/11/polystyrene-not-the-real-culprit-experts-say-ban-of-product-will-not-solve-waste-issues

Boyce, J., Charles, C., & Binkley, B. M. (2008). Consumer perspectives: Take-out packaging and food safety. *British Food Journal*, *110*(8), 819–828. doi:10.1108/00070700810893340

Brown, A., Lamb, C., Brown, M., & Neal, A. (2000). Review of the developmental and reproductive toxicity of styrene. *Regulatory Toxicology and Pharmacology*, *3*(3), 228–247. doi:10.1006/rtph.2000.1406 PMID:11162717

Castle, L. (2006). Chemical migration into food: An overview. In Chemical migration and food contact materials. Woodhead Publishing.

Dillman, D. A. (2007). *Mail and internet surveys: The tailored design method* (2nd ed.). John Willey and Son.

Emma, J., Sherrington, B. A., & Hones, P. A. (2001). The toxicity of styrene monomer. *Adverse Drug Reactions and Toxicological Reviews*, *20*, 9–35. PMID:11395942

Haqim, M. M. (2015). *Knowledge, attitude and practice of styrene relating to usage of polystyrene food packaging among Universiti Putra Malaysia students* [Unpublished doctoral dissertation]. Universiti Putra Malaysia, Serdang, Malaysia.

Jayaraman, K., Haron, H., Sung, G. B., & Lin, S. K. (2011). Consumer reflections on the usage of plastic bags to parcel hot edible items: An empirical study in Malaysia. *Journal of Cleaner Production, 19*(13), 1527–1535. doi:10.1016/j.jclepro.2011.03.019

Kamidi, T. (2015). Styrofoam: The silent killer. *Borneo Post Online*. http://www.theborneopost.com /2015/02/18/styrofoam-the-silent-killer

Khaksar, M. R., & Khansari, M. G. (2009). Determinants of migration monomer styrene from GPPS (General Purpose Polystyrene) and HIPS (High Impact Polystyrene) cups to hot drinks. *Toxicology Mechanisms and Methods, 19*(3), 257–261. doi:10.1080/15376510802510299 PMID:19750020

Marsh, M., & Bugusu, B. (2007). Food packaging: Roles, materials, and environmental issues. *Journal of Food Science, 72*(3), 3. doi:10.1111/j.1750-3841.2007.00301.x PMID:17995809

Neuman, W. L. (2014). *Social research methods: Qualitative and quantitative approaches* (7th ed.). Pearson.

Rennie, D. (1995). Health education models and food hygiene education. *Journal of the Royal Society of Health, 1*(2), 75–79. doi:10.1177/146642409511500203 PMID:7738994

Sheridan, M. (2003). *Packing up profits. Restaurants and institutions*. Retrieved from www.rimag.com/archives/2003/06b/ops1.asp

Teh, W. S. (2016). *Silent killer: Polystyrene containers still widely used despite Ramadan night market ban*. https://malaysiandigest.com/news/618001silent-killer-polystyrene-containers-still-widely-used-despite-ramadan-night-market-ban.html

U.S. National Library of Medicine. (2014). *HSDB: Styrene. Toxicology data network*. https://toxnet.nlm.nih.gov/cgibin/sis/search2/

Chapter 6
Standard Food Safety Practices From Receiving to Cleaning in the Restaurant

Suzita Ramli
Sultan Idris Education University, Malaysia

Jun Xian Wong
Sultan Idris Education University, Malaysia

ABSTRACT

Transmission of hazardous materials could be aggravated by inappropriate handling and storage practices. This results in cross-contamination to foodstuff or cooking utensils. The introduced hazards in the food supply chain might lead to client and reputation loss. The implementation of food safety is necessary to secure safety concerns. All employees should take initiative to be aware and have good attitudes regarding proper hygiene and sanitary practices to assure their product integrity and safety for human consumption. Therefore, this chapter delivered the appropriate and standard food safety protocols to all individuals involved in food storage, preparation, and serving. The scope was structured into (1) identification of hazardous ingredients, (2) purchasing and receiving raw materials, (3) transporting and storage, (4) cooking and reheating, (5) food serving and displaying, (6) leftover storage, and (7) cleaning and sanitation.

DOI: 10.4018/978-1-7998-7415-7.ch006

INTRODUCTION

Food consumption is reflected as a biological need in fulfilling hunger and gaining nutrients. In this modernization and urbanization society, outside catering outlets are mushrooming around the globe due to the limited time required to cook at home and more choices provided in outdoor catering outlets (Souza et al., 2018). However, unhygienic food served to provide the ideal conditions for microbial proliferation and the occurrence of foodborne diseases due to unhygienic food preparation practices (Faridah et al., 2016; Insfran-Rivarola et al., 2020). Foodborne illness imposes a global public health threat due to food quality, economic, and reputation loss of countries. Approximately 2.2 million people died annually due to foodborne or waterborne illnesses (Ncube et al., 2020).

Instead of food safety and public health risk, people are generally more concerned with the service provided, price setting, menu choices (Harris et al., 2018). Unlike home-cooked food, one little mistake did by food handlers can cause a chaotic consequence. Therefore, clientele perception of food safety is crucial to provide consumers with a 'safer' meal. Restaurant owners are responsible to take regulatory initiatives to implement and practice international food safety interventions such as ISO 22000 routinely: 2018, Good Manufacturing practices (GMP), Hazard analysis and critical control point (HACCP), and WHO food safety strategic plan (2013 to 2022) and motivate their subordinates to comply it (Ncube et al., 2020). Table 1 describes categories of restaurant and their respective complement of food safety practices. Bear in mind that high commitment to hygiene practices did not represent high effectiveness in preventing foodborne illness (Mjoka & Selepe 2018). Therefore, impactful food safety training is necessary to be delivered in worldwide restaurants.

This chapter attempts to recommend a complete and standard FSMS to be implemented during food hygiene and sanitation training in offering long-term benefits to the restaurant industries. The scope in this chapter will cover different categories of food contaminants and standard procedures from receiving cleaning and sanitizing in restaurants.

Table 1. Different categories of restaurants establishments and their respective complement level in food safety practices

Restaurant Categories (Provided with examples)	Key Characteristics of Restaurants[a]	General Complement of Food Safety Practices
Full service (Fine dining and casual dining)	• Customers are known as guests. Guests will be provided wide options of food types from the menu • Guests are often served by waiters once they sit down and until they leave • Guests will dine in those particular restaurants and seldom practice takeaway • Takeaway or delivery orders are seldom been practiced	Foods are mostly prepared under hygienic conditions. Most of the food handler staff are educated, well trained, and aware of knowledge regarding food safety and kitchen hygiene (Darko et al., 2015).
Quick service (Fast food settings)	• Fast services provided, long operation hours, low food price, convenient location drive-through service, fast, and efficient customer services • Food formula preparation is always standardized	Most of the employees under are provided with formal hygiene training. Food prepared is considered hygiene and safe to consume (Elobeid et al., 2019).
Quick casual (Pub, café, normal dining restaurants, and street vendors or hawkers)	• A category locates between quick service and full-service restaurant • Dining price will normally be lower than full-service restaurants • Accentuate high food quality but at a reasonable and affordable price • Food only prepared once received customer order • Different restaurants will have their signature or innovative dishes	Food handlers might uneducated, untrained, or unaware of food safety knowledge and hygiene practices. The risk of hazard exposure is very high (Khairuzzaman et al., 2014; Olu-Taiwo et al., 2021).

Superscript a = key characteristics of each category are adapted from Chua et al. (2020)

BACKGROUND

Identification of Hazardous Ingredients in Restaurant Industries

Food Allergens

Food allergens trigger the body's immune system when ingesting and reacting to specific proteins that induce food allergies. Food allergies vary from mild to life-threatening symptoms, including anaphylaxis, lip swelling, and hives (FDA, 2021). Currently, there are no cures for food allergies. Consumers who suffer from food allergies should be more cautious and self-aware in their daily diets (Alvarez & Boye, 2012). In the United Kingdom, allergy individuals prefer dishes containing allergenic ingredients to be listed in restaurant menus (Lieberman et al., 2021).

According to the Food Allergen Labelling and Consumers of Protection Act of 2004 (FALCPA), the eight major food allergens (also known as 'Big 8') consist of egg, milk, peanut, tree nuts, soybean, wheat, fish, and shellfish, as demonstrated in Table 2 (Verhoeckx et al., 2015; FDA, 2018a).

Table 2. The descriptions and common allergens source present in 'Big 8' food

Food Types	Description	Common Food Examples
Egg	The egg is act as a common additive in bakeries, condiments, and sauce	Bread, cake, muffin, pancake, waffles, mayonnaise, custard, and Caesar salad dressing
Milk	Raw milk or ingredients derived from milk are cause the possibility in milk allergic	All types of creams, butter, and cheese
Peanut	Peanut allergy is one of the most common allergies in western countries	Roasted peanuts, boiled peanuts, peanut oils, and peanut butter
Tree nut	A hard shell or fruity bodies in hard shells. The tree nuts are usually consumed in roasted form or applied in confections, pastries, and bakeries	Almond, Brazil nut, cashew, hazelnut, pistachio, and walnut
Soybean	Soybean is very famous in many Asian foods as it will be introduced during cooking	Miso paste, tempeh, tofu, soymilk, and soy sauce
Fish	In several Asian cuisines, seafood, stock, and fish sauces are very crucial as base flavors during the preparation of dishes	Codfish, yellowfin tuna, ocean perch, swordfish, and Atlantic salmon
Shellfish	In several Asian cuisines, seafood, stock, and fish sauces are very crucial as base flavors during the preparation of dishes	Crab, lobster, and shrimp

Source: Alvarez and Boye (2012) and Verhoeckx et al. (2015).

If certain products contain allergenic ingredients, specific labeling is required to protect individuals facing food hypersensitivities (FDA, 2021). Food allergens are categorized into declared allergens and non-declared allergens. Declared allergens are legally required to be labeled to be aware of allergic individuals. Undeclared allergens are unintentionally introduced and mainly found in raw ingredients or cross-contamination during processing or ingredients formulation (Ricci et al., 2017). The undeclared allergen is also known as a 'hidden allergen'. Hidden allergen presents as the technical term or absents in the food ingredient list. For instance, 'textured vegetable protein instead of soy' or 'casein instead of milk' (Gendel, 2012). The generic terms like 'ingredients' or 'spices' are inappropriate and should be avoided when describing the allergenic materials. Specific terms are required for each type of spice in food labeling, such as basil, rosemary, cumin, and parsley, instead of mentioning generic terms without specific within the ingredient list (Pérez, 2013). South Korea had recently identified 30 cases of the prevalence of allergenic substances (almond, egg, milk, nut, and wheat) in candies, chocolates, and snacks, without being labeled (Kim et al., 2021).

Awareness and training to identify food allergens should be introduced and emphasized repeatedly among restaurant food handlers. Among 114 food handlers in Penang, Malaysia, only 1.79% of food staff considered has excellent knowledge in food allergen, whereas 21.37% of respondents have low-risk practices in food allergies (Shafie & Azman, 2015). Food handlers must understand that allergen cross-contamination can occur via food-to-food, hand-to-food contact, interchanging multiple containers, equipment, and utensils without cleaning or washing (Codex Alimentarius, 2020).

Foodborne Pathogens

Foodborne pathogens, especially bacteria, are frequently involved by ingesting contaminated food or microbial toxins to conduct foodborne illness, also known as food poisoning (Ozogul & Hamed, 2017). The foodborne outbreaks are mostly associated with myriad factors such as restaurant policies or food handling practices such as bare hand contact with food or food prepared by sick employees (Hedberg, 2013; Mun, 2020). The restaurant has been reported as one of the potential sites for foodborne infection. The higher likelihood of acquired foodborne diseases is reflected by increased consumption of contaminated food (Angulo & Jones, 2006; Ozogul & Hamed, 2017). There are some common foodborne pathogens detected in restaurant industry as summarized in Table 3 below.

Table 3. Common pathogens found in restaurant industries and their associated food sources and clinical symptoms

Foodborne Pathogens	Associated Food Types	Symptoms
Bacillus cereus	Powdered milk, dairy products, chicken, and meat	Abdominal cramp, diarrheal, nausea, and vomiting
Campylobacter jejuni	Raw milk and undercooked shellfish, poultry, or meat	Abdominal pain, diarrheal, fever, nausea, and vomiting
Clostridium perfringens	Poultry, meat, species, stew, and salmon	Abdominal cramp, diarrheal, and nausea
Shiga-toxin producing *E. coli* (STEC)	Undercooked ground meat or beef, unpasteurized dairy products, dirty water sources, and vegetables (radish, sprouts, and lettuce).	Bacteremia, diarrheal, fever, headache, nausea, vomiting, and typhoid fever
Listeria monocytogenes	Poultry, meat, dairy product, and vegetables	Abdominal pain, fever, headache, and diarrheal
Salmonella spp.	Poultry, meat, and raw milk	Abdominal cramp, diarrheal, nausea, and vomiting

Source: Ozogul & Hamed (2017).

Despite worldwide are coping with the COVID-19 pandemic crisis, the issues of foodborne illness should not be ignored as well. Serious foodborne diseases might cause fatal to individuals. Foodborne pathogens can be introduced at any point from farm to fork supply chain, restaurant or any dining service sites are the most common site for illness occurrence among chain (Al-Ghazali et al., 2020). In most of the developed countries, restaurant industries contributed the highest among food service and preparation facilities, which accounted for 40% of reported outbreaks (Xue & Zhang, 2013; Angelo et al., 2017; Cui et al., 2021).

According to a surveillance report from Centers of Disease Prevention and Control (CDC), 61% of foodborne cases are associated with restaurants. The majority of cases originate from sit-down restaurants (48%), followed by catering and banquet functions (14%), and hotels (8%). The highest outbreak contribution of food sources was meat and poultry (44%), followed by seafood (27%) and vegetables (26%). This finding showed food preparation and serving facilities pose a huge concern in public health (Dewey-Mattia et al., 2018). Another finding supported seafood (26%), meat and poultry (36%), and fresh produce (14%) are common foodborne transmission vehicles by *S. enterica*, *C. botulinum*, *E. coli* O157:H7, *C. jejuni*, *Vibrio*, *Shigella*, and *B. cereus* (Angelo et al., 2017). Similarities of both cases are due to unhygienic practices and lack of food safety knowledge (including slow cooling, insufficient or inadequate temperature or time during reheating, cold or hot holding, inadequate and dirty utensils, and equipment) thus caused the proliferation of foodborne bacteria.

Purchasing and Receiving

Purchasing is mainly done by the procurement department in restaurant authorities or hotel executives. Therefore, selecting the right and reliable supplier is crucial to restaurant management for the sake of food safety. All raw materials need to be procured from reliable sources and reputable suppliers (CFS, 2017). Suppose there are some inquiries regarding sources and food safety practices. In that case, there are necessary to conduct a site inspection or supplier audit on the particular companies or industry plants from time to time. The purpose of site inspection is to ensure all food handling practices comply with the food safety requirements (CFS, 2017; Trafialek, 2019).

Before delivery, the person in charge should instruct the suppliers to ensure all the goods are kept at a specified temperature. Upon receiving, personal inspection shall conduct to the hygiene status of delivery trucks, label (best before and expiry date), packaging, and freshness of food by restaurant authorities or person in charge. In cold delivery of raw supply, the chilling or freezing trucks should maintain below 4°C and -18°C, respectively. Frozen foods should entirely be freeze in the truck. All cold-stored food must be chilled and froze immediately within 10 mins upon receiving and inspecting to be free of evidence of previous temperature abuse during transporting and loading (FDA, 2010; CFS, 2017). The packaging materials types and conditions must be inspected as well. For example, all the fresh leafy greens such as mixed salad, cabbage, endives, iceberg lettuce, and romaine lettuce should be delivered by cold storage trucks and packed in sealed plastic bags then in carton boxes to minimize physical damage of vegetables (Coleman et al., 2013). After delivering all the ordered items, the delivery invoices and receipts should provide accurate and detailed information of supplier and products (company names, address, use by or expiry date) for the record, traceback, or identification purposes (FDA, 2010; FSAI, 2014; FSANZ, 2019a).

The delivery truck must operate in clean, tidy, and without emitting putrid odors. All cold-stored food must be delivered with chiller or freezer trucks. The interior temperature of chillier or freezing trucks must regularly check once delivered goods. The frozen food should be solid without thawing or refreezing signs by observing the large ice crystals or liquids appear below the container bottom (USDA, 2012). During the inspection, the hygienist or restaurant staff needs to probe a clean and sanitized thermometer to ensure the core temperature of the fresh meat, poultry, and fish are kept below 5°C and below 7°C for milk (SFA, 2020). There are some criteria of goods that should be rejected for consumer safety reasons during receiving and purchasing. Conditions that lead to item rejection or not being purchased are tabulated in Table 4.

Table 4. List of food categories that should be rejected or not purchased

Food Categories	Conditions Lead to Reject or Not Purchase
Canned food	Dented, rusty, and swollen cans
Package food	Unsealed, opened, or pored packets
Table eggs	Cracked and dirty eggs shell surfaces
Fruits and vegetables	Shriveled, moldy, bruised, wilted leaves, and blemished appearance
Meat and poultry	• Stale appearance and release of off odors • No excessive moisture amount on meat and poultry surfaces • Make sure the meat is in a hard and frozen state • If the meat is packed in form, no torn, and dripping of meat juices
Seafood	Fish • Red, cloudy fish eyes and slimy flesh • Strong ammonia fishy-smell • The gills and liver are not in red color • The fish edges are found with brown or yellowish discoloration Shellfish • Slippery surface, black spots, mild odor, and shell discoloration are detected on prawn shells • Oysters, mussels, and clamshells should close tightly when touched. The gapping shells indicated shellfish are dead • The tails of lobsters should be curled during handpicked • Crabs should be able to exhibit the leg movements

Sources: FDA (2010) and SFA (2020).

Storage of Fresh Raw Ingredients

Chilling and Freezing

Domestic fridge usage seems easy, but consumers often failed to complement the correct food storage and maintenance practices. Refrigerators served in restaurants could render microbial contamination during stock arranging, unwashed, unpackaged raw uncooked food, and packaging leakage (Madhwal & Sharma, 2017). Temperature display on the refrigerator is critical to determine whether the current temperature is appropriate or not to keep materials. Additionally, the displayed temperature should always be tallied with the actual temperature inside the chiller or freezer (CDE, 2000). Ensure all perishable food products to chill or freeze within I to 2 h at 4°C or below. The foodborne pathogens can multiply quickly in the food danger zone (5 to 60°C) at room temperature for more than 2 h. However, pathogenic microbes required 1 h to multiply if the surrounding temperature is higher than 32°C (Odumeru, 2012; CDC, 2021). Table 5 tabulates all suggested temperature and storage periods to store all perishable foods during chilling and freezing.

Table 5. Suggested temperature and storage periods during chilling and freezing of perishable foods

Food Categories	Chilling (below 4°C)	Freezing (below -18°C)
Mix salads (contain hams, tunas, chickens, and eggs)	3-4 days	Freezing is not allowed
Hot dogs	1 week (opened package) or 2 weeks (unopened package)	1-2 months
Luncheon meat	3-5 days (opened package or in deli sliced form) or 2 weeks (unopened package)	1-2 months
Sausage	1-2 days (raw from chicken, beef, pork, and turkey) 1 week (fully cooked from chicken, beef, pork, and turkey)	1-2 months 1-2 months
Hamburger patties	1-2 days (chicken, ground beef, turkey, lamb, veal, and pork patties)	3-4 months
Steak, chop, roasts	3-5 days	4-12 months
Ham	5-7 days or follow use-by date (cured) 3-5 days (uncured) 2 weeks (whole ham or sliced vacuum-packed with cooked hams)	3-4 months 3-6 months 1-2 months
Poultry products	1-2 months (whole or sliced chicken and turkey)	1 year (whole) 9 months (sliced)

continues on following page

Table 5. Continued

Food Categories	Chilling (below 4°C)	Freezing (below -18°C)
Eggs	3-5 weeks (raw shell eggs) 2-4 days (manually cracked raw eggs) 1 week (hard-boiled eggs)	Freezing is not allowed. If accidentally froze the raw eggs, used them immediately after thawing 12 months (beat the albumen and yolks together before keeping in the freezer) Freezing is not allowed
Fish	1-2 days (freshwater fish) 1-2 days (salmon steak) 1-2 days (fish fillets: cod, mullet, sea trout, ocean perch, sea bass) 3-4 days (cooked fish)	4-6 months 2 months 4-6 months 1-3 months
Shellfish	1-2 days (crabs) 1-2 days (lobsters) 1-2 days (shrimps) 1-2 days (clams, oysters, and scallops)	4 months 6 months 3-6 months 3-4 months

Sources: McCurdy et al. (2009) and SDA et al. (2021).

Still, some perishable ingredients are not suitable neither chilling nor freezing. Tomato cannot be chilled store since it promotes the formation of grainy texture. Besides, banana is not ideal for chilling unless they are fully ripened but need to consume within 2 days. Banana skin will blacken due to chilling injury, or the ripening process of unripe banana would be terminated. Some of the tuber vegetables such as onions, garlic, potatoes, and sweet potatoes are required to keep under dark and cool room temperature conditions (Dales, 2018).

Refrigerator doors must close and be tightly sealed anytime to ensure the coolness of the internal environment and prevent external heat gain into the refrigerator (CDE, 2020; CDC, 2020a). Overcrowd is not allowed during refrigerator shelving to ensure free circulation of cool air around the ingredients. Shelving items are required to follow the first in first out (FIFO) policy by rotating and utilizing the older date labeling first (Dales, 2018; MOHM, 2021). Spillage from liquid and meat drip are frequently found in the refrigerator, however, the spills need to be wiped or cleaned immediately to reduce the chances of *L. monocytogenes* surviving or growth under low temperature (USDA, 2017).

Some emergencies like weather or power outages are unpredictable and may happen any time of year to disrupt the normal usage of the chiller or freezer. We need to decide which ingredients to keep or discard after a long circuit cut since the freshness will be altered without sufficient power supplies to keep the food cold. A freezer container, block ice, dry ice, or cold gel packs should always be prepared and available to transfer as another backup temporary storage space when the chiller or freezer is in down functions (CDC, 2020a). The doors must be closed all-time for freezer and chiller compartments. A short power breakdown will not affect the freshness of the frozen foods. However, the perishable food must be discarded after being exposed to an unrefrigerated condition for more than 4 h continuously, altered color, changed texture, and internal temperature of food is above 4°C (Harris, 2002; USDA, 2021). After the power restore, the food temperature must be measured again and discarded any ingredients found above 4°C. It is strictly not allowed to taste the food to determine its safety or freshness during or after circuit break down. This is because the unrefrigerated temperature will provide an ideal condition for microorganisms to divide in the food temperature danger zone. Table 6 lists down all actions required to be taken against high-risk items after a short-term power outage in the freezer and chiller.

Table 6. Action required to respond during a short-term power outage in high-risk foods

Food Categories	Actions
Partially cooked poultry and meat	Subjected the partially cooked poultry or meat into high temperature reheating to achieve the desired internal temperature (at least 74°C), or Discard the ingredients if in doubt of cooking and consuming or kept for more than 2 h in food temperature danger zone.
Fully cooked food	Chill back the food after power is restored, or Discard the ingredients if kept for more than 2 h in food temperature danger zone.
Partially cooked vegetables	The ingredients need to be eaten as soon as possible, or Heated the ingredients until they become fully cooked using the grill or stove, or Chilled back into the chiller immediately once the power is restored.
Fresh-cut fruits and vegetables	Discard the ingredients if kept for more than 2 h in the temperature danger zone.
Eggs (with or without shells), milk (all types), cheeses (all types), fishes (all types), and shellfish (all types)	Discard the ingredients if kept for more than 2 h in the temperature danger zone.

Sources: Harris (2002) and USDA (2021).

Preparation and Cooking

Personal Hygiene Practices, Knowledge, and Attitudes in Kitchens

GMP and HACCP provide a series of guidelines and knowledge for all food handlers to comply in their working area. However, lacking knowledge and training causing majorities of people did not comply with the listed protocols given in GMP and HACCP. For example, most of the kitchen staff (70.4%) did not have idea about proper food handling method, training course, and ppm levels of chlorine to be applied at fruits and vegetables as stated in GMP and HACCP in restaurants (Al-Ghazali et al., 2020).

A high level of personal hygiene in grooming and behavior is revealed as an important consideration to keep the food safe. Poor awareness of personal hygiene has been identified as one of the main risks in food contamination or foodborne illness (Tan et al., 2013). The grooming standard must be suitable, clean, and protective to prevent unnecessary cross-contamination to raw ingredients or RTE food. All the chef must tie their hair and be covered with hair net or hat. All personal accessories such as watches, jewelry, and hairband are not allowed during food handling and preparation. Individuals need to avoid touching face and hair while doing food preparation and cooking (FSA, 2018). Furthermore, drinking, smoking, or injured people with cut wounds are prohibited from the food preparation areas.

Handwashing Method

Food contamination can occur anytime during food preparation or serving via hand contact, especially with physical and microbiological hazards. It poses public health risk if consumers are taking contaminated food unconsciously. Proper handwashing methods can efficiently reduce the transmission of pathogens from bare hands to food (Green et al., 2006). In this regard, a handwashing station could be so crucial in domestic kitchens. A handwashing station is a dedicated area that provided soaps and water for handwashing. The location of the handwashing station should be convenient for food handlers, and it can either be established near to washroom or food preparation areas (Hulland et al., 2013). Handwashing areas must restrict only to handwashing without washing food or utensils. A complete handwashing station should consist of liquid soaps dispensers, consistent of water supply (preferably warm water), hands-free water taps (foot-operated pedal or lever arm type), hand scrub with alcohol, hand drying facilities (clean paper towels or electric dryer), and enclosed trash cans (FSAI, 2017; APEC & FSCF PTIN, 2018; Knight et al., 2020). Handwashing should be done before, during, and after food preparation, handling food equipment, eating, or coming back from the washroom. During handwashing, both hands need to undergo scrubbing for 20 s, accompanied by warm running water and soaps must be used. CDC has been suggested 20 s of handwashing can be done by humming the "Happy birthday" song twice from the beginning till the end of washing. Finally, a hand sanitizer must be prepared at handwashing areas. Similarly, 20 s is required by rubbing sanitary gel over the surfaces of hands, palms, and fingers (Coorey et al., 2018; CDC, 2020b; CDC, 2021).

Handwashing is the most easily forgotten or wrongly practiced step during food handling. Inadequate handwashing is the primary contributor to foodborne outbreaks associated with restaurants. Based on a study conducted in Jiangsu Providence, China, less frequent handwashing practice among 453 kitchen chefs is identified, either after leaving their work temporarily or before leaving the kitchen. Respondents are found seldom using kitchen towels or handwashing areas after touching used cutleries (Cui et al., 2021).

Wearing Gloves

Even though handwashing is vital before food preparation, but it is impossible to eliminate all microbes from both hands. Therefore, the FDA Food Code has specified the necessity for food handlers to wear disposable gloves during the handling of non-RTE food (Green et al., 2006). Disposable hand gloves provided an efficient barrier between bare hands and food to prevent cross-contamination. However, a standard handwashing procedure is required before putting on disposable gloves. Hand gloves are necessary to change frequently, especially during non-food-related cleaning, such as cleaning trash or emptying dustbins (FSAI, 2017). Besides, different gloves should be used during handling raw, cooked, or RTE food (SFA, 2020).

Table 7. The types of color codes in cutting boards and suitable materials to be used on it

Color Codes	Ingredient Types
Red	Raw meat
Brown	Cooked meat
Yellow	Poultry
Blue	Seafood
White	Dairy and bakery product
Green	Fruit and vegetable

Source: Mallett (2017).

Cutting

Before start cutting, it is advisable to clean cutting boards and knife with hot soapy water after cutting one food before switching to another ingredient (Odumeru, 2012; CDC, 2021). An uncleaned chopping board will serve as a cross-contamination vehicle to other ingredients, especially handle the ready-to-eat products (RTE) on raw ingredients chopping board (FSAI, 2014). In commercial kitchens, the cutting board should use plastic-based or non-porous materials type. The benefits of using a plastic chopping board are easily clean and sanitize, non-absorbent, non-easily contaminated and waterproof. According to HACCP International guidelines, cutting boards must be assigned into different colors to handle specific ingredients to prevent cross-contamination (Mallett, 2017). Table 7 has presented a series of color codes in cutting boards and respective food ingredients that are suitable to be used on it.

Cooking

Chefs should never judge food texture and color to determine whether food is raw or cooked. It is unreliable and inaccurate without measuring the food's internal temperature. Food is considered safe to consume when the desired internal temperature is obtained, as demonstrated in Table 8. Pathogenic microorganisms only can be eliminated when sufficient core temperature of food is achieved (USDA, 2017; CDC, 2021). A thermometer probe is recommended to insert into the food central part to ensure they are fully cooked. The probe should calibrate time by time to ensure the instrument functions optimally (Brown et al., 2012).

Table 8. The internal temperature required to achieve fully cooked in selective food ingredient lists.

Food	Temperature (°C)
Whole or cut poultry (ground chicken or turkey)	82
Ground beef, veal, lamb, and pork	71
Ground chicken and turkey Pre-cooked ham Casseroles (mixture of chopped meat and vegetables)	73-74
Whole beef, pork, lamb, veal cuts, raw fresh ham, roasts, chops, and steaks	63
Fin fishes	63°C or cooked until the flesh turns opaque, and easily separated when using a fork
Crustaceans (crabs, shrimps, and lobsters) and scallops	Cooked until the flesh turns pearly white or opaque
Clams, mussels, and oysters	Cooked until the shells are opening
Shelled eggs	Cooked until the albumen and yolk turn firm

Sources: Odumeru (2012) and CDC (2021).

Food Serving and Displaying

This section describes standard safety guidelines that should comply during food serving or buffet displaying either in restaurants or hotel settings. A two-hour rule is very crucial during buffet serving. Any dishes left beyond two hours on buffet counters should be discarded. The buffet holding period needs to be reduced to one hour if the temperature of the holding warmer at above 32°C (FDA, 2018b). Ensuring all the cooked or RTE food will not be displayed too early before the buffet session is start. It is advisable to seal or wrapped the food container before serving or on the way to be delivered to the buffet counter. Different sets of utensils (tong, fork, or spoon) will be required for different displayed food to minimize the probability of cross-contamination (CFS, 2017). Buffet can be categorized into hot and cold items. 'Keeping hot food at hot condition' and 'keeping cold food at cold condition' is important to prevent food from being kept in the food danger zone (4 to 60°C) (FDA, 1999; CDC, 2020). In the serving of high-risk food such as salad, seafood, and sushi, a minimum quantity of high-risk food needs to be displayed on buffet counters. Recook or top-up the finished food items can prevent temperature abuse and minimize the high-risk food standing too long at displayed temperature (FSAI, 2014).

Displaying of Hot and Cold Buffet Dishes

The temperature of the hot buffet counter should maintain at least 60°C or above during displaying. Eggs or egg-based dishes can be chilled, stored, and reheated to 74°C before serving (FDA, 1999; CDC, 2020c). In non-buffet restaurants, the chilled dishes should be refrigerated before serving. During buffet cold holding, ices are required to retain the chill condition of high-risk dishes, including sashimi, raw oysters, RTE salad, and display not more than two hours. These foods are principal sources of foodborne pathogens such as *Staphylococcus aureus*, *Vibrio,* and *Salmonella* (FDA, 1999). During salad-bar display, all fresh-cut vegetables need to be held at less than 5°C. The chilled or frozen food should transport with pre-cooled insulated vehicles or containers (FDA, 2010; FSAI, 2017).

Leftover Storage

Storing Opened Ingredients

Generally, the unopened ingredients and products can be stored according to the time frame of best before or expiry date (primary shelf life) specified by manufacturers. However, the primary shelf life is no longer accurate as the package has been opened.

All leftover food should be marked with the desired date to be disposed of. Secondary shelf life is defined as the period once the container or packaging is opened, thus might inducing a sharp acceleration in the product deterioration (Brown et al., 2012; Nicoli & Calligaris, 2013; Dales, 2018). Table 9 list-out the secondary shelf life of food ingredients and the time frame given to keep the opened products.

Uncooked raw materials like the whole raw meat, poultry, or fish should be stored in the freezer. It is recommended to cut large portions into smaller parts to facilitate the frozen process within a short time. All the cut portions need to wrap or cover with plastic film or freezer bags (MOHM, 2021).

Table 9. The examples of secondary shelf life of food ingredients once it opened

Categories	Types	Opened Storage Periods by Refrigeration
Condiments and sauces	Ketchup and chili sauces	4 months
	Pickles and salad dressing (commercial)	1-3 months
	Mayonnaise (commercial)	2 months
	Spaghetti sauce	4 days
	Black pickled olives	2 weeks
	Oils (olive and vegetable)	3-5 months
Dried materials	Baking powder	3-6 months
	Flour (white or wheat)	6-8 months
	Icing sugar	2-3 weeks
	Sugar (brown, white, and confectioners)	18-24 months
	Grounded spices	24-36 months
Canned foods	Fish (salmon, tuna, and sardine), meat (beef, chicken, turkey, and pork), soups, and vegetables	3-4 days
	Fruits (flesh and juices)	5-7 days
Baking materials	Cakes and muffins (commercial)	7-10 days
	Bread (pans, flats, rolls, and buns)	2-3 weeks
	Cookies (soft and crispy)	8-12 months (frozen)
	Tortilla (corn and flour)	6-8 months
Dairy products	Butter	1-2 months
	Margarine	6 months
	Cheese	6 months (hard and block) 1 month (shredded)
	Yogurt	1-2 weeks
	Milk (plain or flavored)	1 week

continues on following page

Table 9. Continued

Categories	Types	Opened Storage Periods by Refrigeration
Prepared or cooked foods	Pasta	3-5 days
	Rice	4-6 days
	Fresh cut fruits	4 days
	Fresh cut vegetables or mix-salad	3-5 days
	Leftovers (with meat, poultry, fish, egg, cooked vegetables), meats served with broth or gravy, salads (mix with meat, poultry, fish, and egg), soup, and stew	3-4 days
Meat, fish, and shellfish (fresh)	Roasts, steaks, and chops (beef, pork, and lamb)	3-5 days
	Ground meats (beef, lamb, poultry, and pork), fatty fish (catfish, salmon, tuna, and mackerel), lean fish (cod, rockfish, sea trout, and others), whole cut poultry, raw scallops, and shrimps	1-2 days
	Raw crab (legs) and raw lobsters (tails)	3-5 days
Meat, fish, and shellfish (cooked, processed, or smoked)	Bacon, hams, hotdogs	1 week
	Sausage	1-2 days (raw), 1 week (smoked and cooked)

Sources: Andress & Harrison (2011) and Dales (2018).

Cooling of Hot Food

Food cooling practices are normally well-practiced by most of the food premises before undergoing chilling on pre-cooked food, yet appropriate cooling procedures are poorly implemented by individuals (Brown et al., 2012). According to the Ministry of Health, Malaysia, hot prepared food cannot directly keep in the refrigerator as heat vapor generated from the dishes will subsequently increase the internal temperature and humidity of the fridge (MOHM, 2021). Slow cooling leads to the risk of food safety issues in retail foodservice operations. From 1998 to 2008, the CDC reported 504 outbreaks associated with catering establishments caused by hot food undergoing a prolonged cooling period (CDC, 2020c). For instance, *Clostridium perfringens* and *Bacillus cereus* are major pathogens to germinate spores in improper cooling food (Schaffner et al., 2015). According to the FDA Food Code, the whole cooling process needs to cool down below 5°C within 2 h, provided with a total two-stage cooling method to be applied to monitor rapid cooling of hot prepared food. Internal food temperature must achieve from 57.2-21.2°C within 2 h, and followed by 21.1-5°C within 4 h or less. Periods of hot food cooling should minimize as the temperature danger zone ranges from 5-57.2°C. Foodborne pathogens can increase their counts within the temperature danger zone. Therefore, monitoring time and temperature should be conducted on a routine basis (Roberts et al., 2013; FSANZ, 2019).

Food Code and United States Department of Agriculture (USDA) recommend that multiples cooling methods can be conducted simultaneously to ensure food temperature can be lowered below 5°C within 6 h are as follow: (1) placing the large pot of cooked sauce, stew, or soup inside the shallow pans, inserting the pans in the ice water bath, then refrigerating at maximum holding temperature of 5°C, (2) separating whole food (e.g. large cut of poultry or meat) into thinner or smaller portion and wrapped separately before refrigerating at maximum holding temperature of 5°C, (3) stirring the food in a shallow pan or container placed in the water ice bath, and (4) using the rapid cooling equipment such as ice wands and blast chillers. Also, Food Code, and USDA suggest the food pans or containers can leave uncovered during cooling for well ventilated during cooling. It can leave uncovered or loosely covered if it is necessary to avoid food from overhead contamination. Other than the commercialized refrigerator, the cooling standards are appropriate with a walk-in chiller or walk-in freezer as well. Last but not least, do remind that the cooling food should not leave at temperature danger zone for more than 2 h. The person in charge of food service establishments (manager, sous chef, or bar captain) must ensure all food is being appropriately cooled via routine monitoring of food internal temperature (Brown et al., 2012; Haraminac, 2017).

Reheating of Chilled Food

Many people did not know the correct method of food reheating. A previous study in Saudi Arabia observed the restaurant supervisors had demonstrated poor knowledge in desired hot food temperature (14/97; 14.4%), temperature danger zone (15/97; 15.5%), and correct food cooling method after cooking (45/97; 46.4%) (Al-Mohaithef et al., 2021). Another research discovered some respondents (270/1178; 22.9%) were not aware that cooked food is required to be reheated again before being consumed (Low et al., 2016). Cooled food needs to be reheated until steam vapor is released and achieve at least 70°C and hold at least 2 min at this temperature. In the stove usage, the pans or woks need to be located at the stovetop and heated thoroughly until the internal temperature of food is reached 74°C. Stirring, rotating, and covering is necessary during microwave reheating until the food temperature is reached at least 74°C. Slow cooker and steam tables are not recommended during reheating of foods since they are letting the foods for standing too long in the food danger zone (USDA, 2019).

Cleaning and Sanitation

RTE products like fruits and vegetables are mostly lacking decontamination and preservation treatments before undergoing packaging, transporting, or distributing (Wong et al., 2021a; Wong et al., 2021b). Therefore, cleaning is crucial in pre-cooking, however, cleaning with untreated water or without sanitized are strictly unencouraged as non-treated water can be transmission vehicles of waterborne pathogens. Studies showed that cleaning with one-cycle used water (such as groundwater, well water, tap water, river water, and wastewater) cannot reduce microbial loads effectively (USDA, 1998; Balali et al., 2020). Water may serve as decontamination but also cause recontamination. A previous study in Malaysia, the respondents provided the wrong answer (1123/1178; 95.3%) based on the question 'Washing fruits and vegetables thoroughly with tap water is necessary to prevent foodborne diseases' (Low et al., 2016). Therefore, filtered water or sanitized water with chlorine and ozone treatments are suggested to prevent recontamination and decontamination (López-Gálvez et al., 2021). Das et al. (2019) compared washing vegetables by tap water and filter water reduced 0.85 to 2.05 and 2.38 to 3.36 log CFU/g of microbial loads, respectively. In general, 4 to 7 log CFU/g of microbiota can be retained without thoroughly cleaning. In the same study, washing with chlorine water (200 ppm) achieved an additional microbial reduction (2 log CFU/g).

Three essential steps to conduct sequencing cleaning and sanitation; (1) washing, (2) sanitizing, and (3) drying. All the cooking utensils are required to soak in warm soapy water. The utensils should be visibly clean and dry after washing. Deep washing and cleaning could help to reduce cross-contamination of food allergens while preparing food for allergic people (Ricci et al., 2017; FSANZ, 2019b). Some restaurants have three compartment sinks, also known as a three-step washing system or pot washing station (PWT) dedicated to cleaning, washing, and sanitizing the dirtied pots, cooking equipment, and utensils. The first compartment is mainly for detergent cleaning or above 43°C, following by rinse cleaning with water (second compartment) and sanitizing using sanitizing solution or hot water immersing at above 77°C for 30 s (third compartment). The water supply for each compartment must be visibly clean and warm enough to achieve the desired temperature as specified earlier (FDA, 2013; USDA, 2017).

CONCLUSION

To summarize, the food safety training program is significant to ensure all massages given are successfully delivered to all food handlers, servers, and management executives. The purpose is to educate all people regarding the food safety knowledge and guideline provided. Based on this chapter, all standard procedures are important from receiving to cleaning, to prepare a portion of safe food for the consumers. It is undeniable that it will be tedious for all food handlers to conduct all the so-called implemented standard protocols. However, there are no ways for us to minimize the public health risk if we are compromised to follow the food handling practices. As the saying goes, no man is on the island. Each individual has responsibility in the practicing and participating of proper food handling manner in the restaurant industries. In should contribute their effort in practicing food hygiene practices and build awareness in their mindset. In short, delivering is heavily dependent on people's attitude and how they practice standard appropriate hygiene culture in their working environment.

ACKNOWLEDGMENT

This work was supported by the Fundamental Research Grants Scheme (FRGS-RACER) (2019-0169-103-62(RACER/1/2019/STG03/UPSI//1) provided by the Ministry of Higher Education of Malaysia. The authors would like to extend their gratitude to Sultan Idris Education University (UPSI) for managing the grant.

REFERENCES

Al-Ghazali, M., Al-Bulushi, I., Al-Subhi, L., Rahman, M. S., & Al-Rawahi, A. (2020). Food safety knowledge and hygienic practices among different groups of restaurants in Muscat, Oman. *International Journal of Food Sciences*, *2020*, 1–8. Advance online publication. doi:10.1155/2020/8872981 PMID:33415137

Al-Mohaithef, M., Abidi, S. T., Javed, N. B., Alruwaili, M., & Abdelwahed, A. Y. (2021). Knowledge of safe food temperature among restaurant supervisors in Dammam, Saudi Arabia. *Journal of Food Quality*, *2021*, 1–8. Advance online publication. doi:10.1155/2021/2231371

Alimentarius, C. (2020). *Code of practice on food allergen management for food business operators*. Codex Alimentarius International Food Standards. https://www.fao.org/fao-who-codexalimentarius/shproxy/en/?lnk=1&url=https%253A%252F%252Fworkspace.fao.org%252Fsites%252Fcodex%252FStandards%252FCXC%2B80-2020%252FCXC_080e.pdf

Alvarez, P. A., & Boye, J. I. (2012). Food production and processing considerations of allergenic food ingredients: A review. *The Journal of Allergy*. PMID:22187573

Andress, E. L., & Harisson, J. A. (2011). *Food storage for safety and quality*. University of Georgia. https://nchfp.uga.edu/how/store/UGA_foodstorage_2011.pdf

Angelo, K. M., Nisler, A. L., Hall, A. J., Brown, L. G., & Gould, L. H. (2017). Epidemiology of restaurant-associated foodborne disease outbreaks, United States, 1998-2013. *Epidemiology and Infection*, *145*(3), 523–534. doi:10.1017/S0950268816002314 PMID:27751201

Angulo, F. J., Jones, T. F., & Angulo, F. J. (2006). Eating in restaurants: A risk factor for foodborne disease? *Clinical Infectious Diseases*, *43*(10), 1324–1328. doi:10.1086/508540 PMID:17051501

APE Cooperation (APEC) & Food Safety Cooperation Forum's Partnership Training Institute Network (FSCF PTIN). (2018). *Improving Food Safety through Hand Washing and Drying Capacity Building*. http://fscf-ptin.apec.org/docs/training/Handwash_FN-June-2018.pdf

Balali, G. I., Yar, D. D., Afua Dela, V. G., & Adjei-Kusi, P. (2020). Microbial contamination, an increasing threat to the consumption of fresh fruits and vegetables in today's world. *International Journal of Microbiology*, *2020*, 1–13. Advance online publication. doi:10.1155/2020/3029295 PMID:32565813

Brown, L. G., Ripley, D., Blade, H., Reimann, D., Everstine, K., Nicholas, D., Egan, J., Koktavy, N., & Quilliam, D. N. (2012). Restaurant food cooling practices. *Journal of Food Protection*, *75*(12), 2172–2178. doi:10.4315/0362-028X.JFP-12-256 PMID:23212014

California Department of Education (CDE). (2000). *Proper storage temperatures for USDA commodities*. https://www.cde.ca.gov/ls/nu/fd/mb00404.asp

Center for Food Safety (CFS). (2017). *HACCP system: Seven principles of HACCP system*. https://www.cfs.gov.hk/english/programme/programme_haccp/programme_haccp_7requirement.html

Centers for Disease Control and Prevention (CDC). (2018). *Surveillance for Foodborne Disease Outbreaks, United States, 2016, Annual Report*. U.S. Department of Health and Human Services, CDC.

Centers for Disease Control and Prevention (CDC). (2020a). *Food safety for power outages*. CDC.

Centers for Disease Control and Prevention (CDC). (2020b). *Four steps to food safety: Clean, separate, cook, chill*. https://www.cdc.gov/foodsafety/keep-food-safe.html

Centers for Disease Control and Prevention (CDC). (2020c). *Food cooling practice improvements: Key takeaways from 3 food safety reports.* https://www.cdc.gov/nceh/ehs/ehsnet/plain_language/food-cooling-improvements.html

Centers for Disease Control and Prevention (CDC). (2021). *When and how to wash your hands.* Handwashing: Clean hands save lives. https://www.cdc.gov/handwashing/when-how-handwashing.html

Chua, B. L., Karim, S., Lee, S., & Han, H. (2020). Customer restaurant choice: An empirical analysis of restaurant types and eating-out occasions. *International Journal of Environmental Research and Public Health, 17*(17), 1–23. doi:10.3390/ijerph17176276 PMID:32872267

Coleman, E., Delea, K., Everstine, K., Reimann, D., & Ripley, D. (2013). Handling practices of fresh leafy greens in restaurants: Receiving and training. *Journal of Food Protection, 76*(12), 2126–2131. doi:10.4315/0362-028X.JFP-13-127 PMID:24290691

Coorey, R., Ng, D. S. H., Jayamanne, V. S., Buys, E. M., Munyard, S., Mousley, C. J., Njage, P. M. K., & Dykes, G. A. (2018). The impact of cooling rate on the safety of food products as affected by food containers. *Comprehensive Reviews in Food Science and Food Safety, 17*(4), 827–840. doi:10.1111/1541-4337.12357 PMID:33350119

Cui, B., Li, S. Y., Wang, L. D. L., Chen, X., Ke, J., & Tian, Y. (2021). Hand hygiene knowledge and self-reported hand washing behaviors among restaurant kitchen chefs in Jiangsu Province, China. *International Journal of Environmental Research and Public Health, 18*(4), 1–14. doi:10.3390/ijerph18042149 PMID:33671843

Dales, A. (2018). *A shelf life guide.* Los Angeles Regional Food Bank. https://www.cerritos.edu/basic-needs/_includes/docs/ShelfLifeGuide.pdf

Darko, S., Mills-Robertson, F. C., & Wireko-Manu, F. D. (2015). Evaluation of some hotel kitchen staff on their knowledge on food safety and kitchen hygiene in the Kumasi Metropolis. *International Food Research Journal, 22*(6), 2664–2669.

Das, A. K., Sultana, Z., Kabir, A., & Kabir, M. S. (2019). Effect of washing on reducing bacterial loads in common vegetables sold in Dhaka City. *Bangladesh Journal of Microbiology, 35*(2), 96–101. doi:10.3329/bjm.v35i2.42637

Dewey-Mattia, D., Manikonda, K., Hall, A. J., Wise, M. E., & Crowe, S. J. (2018). Surveillance for foodborne disease outbreaks - United States, 2009-2015. *MMWR. Surveillance Summaries, 67*(10), 1–11. Advance online publication. doi:10.15585/mmwr.ss6710a1 PMID:30048426

Elobeid, T., Savvaidis, I., & Ganji, V. (2019). Impact of food safety training on the knowledge, practice, and attitudes of food handlers working in fast-food restaurants. *British Food Journal, 121*(4), 937–949. doi:10.1108/BFJ-01-2019-0066

Faridah, H. I., Chemah, T. C., Rosmaliza, M., & Norhayati, M. Y. (2016). Food safety knowledge and personal hygiene practices amongst mobile food handlers in Shah Alam, Selangor. *Procedia: Social and Behavioral Sciences, 222*, 290–298. doi:10.1016/j.sbspro.2016.05.162

FDA. (1998). *Guide to minimize microbial food safety hazards for fresh fruits and vegetables*. Center for Food Safety and Applied Nutrition.

Food and Drug Administration (FDA). (2010). *Program information manual retail food protection: recommendations for the temperature control of cut leafy greens during storage and display in retail food establishments*. https://www.fda.gov/food/retail-food-industryregulatory-assistance-training/ program-information-manual-retail-food-protection-recommendations-temperature-control-cut-leafy #prop

Food and Drug Administration (FDA). (2012). *Food allergies*. https://www.fda.gov/food/food-labeling-nutrition/food-allergies

Food and Drug Administration (FDA). (2013). *Food Code*. Public Health Service of Food and Drug Administration.

Food and Drug Administration (FDA). (2018). *Food allergen labeling and consumer protection act of 2004 (FALCPA)*. https://www.fda.gov/food/food-allergensgluten-free-guidance-documents-regulatory-information/food-allergen-labeling-and-consumer-protection-act-2004-falcpa

Food and Drug Administration (FDA). (2018). *Serving up safe buffets*. https://www.fda.gov/food/buy-store-serve-safe-food/serving-safe-buffets

Food Safety Authority of Ireland. (2017). *Hand washing and food safety*. https://www.fsai.ie/faq/hand_washing.html

Food Safety Authority of Ireland (FSAI). (2014). Country markets: Guide to good hygiene practice. Food Safety Authority of Ireland.

Food Safety of Australia and New Zealand (FSANZ). (2019). *Preparing and cook food*. https://www.foodstandards.gov.au/consumer/safety/faqsafety/pages/foodsafetyfactsheets/charitiesandcommunityorganisationsfactsheets/preparingandcookingf1479.aspx

Food Standards Agency. (2018). *Personal hygiene: Guidance on what you and your staff must do when handling food.* https://www.food.gov.uk/business-guidance/personal-hygiene

Food Standards Australia New Zealand (FSAI). (2019). *Receiving food safely.* https://www.foodstandards.gov.au/consumer/safety/faqsafety/pages/foodsafetyfactsheets/receivingfoodsafely.aspx

Gendel, S. M. (2012). Comparison of international food allergen labeling regulations. *Regulatory Toxicology and Pharmacology, 63*(2), 279–285. doi:10.1016/j.yrtph.2012.04.007 PMID:22565206

Green, L. R., Selman, C. A., Radke, V., Ripley, D., Mack, J. C., Reimann, D. W., Stigger, T., Motsinger, M., & Bushnell, L. (2006). Food worker hand washing practices: An observation study. *Journal of Food Protection, 69*(10), 2417–2423. doi:10.4315/0362-028X-69.10.2417 PMID:17066921

Haraminac, E. (2017). Cooling hot food, do it right to prevent bacterial growth. *Michigan State University Extension.* http://msue.anr.msu.edu/news/cooling_hot_food_do_it_right_to_prevent_bacterial_growth

Harris, K. J., Ali, F., & Ryu, K. (2018). Foodborne illness outbreaks in restaurants and patrons' propensity to return. *International Journal of Contemporary Hospitality Management, 30*(3), 1273–1292. doi:10.1108/IJCHM-12-2016-0672

Harris, L. J. (2002). *Guidelines for food safety during short-term power outages: Consumer fact sheet.* University of California, Department of agriculture and natural resources. doi:10.3733/ucanr.7264

Hedberg, C. W. (2013). Explaining the risk of foodborne illness associated with restaurants: The environmental health specialists' network (ehs-net). *Journal of Food Protection, 76*(12), 2124–2125. doi:10.4315/0362-028X.JFP-13-270 PMID:24290690

Hulland, K. R. S., Leontsini, E., Dreibelbis, R., Unicomb, L., Afroz, A., Dutta, N. C., Nizame, F. A., Luby, S. P., Ram, P. K., & Winch, P. J. (2013). Designing a handwashing station for infrastructure-restricted communities in Bangladesh using the integrated behavioural model for water, sanitation and hygiene interventions (IBM-WASH). *BMC Public Health, 13*(1), 877. Advance online publication. doi:10.1186/1471-2458-13-877 PMID:24060247

Insfran-Rivarola, A., Tlapa, D., Limon-Romero, J., Baez-Lopez, Y., Miranda-Ackerman, M., Arredondo-Soto, K., & Ontiveros, S. (2020). A systematic review and meta-analysis of the effects of food safety and hygiene training on food handlers. *Foods*, *9*(9), 1169. Advance online publication. doi:10.3390/foods9091169 PMID:32854221

Khairuzzaman, M., Chowdhury, F. M., Zaman, S., Al Mamun, A., & Bari, M. L. (2014). Food safety challenges towards safe, healthy, and nutritious street foods in Bangladesh. *International Journal of Food Sciences*, *2014*, 1–9. Advance online publication. doi:10.1155/2014/483519 PMID:26904635

Kim, K.-S., Song, S.-M., Kwon, S.-H., Jang, S.-E., Lee, B.-M., Kim, M.-H., Han, Y.-S., Hur, M.-J., & Kwon, M.-J. (2021). A Survey on the actual condition of products not labelled with allergens. *Journal of Food Hygiene and Safety*, *36*(3), 257–263. doi:10.13103/JFHS.2021.36.3.257

Knight, J. L. K., Joana Da Cunha Forte, G. M., & Gautam, O. P. (2020). *Technical Guide for handwashing facilities in public places and buildings*. WaterAid. https://washmatters.wateraid.org /sites/g/files/jkxoof256/files/technical-guide-for-handwashing-facilities-in-public-places-and-buildings.pdf

Lieberman, J. A., Gupta, R. S., Knibb, R. C., Haselkorn, T., Tilles, S., Mack, D. P., & Pouessel, G. (2021). The global burden of illness of peanut allergy: A comprehensive literature review. *Allergy: European Journal of Allergy and Clinical Immunology*, *76*(5), 1367–1384. doi:10.1111/all.14666 PMID:33216994

López-Gálvez, F., Gómez, P. A., Artés, F., Artés-Hernández, F., & Aguayo, E. (2021). Interactions between microbial food safety and environmental sustainability in the fresh produce supply chain. *Foods*, *10*(7), 1655. Advance online publication. doi:10.3390/foods10071655 PMID:34359525

Low, W. Y., Jani, R., Halim, H. A., Alias, A. A., & Moy, F. M. (2016). Determinants of food hygiene knowledge among youths: A cross-sectional online study. *Food Control*, *59*, 88–93. doi:10.1016/j.foodcont.2015.04.032

Madhwal, S., & Sharma, S. (2017). Food Safety Issues and Risk Associated with Refrigerated Foods. *International Journal of Current Microbiology and Applied Sciences*, *6*(12), 4196–4203. doi:10.20546/ijcmas.2017.612.482

Mallett, R. (2017). *HACCP International Food Safety Bulletin*. HACCP Europe., doi:10.18502/jfsh.v5i4.5700

McCurdy, S., Peutz, J., & Wittman, G. (2009). *Storing food for safety and quality*. Pacific Northwest Extension Publication. https://catalog.extension.oregonstate.edu/sites/catalog/files/project/pdf/ pnw612.pdf

Ministry of Health Malaysia (MOHM). (2021). *Guide to Safe Food Storage – Portal My Health*. http://www.myhealth.gov.my/en/guide-to-safe-food-storage/

Mjoka, J., & Selepe, M. (2018). Assessment of food hygiene knowledge and practices among food handlers in selected hotels around uMhlathuze Area. *African Journal of Hospitality, Tourism and Leisure*, *7*(4).

Mun, S. G. (2020). The effects of ambient temperature changes on foodborne illness outbreaks associated with the restaurant industry. *International Journal of Hospitality Management*, *85*, 102432. Advance online publication. doi:10.1016/j.ijhm.2019.102432

Ncube, F., Kanda, A., Chijokwe, M., Mabaya, G., & Nyamugure, T. (2020). Food safety knowledge, attitudes and practices of restaurant food handlers in a lower-middle-income country. *Food Science & Nutrition*, *8*(3), 1677–1687. doi:10.1002/fsn3.1454 PMID:32180975

Nicoli, M. C., & Calligaris, S. (2018). Secondary shelf life: An underestimated issue. *Food Engineering Reviews*, *10*(2), 57–65. doi:10.100712393-018-9173-2

Odumeru, J. A. (2012). Microbial safety of food and food products. In B. K. Simpson (Ed.), *Food Biochemistry and Food Processing* (pp. 785–797). Wiley Blackwell. doi:10.1002/9781118308035.ch41

Olu-Taiwo, M., De-Graft, B. M., & Forson, A. O. (2021). Microbial quality of sliced pawpaw (*Carica papaya*) and watermelon (*Citrullus lanatus*) sold on some streets of Accra Metropolis, Ghana. *International Journal of Microbiology*, *2021*, 1–8. Advance online publication. doi:10.1155/2021/6695957 PMID:33574850

Ozogul, F., & Hamed, I. (2018). The importance of lactic acid bacteria for the prevention of bacterial growth and their biogenic amines formation: A review. *Critical Reviews in Food Science and Nutrition*, *58*(10), 1660–1670. doi:10.1080/10408398.2016.1277972 PMID:28128651

Pérez, J. S. (2013). *Guidance on Food Allergen Management for Food Manufacturers*. Food Drink Europe.

Ricci, A., Chemaly, M., Davies, R., Fernández Escámez, P. S., Girones, R., Herman, L., Lindqvist, R., Nørrung, B., Robertson, L., Ru, G., Simmons, M., Skandamis, P., Snary, E., Speybroeck, N., Ter Kuile, B., Threlfall, J., Wahlström, H., Allende, A., Barregård, L., & Bolton, D. (2017). Hazard analysis approaches for certain small retail establishments in view of the application of their food safety management systems. *EFSA Journal*, *15*(3). Advance online publication. doi:10.2903/j.efsa.2017.4697 PMID:32625423

Roberts, K. R., Olds, D. A., Shanklin, C., Sauer, K., & Sneed, J. (2013). Cooling of foods in retail foodservice operations. *Food Protection Trends*, *33*(1), 27–31.

Schaffner, D. W., Brown, L. G., Ripley, D., Reimann, D., Koktavy, N., Blade, H., & Nicholas, D. (2015). Quantitative data analysis to determine best food cooling practices in U.S. restaurants. *Journal of Food Protection*, *78*(4), 778–783. doi:10.4315/0362-028X.JFP-14-252 PMID:25836405

Shafie, A. A., & Azman, A. W. (2015). Assessment of knowledge, attitude and practice of food allergies among food handlers in the state of Penang, Malaysia. *Public Health*, *129*(9), 1278–1284. doi:10.1016/j.puhe.2015.03.016 PMID:25931434

Singapore Food Agency (SFA). (2020). *Guidelines on food safety & hygiene practices for residents preparing food under the HDB/URA's home-based small-scale business scheme*. Author.

Souza, C., Azevedo, P. R. M., & Seabra, L. M. A. J. (2018). Food safety in Brazilian popular public restaurants: Food handlers' knowledge and practices. *Journal of Food Safety*, *38*(5), e12512. Advance online publication. doi:10.1111/jfs.12512 PMID:30449912

Tan, S. L., Cheng, P. L., Soon, H. K., Ghazali, H., & Mahyudin, N. A. (2013). A qualitative study on personal hygiene knowledge and practices among food handlers at selected primary schools in Klang Valley area, Selangor, Malaysia. *International Food Research Journal*, *20*(1), 71–76.

Trafialek, J. (2019). The role of suppliers of raw materials in ensuring food safety. *Global Journal of Nutrition & Food Science*, *1*(4). Advance online publication. doi:10.33552/GJNFS.2019.01.000517

United States Department of Agriculture. (2012). Food safety plan HACCP-based standard operating procedures. Mississippi: National Food Service Management Institute.

United States Department of Agriculture. (2017). *Refrigerator thermometers - cold facts about food safety*. https://www.fda.gov/food/buy-store-serve-safe-food/refrigerator-thermometers-cold-facts-about-food-safety

United States Department of Agriculture. (2019). *What methods of reheating food are safe?* https://ask.usda.gov/s/article/What-methods-of-reheating-food-are-safe

United States Department of Agriculture. (2019). *Cold food storage chart*. https://www.foodsafety.gov/food-safety-charts/cold-food-storage-charts

United States Department of Agriculture. (2021). *Food safety during power outage*. https://www.foodsafety.gov/food-safety-charts/food-safety-during-power-outage

Verhoeckx, K. C. M., Vissers, Y. M., Baumert, J. L., Faludi, R., Feys, M., Flanagan, S., Herouet-Guicheney, C., Holzhauser, T., Shimojo, R., van der Bolt, N., Wichers, H., & Kimber, I. (2015). Food processing and allergenicity. *Food and Chemical Toxicology*, *80*, 223–240. doi:10.1016/j.fct.2015.03.005 PMID:25778347

Wong, J. X., Ramli, S., & Chen, S. N. (2021a). Assessment of toxic effect of *Centella asiatica* extract and its application as natural preservative in fresh-cut mango, pear and cabbage. *Journal of Food Processing and Preservation*, *45*(10), e15824. doi:10.1111/jfpp.15824

Wong, J. X., Ramli, S., Desa, S., & Chen, S. N. (2021b). Use of *Centella asiatica* extract in reducing microbial contamination and browning effect in fresh cut fruits and vegetables during storage: A potential alternative of synthetic preservatives. *Lebensmittel-Wissenschaft + Technologie*, *151*, 112229. Advance online publication. doi:10.1016/j.lwt.2021.112229

Xue, J., & Zhang, W. (2013). Understanding China's food safety problem: An analysis of 2387 incidents of acute foodborne illness. *Food Control*, *30*(1), 311–317. doi:10.1016/j.foodcont.2012.07.024

KEY TERMS AND DEFINITIONS

Cross-Contamination: The transferring of any microorganisms or foreign substances from one medium to another medium. The foreign substances are including food allergens, microbial toxins, chemicals, and microbial cells.

Food Code: The Food Code is a model that representing the best advice from FDA for a consistent set of provisions addressing the safety and protection of food sold in retail and in foodservice. Besides, it is a systematized basic standard to protect public health relating to food, beverages, raw materials, utensils, and equipment use and domestic consumption. At last, it will ensure all the foods to be safe prepared, without unadulterated and accurately served to consumers

Food Danger Zone: The temperature range allows the foodborne bacteria to grow faster from 5 to 60°C and doubling the bacterial population as more quickly as 20 minutes.

Hazard Analysis and Critical Control Point (HACCP): HACCP is an international food safety management system that emphasizing on the food safety by analyzing and monitoring the physical, chemical, and biological hazards from raw ingredients production, food purchasing, and handling until to the finalized product manufacturing, distribution, and human consumption.

Internal Food Temperature: The core temperature inside the central part of the cooked food. The internal food temperature can be obtained using the temperature probe by inserting it into the central part of food.

Partially Cooked Food: The food has been cooked in halfway without entirely cooked. It will be cooked again later on before serving to the guest.

Perishable Food: Perishable food refers to any edible food or raw ingredients that are highly susceptible to food spoilage and become unfit for human consumptions. They usually have a shorter shelf life and most probably need to store in a chilling or freezing environment to prolong their storage shelf life.

Primary Shelf Life: The time elapsed between the manufacture and packaging of a product and the point at which the product first becomes unacceptable under specified environmental parameters.

Ready-to-Eat (RTE): The RTE food can be directly consumed without the need to further cooking or reheat. It can be directly serving to consumers. The examples of RTE food consist of sandwiches, cheese, and smoked fish. Some of the RTE food can even be served cold in advance such as sashimi and sushi.

Secondary Shelf Life: The period when the original food packaging or containers had opened by consumers and lead to a sudden sharp acceleration of alteration in the product quality.

Two-Hour Rule: The time frame provided to display any perishable items that are left without cold or hot holding on the buffet counter under room temperature conditions. After 2 hours, the perishable ingredients need to be discarded.

Chapter 7
A Systematic Literature Review of Assessing the Level of Food Safety Knowledge Among Food Handlers in Malaysia

Genevie Eleanor Ruby
Universiti Putra Malaysia, Malaysia

Ungku Fatimah Ungku Zainal Abidin
Universiti Putra Malaysia, Malaysia

ABSTRACT

Being knowledgeable about food safety is one of the strategies to address food-borne diseases (FBD). The systematic review was focuses on food safety knowledge and the respective interventions. Generally, numerous relevant studies have been done to determine the level of food safety knowledge among food handlers, but studies from a Malaysian perspective were limited. Therefore, the present study reviewed a number of previous studies regarding level of food safety knowledge and type of interventions that have been done among various categories of food handlers in Malaysia. For the review purpose, preferred reporting items for systematic reviews and meta-analyses (PRISMA) was adopted based on Science Direct, Scopus, and Google Scholar databases. A total of 22 resulted from the searching and were analyzed systematically. The review of food safety knowledge was divided into three themes consisting of food handlers at premises, consumers, and students. The results of this review have identified the knowledge gap of food handlers, and the authors provide recommendations for future food safety education.

DOI: 10.4018/978-1-7998-7415-7.ch007

INTRODUCTION

The food and waterborne diseases in Malaysia include cholera, dysentery, food poisoning, Hepatitis A, and typhoid (Ministry of Health Malaysia, 2017). Recently, several cases of food poisoning and even some deaths have been reported. For example, Malaysia was shocked by food poisoning because consuming "*Laksa*" contributed to 34 cases, including two deaths in Perak, 25 cases in Kedah, and 24 cases in Selangor (The Malaysian National News Agency, 2018). In addition, 36 students from *tahfiz* center in Kedah (Noorazura Abdul Rahman, 2019), 49 primary school students in Keramat (New Straits Times, 2019), and 110 college students in Kajang (The Malaysian National News Agency, 2019) was rushed to hospital after experiencing symptoms such as vomiting, stomach ache, and diarrhea. All of them have been confirmed suffering from food poisoning. A recent case of food poisoning was reported at the training center in Kuala Lumpur involving a total of 137 kindergarten teachers after consuming "*ayam masak lemak*" dish (The Malaysian National News Agency, 2020).

Clinical reports of Malaysian diarrheal patients revealed that non-typhoid *Salmonella* (57%) become among the most significant pathogens isolated from their stool samples, followed by enteropathogenic *Escherichia coli* (14%), *Shigella* (11%), *Campylobacter* (5%), and *Aeromonas* (5%) (Dewanti-Hariyadi & Gitapratiwi, 2014). All of these pathogens can be found in common foods, for example, *Salmonella* discovered in meat, especially buffalo (Saira Banu, Humairak, Zakiah, & Siti Adila, 2019), or leafy vegetables such as cabbages and cucumbers (Saw et al., 2020).

Poor personal hygiene among food handlers was also a significant cause for the transmission of the pathogen into food. A microbiological study by Nasrolahei, Mirshafiee, Kholdi, Salehian, & Nasrolahei (2017) revealed that *Staphylococcus aureus* and *Escherichia coli* were prevalent in the fingernails of butchers, fast food handlers, and school cafeterias staff. Furthermore, during the typhoid outbreaks in Kelantan, Malaysia, food handlers also identified as the chronic carriers for *Salmonella* Typhi that were detected from their stool samples (Chua et al., 2015).

Pathogens transferred from the unhygienic hands of food handlers may continue to survive in food that is not stored and cooked at a recommended temperature. In the case of consuming raw or uncooked eggs that lead to the infection of *Salmonella enterica*, it could be prevented by immersed the half-boiled egg in freshly boiled water for at least 15 min (Mohamad Fithri et al., 2018). Regarding this matter, food handlers also recommended cooking the egg at a temperature more than 100 °C for 8 min and low storage temperature in the refrigerator to be practiced at home and other food premises to minimize the risk of food poisoning caused by *Staphylococcus* (Sánchez, Neira, Laca, Laca, & Díaz, 2019). Therefore, to ensure that food is safe to consume, the food handlers should adopt food safety practices. Numerous studies abroad indicated that a good food safety knowledge level promotes better food handlers' food safety practices from various premises (Al-kandari, Al-abdeen, & Sidhu, 2019; Alqurashi, Priyadarshini, & Jaiswal, 2019; Omemu & Aderoju, 2008; Unusan, 2007).

To date, limited study has synthesized the literature using systematic review methods to assess the food safety knowledge of food handlers in Malaysia. The increment in the incidence rate of 7.87 per 100,000 population for food poisoning cases in 2017 compared to 2016 indicated the importance of having knowledge about food safety among food handlers (Ministry of Health Malaysia, 2016; 2017). This synthesis will identify the food safety knowledge gaps among food handlers and develop food safety intervention programs to minimize the risk of food borne diseases (FBD). Therefore, the current study aimed to conduct a systematic review using standardized and structured techniques to identify, characterize, and summarize the existing literature investigating the food safety knowledge among food handlers and interventions implemented to improve their knowledge. In addition, this study also will examine the trends in research related to food safety knowledge. Results from this systematic review may emphasize the important areas that could be targeted for future education and assist future researchers interested in studies regarding food safety knowledge in Malaysia.

METHODS AND MATERIALS

Search Strategy

As systematic literature review (SLR) will help summarize the previous literature and explain differences among the same questions of the studies; therefore, scientific strategies and a rigorous method to minimize its bias (Cook, Mulrow, & Haynes, 1997). The Preferred Reporting Items for Systematic Reviews and Meta-Analyzes (PRISMA) has been applied to enable this SLR to be more comprehensive, explain the method in more detail and take into account the quality differences between the studies (Moher, Liberati, Tetzlaff, Altman, & The Prisma Group, 2009). The searching was done in three academic databases such as Science Direct, Scopus, and Google Scholar by using the search strings ("food safety knowledge" OR "food safety education" OR "food safety awareness" OR "food safety assessment*" OR "food safety concern*" OR "food hygiene knowledge" OR "food hygiene education" OR "food hygiene awareness" OR "food hygiene assessment*" OR "food hygiene concerns*" AND "Malaysia"). A total of 40 articles were retrieved in the initial stage of the systematic review process.

Screening Process

A total of 6 duplicate articles were removed before initiated the screening process. Then, a total of 34 articles were screened in accordance with the inclusion and exclusion criteria that have been determined. The first inclusion criteria where the type of literature must be research articles because it contains empirical data. Therefore, secondary sources such as conference proceedings, theses, books, or book chapters were excluded from this review. Only studies conducted in Malaysia included, in line with the study objectives. In addition, this review only focused on articles published in English, and there was no limit on the publication date due to the known gap in literature specific to food safety knowledge among food handlers in Malaysia. The earliest record found was published in 2002, and the latest was published in 2020.

Eligibility

A total of 29 articles were prepared for the eligibility process. During this process, the articles were examined its titles, abstracts, and main contents to ensure that they fulfilled the inclusion criteria and met the current review's objectives. Consequently, four articles were excluded because they do not focus on Malaysian food handlers and are not appropriate to the objectives. A total of 22 remaining articles to be analysed was indicated in Figure 1.

RESULTS AND DISCUSSIONS

General Characteristics of the Studies in the Review

The analysis produced a total of 22 articles related to food safety knowledge in Malaysia. As shown in Table 1, a study on food safety knowledge is focused on food handlers involved in businesses that operated at the premises, including schools, food courts, colleges, universities, and home-based (n = 13), home (n = 4), and students (n = 4). In addition, there was also a study that combined food handlers in the cafeteria and students. Most of the studies (n = 20) have been done in West Malaysia, while only two (n=2) analyses were carried out in East Malaysia.

The discussion in this section is about the level of knowledge of food safety based on the category of food handlers.

Figure 1. Flow diagram of the literature review process

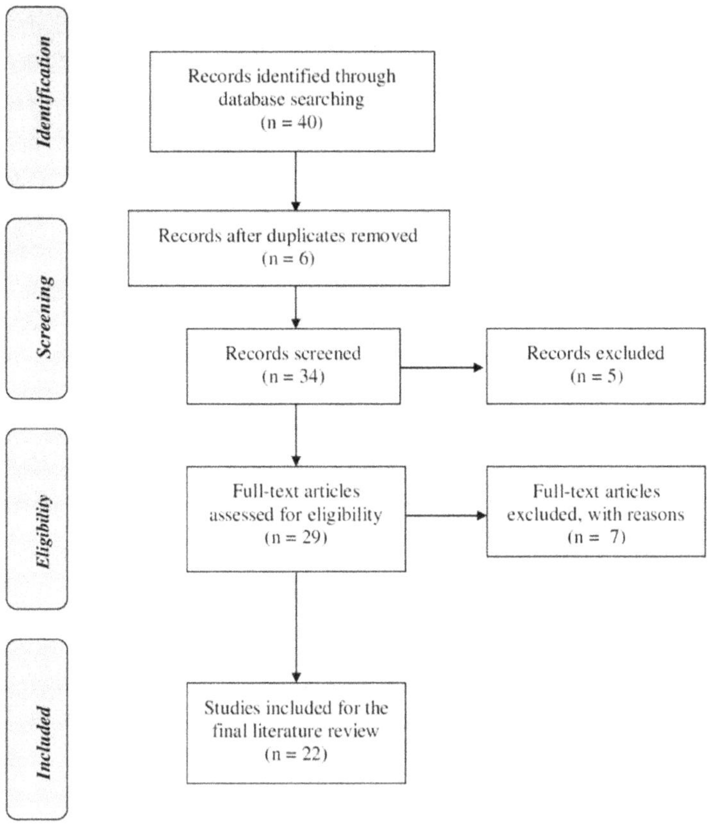

Table 1. Summary of food safety knowledge studies in Malaysia

Author	Location	Type of respondent	Number of respondents	Data collection method	
				Questionnaire	Interview
Mohd Zain & Naing (2002)	Kota Bharu, Kelantan	Food handlers on-premises	430	/	
Nee & Abdullah Sani (2011)	Universiti Kebangsaan Malaysia	Food handlers in residential colleges and canteens	65	/	
Abdul-Mutalib et al. (2012)	Kuala Pilah, Negeri Sembilan	Food handlers on premises	64	/	
Tan, Cheng, Soon, Ghazali, & Mahyudin (2013)	Klang Valley, Selangor	Food handlers in primary school	25		/
Tan, Fatimah, Muhammad Shahrim, Lee, & Nor Ainy (2013)	Hulu Langat, Selangor	Food handlers in primary school	85	/	

continues on following page

Table 1. Continued

Author	Location	Type of respondent	Number of respondents	Data collection method	
				Questionnaire	Interview
Rosnani, Son, Mohhidin, Toh, & Chai (2014)	Putrajaya	Food handlers in restaurants	127	/	
Mohd Firdaus Siau, Son, Mohhiddin, Toh, & Chai (2015)	Putrajaya	Food handlers in food courts	274	/	
Lee, Abdul Halim, Thong, & Chai (2017)	Kuala Lumpur	Food handlers at canteens in the university	67	/	
Mustaffa, Rahman, Hassim, & Ngadi (2017)	Peninsular Malaysia	Food handlers in university cafeterias	30	/	
Siti Nurul Ain, Sahilah, & Razalee (2018)	Kuala Pilah, Negeri Sembilan	Food handlers on-premises	134	/	
Asmawi et al. (2018)	Petaling Jaya	Food handlers in food courts	108	/	
Dora-Liyana, Mahyudin, Ismail-Fitry, Ahmad-Zaki, & Rasiyuddin (2018)	Northern Region of Malaysia	Food handlers in boarding schools	134	/	
Mohd Yusof, Rahman, & Haque (2018)	Universiti Pertahanan Nasional Malaysia	Food handlers in cafeterias	53	/	
		Dietetic students	53		
Mohamad Hapiz (2019)	Sungai Petani, Kedah	Food handlers on-premises	322	/	
Nur Izyan et al. (2019)	Klang Valley, Selangor	Home-based food providers	111	/	
Lim, Chye, Mohd Rosni, Norazah, & Lee (2016)	Semporna, Sabah	Household food preparers	250	/	
Balaganesan & Siti Nur Afifah (2019)	Selayang, Selangor	Housewives	200	/	
Ruby, Ungku Zainal Abidin, Lihan, Jambari, & Radu (2019)	Sibu, Sarawak	Adult consumers	623	/	
Norazmir et al. (2012)	Johor Bahru, Johor	Upper secondary school students	399	/	
Mohd Halim, Mazni, Mohd Shazali, & Norazmir (2016)	Universiti Teknologi Mara, Puncak Alam	Students of a culinary degree program	18		/
Abdul Nazer, Angeline, Sunil Kumar, & Nazer Zulfika (2018)	Kedah	Private university students	869	/	
Syahira, Huda, & Mohd Rafee (2019)	Hulu Langat, Selangor	Form four secondary school students	575	/	

Food Safety Knowledge among Food Handlers

The level of knowledge of food handlers on food safety is mostly measured in percentage except for a study by Mohamad Hapiz (2019), which uses mean. Various ranges were used to interpret the level of food handler knowledge. The studies by Abdul-Mutalib et al. (2012), Asmawi et al. (2018), Dora-Liyana, Mahyudin, Ismail-Fitry, Ahmad-Zaki, & Rasiyuddin (2018), Nee & Abdullah Sani (2011), Nur Izyan et al. (2019) and Tan, Fatimah, Muhammad Shahrim, Lee, & Nor Ainy (2013) set a score of less than 50% to show that the food handlers have poor knowledge in food safety. Meanwhile, Mohd Yusof, Rahman, & Haque (2018) and Siti Nurul Ain, Sahilah, & Razalee (2018) estimated the percentage of a total score less than 60% and 75% as poor, respectively. In the study by Rosnani, Son, Mohhidin, Toh, & Chai (2014), food handlers who scored lower than 85% were considered less knowledgeable and this is among the highest score recorded.

Most studies found that food handlers are knowledgeable in maintaining the hygiene of the food, which can be prevented by controlling cross-contamination. The food handlers agreed that food contamination can be controlled by washing hands before handling food (Abdul-Mutalib et al., 2012; Lee, Abdul Halim, Thong, & Chai, 2017; Mustaffa, Rahman, Hassim, & Ngadi, 2017; Nee & Abdullah Sani, 2011; Tan et al., 2013), wearing proper attires including gloves, masks or caps (Abdul-Mutalib et al., 2012; Tan et al., 2013; Tan, Fatimah, Muhammad Shahrim, Lee, & Nor Ainy, 2013) and using clean food utensil during food preparation (Siti Nurul Ain et al., 2018).

However, it was indicated that even though the food handlers were knowledgeable regarding hand hygiene, they still missed the proper steps when washing hands (Dora-Liyana et al., 2018; Tan et al., 2013). The transmissions of pathogens potentially occur if hands do not completely wash and will contaminate the food. While the study by Nur Izyan et al. (2019) found the food handlers comply with one of the procedures before handling food: the use of warm water and soap for handwashing.

In addition, the food handlers also have adequate knowledge regarding the prevention and transmission of FBD (Mohd Zain & Naing, 2002). This fact was reinforced by the findings of a study by Mohd Yusof et al. (2018) indicating food handlers familiar with FBD and identified *Salmonella* as the main cause. The presence of a suitable temperature may accelerate the growth of pathogens. Thus, the food handlers also agreed that proper food storage and recommended temperature for refrigerator or freezer were among the essential measures to control FBD (Mohd Firdaus Siau, Son, Mohhiddin, Toh, & Chai, 2015).

The food handlers are also knowledgeable regarding the correct methods to defrost food especially meat (Asmawi et al., 2018) and have awareness to check the quality of raw food (Mohamad Hapiz, 2019). These matters can prevent food from spoilage due to pathogens action while preserving freshness to ensure it is safe to be consumed.

The studies also found that food handlers were lack of knowledge regarding the temperature control that included the safe temperature of cooked food (Rosnani et al., 2014; Mohd Firdaus Siauet al., 2015; Asmawi et al., 2018; Nur Izyan et al., 2019), temperature for food storage (Nee & Abdullah Sani, 2011; Mohamad Hapiz, 2019) and food thawing (Dora-Liyana et al., 2018). To ensure the food such as meat, poultry and seafood safe for consumption, it must be cooked at the temperature of 70°C, stored at a temperature above 60°C after cooked and no food thawing at room temperature (Ministry of Health Malaysia, 2014). All this measure was to inhibit the growth of pathogens that can spoil food.

Apart from that, food handlers also lack of knowledge related to food poisoning. They were identified did not know about the pathogens that caused food poisoning such as *Staphylococcus* or *Campylobacter* (Mohd Yusof et al., 2018; Mustaffa et al., 2017; Nee & Abdullah Sani, 2011) and its symptoms (Mohd Zain & Naing, 2002).

Another critical knowledge among food handlers was about cross contamination. Most of the food handlers demonstrated low scores related to the use of gloves when handling food (Abdul-Mutalib et al., 2012), cleaning cutting boards after use (Lee et al., 2017) and selection of the plastic cutting boards that are easier to clean and less risk for the growth of microorganisms (Siti Nurul Ain et al., 2018).

Food Safety Knowledge among Consumers

Those involved in food preparation at home and not professional food handlers were defined as consumers (Nesbitt et al., 2014). FBD also has the potential to occur in the home environment. Therefore, the consumers should also be knowledgeable in food safety. In this review, only three studies determined the level of food safety among consumers in Malaysia. A study by Lim, Chye, Mohd Rosni, Norazah, & Lee (2016) found that consumers agreed that it was important to maintain kitchenware hygiene for food safety (mean = 14.64) even though they assumed that the use of water only to wash hand is sufficient to prevent FBD (mean = 12.00). However, the consumers realized that they must wash their hands before preparing food, after using toilets and after handling raw meats and poultry (99.5%) but lack knowledge in terms of temperature control (29.5%) (Ruby, Ungku Zainal Abidin, Lihan, Jambari, & Radu, 2019). On the other hand, the study by Balaganesan & Siti Nur Afifah (2019) on housewives indicated a good knowledge level in clean concept focusing on personal hygiene (89.5%). Still, the level was unsatisfactory for the chill (9%), separate (6.5%), and cooking concept (33.5%).

Food Safety Knowledge among Students

Students should also have knowledge about food safety as they are prospective food handlers in the future, whether at home or industry level. The study on the level of food safety knowledge among students in Malaysia has been done in secondary school and higher education institutions. The majority of the upper secondary school students notified that they must wash their hands after sneezing (92.0%) but failed to identify the risky food that can potentially cause food poisoning (4.5%) and did not know that chilling or freezing could eliminate pathogens (25.6%) (Norazmir et al., 2012; Syahira, Huda, & Mohd Rafee, 2019).

The interview with culinary degree students indicated that they could explain the source of physical and chemical hazards. However, the knowledge of biological hazards was limited as they are still unclear about cross-contamination (Mohd Halim, Mazni, Mohd Shazali, & Norazmir, 2016). Meanwhile, the students in private and public universities were also knowledgeable in recognize FBD symptoms and identified *Salmonella*, *Staphylococcus aureus* and *Listeria* as its source (90.6%) but they got a low score in storage method and failed to identify *Campylobacter* as one of the pathogens that can cause FBD (47.2%) (Abdul Nazer, Angeline, Sunil Kumar, & Nazer Zulfika, 2018; Mohd Yusof et al., 2018).

Interventions

Previous studies also suggest interventions to enhance the food handler's knowledge in respect of food safety. According to Food Hygiene Regulations 2009, the food handlers at the premises are required to undergo the Food Handler Training Course organized by Food Handler Training School accredited by Ministry of Health Malaysia. Therefore, researchers suggested that continuous training should be held to improve the knowledge of food handlers and also regular check-ups to monitor their health status (Abdul-Mutalib et al., 2012; Mohd Zain & Naing, 2002; Mustaffa et al., 2017; Nur Izyan et al., 2019; Siti Nurul Ain et al., 2018; Tan et al., 2013).

To further strengthen the food handler's knowledge, it should also be focusing the hands-on activities instead of just delivering the theory and the test at the end of the training can be considered (Asmawi et al., 2018; Tan et al., 2013). Since the training needs to be emphasized, the curriculum and teaching methods must be reviewed including the usage of appropriate language and illustration module due to various backgrounds of food handlers in Malaysia (Lee et al., 2017 ; Rosnani et al., 2014). In the meantime, the government may also introduce a better system and gazette a new law in regards to reducing FBD outbreaks (Mohamad Hapiz, 2019).

Compared with the food handlers at premises, there were no specific regulations relating to consumers who handle food at home. Therefore, local authorities play an essential role in delivering important information regarding safe food handling (Lim et al., 2016). FightBac! can be used as the basis of the information presented to enhance the consumer's understanding of the concept of clean, chill, separate, and cook, which also can be shared through the mass media (Balaganesan & Siti Nur Afifah, 2019). In addition, the '5 Keys to Safer Food' recommended by the World Health Organization also can be used as a guide to educate consumers at home (Ruby et al., 2019).

For students, not only researchers encourage continuing education, but teachers or lecturers need to be more competent in delivering knowledge on food safety and having adequate infrastructure (Abdul Nazer, et al., 2018; Mohd Halim et al., 2016; Norazmir et al., 2012). Food safety education should also take into account the diverse background of students, incorporates with mainstream subjects, and most importantly, parents should be role models to their children when at home (Syahira et al., 2019).

DISCUSSIONS

The main purpose of this review is to examine the level of food safety knowledge in Malaysia and its interventions based on the category of food handlers. Simultaneously, a systematic literature review is important for the researcher to understand the pattern of previous studies and the derived outcomes can be implemented in the future. From this review, some studies patterns of food safety knowledge level were discovered.

The level of food safety knowledge studies in Malaysia focused on three main groups: food handlers at premises, food handlers at home, and senior students. While abroad, the study of food safety knowledge was also carried out on other groups such as street food vendors (e.g., Samapundo, Cam Thanh, Xhaferi, & Devlieghere, 2016; Madaki & Bavorova, 2019), primary school students (e.g., Tutu, Hushie, Asante, & Egyakwa-Amusah, 2020), hotel staff (e.g., Baser, Ture, Abubakirova, Sanlier, & Cil, 2017) and meat handler (e.g., Tegegne & Phyo, 2017). Thus, it was recommended that the food safety knowledge be determined to all food handlers since each group has various backgrounds and environments of food handling.

Based on the number of studies, it obviously showed that previous researchers were more focused on food handlers at premises and most limited studies have been done on consumers. This matter should be given due attention as it was reported that 42% of food poisoning cases were due to negligence of food handling at home (World Health Organization, 2002). While Redmond & Griffith (2003) also estimated that almost 87% of the FBD occurrence is contributed by food mishandling at home.

Consumers who handle food at home should be aware of food safety measures. They play the main role in providing food for the family members, including vulnerable groups such as children or the elderly.

Most researchers tend to conduct research using the questionnaire as a tool to collect data rather than qualitative research interviews. Methods of data collection through the questionnaire become popular because it can reach a larger target population that will produce greater statistical power, gain a high response rate, and be effective in saving time and financial (Jones, Baxter, & Khanduja, 2013). However, qualitative studies may also be practiced as this method may provide more in-depth information about a study and understand the various dimensions of the problem instead of concerning the quantified aspects. The researchers may adopt methods such as case studies, focus groups, and field research that are frequently adopted in qualitative methodology (Queirós, Faria, & Almeida, 2017).

Even though the food handlers were knowledgeable in food safety, two knowledge gaps were identified: temperature control and cross-contamination. The knowledge in temperature included storage, cooking and refrigerator temperatures. All were aware that the use of inaccurate temperature would promote the growth of microorganisms. In the meantime, cross contamination has potentially transferred the microorganisms from any source to food and contaminate it. Previous studies also have shown that food handlers were lack of knowledge regarding temperature control (Al-Kandari et al., 2019; Okour, Alzein, Saadeh, & Alfaqih, 2020). Temperature control was related to knowledge about the use of thermometers. Lack of knowledge about the importance of thermometers among consumers may be due to several perceived barriers such as inconvenient or impractical, lack of time, the difficulty of thermometer usage and cost (Elshahat, Woodside, & Mckinley, 2019). Apart from that, it was found that some food premises did not supply the thermometer for use by food handlers (Asim et al., 2019).

Knowledge regarding refrigerator's temperature, which is to be maintained below 4°C also crucial to ensure the food can safely consumed or it will increase the risk of accelerating the growth of microorganisms. The microbiological test of the home refrigerator confirmed *Listeria monocytogenes* inside the refrigerator with temperature 5–63 °C (Dumitrașcu et al., 2020). Technological developments can be considered in addressing the food handler negligence in using the thermometer in the refrigerator. The manufacturer may design refrigerators that are readily installed with a sensor that monitor the temperature in the refrigerator or lowers the temperature of food stored in the refrigerator door shelves (Katiyo, de Kock, Coorey, & Buys, 2019; Rodriguez-Martinez et al., 2020).

Another problem among food handlers is a lack of awareness about using the thermometer when cooking, especially involving food such as meat or poultry. A thermometer is used to check the doneness of food because eating raw or undercooked food may increase the risk of contracting the infection of *Salmonella* and *Campylobacter*. However, less emphasis on food handlers about using the thermometer as a method to determine a food cooked, for example, based on length of cooking time, checked using equipment such as a knife or toothpick and looked for meat juice (Kosa, Cates, Godwin, & Chabers IV, 2017). Only a handful of food handlers used the thermometer to check their food. Then this contributes to the lack of knowledge about the needs of its usage for food safety.

On the other hand, food handlers should realize that raw materials, contact surfaces and cooking utensils may promote cross-contamination (Parry-Hanson Kunadu, Ofosu, Aboagye, & Tano-Debrah, 2016). Cross-contamination can also be controlled if the food handlers wear proper attire during handling food, use separate cutting boards for different types of food and ensure that the knife and cutting board always sanitize (Rebouças et al., 2017; Taha et al., 2020). In the meantime, Garayoa, Abundancia, Díez-Leturia, & Vitas (2017) also suggested that those involved in food service apply disinfectants to eliminate microorganisms from food contact surfaces such as serving trays and cutting boards. Thus, all this effort may minimize the number of microorganisms from being transmitted into food.

The studies that have been carried out in Malaysia also suggested that intervention programs be developed to improve knowledge regarding food safety for food handlers. However, these interventions focus on food handlers operating at food premises operators compared to other food handlers. A systematic review by Young, Waddell, Wilhelm, & Greig (2020) also agreed that the interventions were most frequently targeted on food handlers in educational institutions and restaurants with personal hygiene and prevention of cross-contamination became the most covering contents in the training session.

Other food handlers, such as consumers and students who are not involved in food safety training, should be targeted in the educational program. Social media may use as an effective medium to distribute any information regarding food safety. For example, the authorities should more actively promote The Official Facebook Page of Food Safety and Quality Programme to disseminate the food safety information among social media users. This method received positive feedback from the public after each registered article in the portal received an average of 13,920 readers, while every posting recorded an average of 131,074 readers (Food Safety and Quality Division, 2018).

For students, the knowledge about food safety may be incorporated into the formal curriculum by educational materials such as posters or modules. This teaching aid should be student-centered to ensure its effectiveness. Lee, Jeong, Ko, Park, & Ko (2016) have developed a textbook and poster under the theme of foodborne illness that required the students to explain, differentiate, and ways to prevent foodborne illness based on social cognitive theory supplementing with hands-on activities.

Take home worksheets that were administered based on the positive deviance model also enhanced the student's knowledge by applying the theories they learned in class after school sessions (Whited, Feng, & Bruhn, 2019). In addition, the application of simulation games on tablet personal computer could be implemented to boost the student's motivation to learn food safety knowledge (Yu, Sirsat, & Madera, 2018). It would also be recommended to develop the food safety curricular based on other behavior theory, such as introspection-based, behavioristic, and sociological.

CONCLUSION

This systematic review was able to identify the pattern of previous studies related to food safety knowledge focusing on premises food handlers, consumers, and students in Malaysia. Outcomes of the analysis indicated that further research needs to be done in certain areas. The research gap recommended that future research about food safety knowledge in Malaysia could widen the target groups, vary the study methods and determine a depth study regarding temperature control and cross-contamination. Furthermore, this review also recommended interventions and education campaigns to improve the knowledge among consumers and students that were least involved in food safety training.

ACKNOWLEDGMENT

We would like to thank Dr. Thavamaran Kanesan and his team for providing their professional proofreading service. This research received no external funding. No conflict of interest among all the authors.

REFERENCES

Abdul-Mutalib, N.-A., Abdul-Rashid, M.-F., Mustafa, S., Amin-Nordin, S., Awang Hamat, R., & Osman, M. (2012). Knowledge, attitude and practices regarding food hygiene and sanitation of food handlers in Kuala Pilah, Malaysia. *Food Control*, *27*(2), 289–293. doi:10.1016/j.foodcont.2012.04.001

Abdul Nazer, A., Angeline, F. W., Sunil Kumar, P., & Nazer Zulfika, A. (2018). A KAP Study on Food Safety and Hygiene Among Private University Students in Kedah State, Malaysia. *Journal of Natural Remedies*, *18*(3), 113–121. doi:10.18311/jnr/2018/22289

Abdul Rahman, N. (2019, May 15). 36 tahfiz students down with food poisoning. *New Straits Times*.

Al-Kandari, D., Al-abdeen, J., & Sidhu, J. (2019). Food safety knowledge, attitudes and practices of food handlers in restaurants in Kuwait. *Food Control, 103*, 103–110. doi:10.1016/j.foodcont.2019.03.040

Asim, H. S., Elnemr, I., Goktepe, I., Feng, H., Park, H. K., Alzeyara, S., AlHajri, M., & Kushad, M. (2019). Assessing safe food handling knowledge and practices of food service managers in Doha, Qatar. *Food Science & Technology International, 25*(5), 440–448. doi:10.1177/1082013219830843 PMID:30791700

Asmawi, U. M. M., Norehan, A. A., Salikin, K., Rosdi, N. A. S., Munir, N. A. T. A., Basri, N. B. M., ... Nor, N. M. (2018). An Assessment of Knowledge, Attitudes and Practices in Food Safety Among Food Handlers Engaged in Food Courts. *Current Research in Nutrition and Food Science, 06*(2), 346–353. doi:10.12944/CRNFSJ.6.2.09

Balaganesan, S., & Siti Nur Afifah, J. (2019). A Survey On Food Handling Knowledge And Practices Among Housewives In Selayang, Selangor. *MAEH Journal of Environmental Health, 1*(1), 38–43.

Baser, F., Ture, H., Abubakirova, A., Sanlier, N., & Cil, B. (2017). Structural modeling of the relationship among food safety knowledge, attitude and behavior of hotel staff in Turkey. *Food Control, 73*, 438–444. doi:10.1016/j.foodcont.2016.08.032

Chua, A. L., Aziah, I., Balaram, P., Bhuvanendran, S., Anthony, A. A., Mohmad, S. N., Nasir, N. M., Hassan, H., Naim, R., Meran, L. P., Hussin, H. M., & Ismail, A. (2015). Identification of Carriers Among Individuals Recruited in the Typhoid Registry in Malaysia Using Stool Culture, Polymerase Chain Reaction, and Dot Enzyme Immunoassay as Detection Tools. *Asia-Pacific Journal of Public Health, 27*(2), NP2740–NP2748. doi:10.1177/1010539512458521 PMID:23000800

Cook, D. J., Mulrow, C. D., & Haynes, R. B. (1997). Systematic Reviews: Synthesis of Best Evidence for Clinical Decisions. *Annals of Internal Medicine, 126*(5), 376–380. doi:10.7326/0003-4819-126-5-199703010-00006 PMID:9054282

Dewanti-Hariyadi, R., & Gitapratiwi, D. (2014). Prevalence of Foodborne Diseases in South East and Central Asia. In Encyclopedia of Food Safety (Vol. 1, pp. 287–294). Academic Press.

Dora-Liyana, A. L., Mahyudin, N. A., Ismail-Fitry, M. R., Ahmad-Zaki, A., & Rasiyuddin, H. (2018). Food Safety and Hygiene Knowledge, Attitude and Practices among Food Handlers at Boarding Schools in the Northern Region of Malaysia. *International Journal of Academic Research in Business & Social Sciences*, *8*(17), 238–266.

Dumitraşcu, L., Ioana, A., Neagu, C., Didier, P., Maître, I., Nguyen-the, C., ... Borda, D. (2020). Time-temperature profiles and Listeria monocytogenes presence in refrigerators from households with vulnerable consumers. *Food Control*, *111*, 107078. doi:10.1016/j.foodcont.2019.107078

Elshahat, S., Woodside, J. V., & Mckinley, M. C. (2019). Meat thermometer usage amongst European and North American consumers: A scoping review. *Food Control*, *106*, 106692. doi:10.1016/j.foodcont.2019.06.018

Food Safety and Quality Division. (2018). *Annual Report Food Safety And Quality 2018*. Author.

Garayoa, R., Abundancia, C., Díez-Leturia, M., & Vitas, A. I. (2017). Essential tools for food safety surveillance in catering services: On-site inspections and control of high risk cross-contamination surfaces. *Food Control*, *75*, 48–54. doi:10.1016/j.foodcont.2016.12.032

Jones, T. L., Baxter, M. A. J., & Khanduja, V. (2013). A quick guide to survey research. *Annals of the Royal College of Surgeons of England*, *95*(1), 5–7. doi:10.1308/003588413X13511609956372 PMID:23317709

Katiyo, W., De Kock, H. L., Coorey, R., & Buys, E. M. (2019). Assessment of safety risks associated with handling chicken as based on practices and knowledge of a group of South African consumers. *Food Control*, *101*, 104–111. doi:10.1016/j.foodcont.2019.02.027

Kosa, K. M., Cates, S. C., Godwin, S., & Chabers, I. V. E. (2017). Barriers to Using a Food Thermometer When Cooking Poultry at Home: Results from a National Survey. *Food Protection Trends*, *37*(2), 116–125.

Lee, H. K., Abdul Halim, H., Thong, K. L., & Chai, L. C. (2017). Assessment of Food Safety Knowledge, Attitude, Self-Reported Practices, and Microbiological Hand Hygiene of Food Handlers. *International Journal of Environmental Research and Public Health*, *14*(1), 55–68. doi:10.3390/ijerph14010055 PMID:28098788

Lee, J., Jeong, S., Ko, G., Park, H., & Ko, Y. (2016). Development of a Food Safety and Nutrition Education Program for Adolescents by Applying Social Cognitive Theory. *Osong Public Health and Research Perspectives*, *7*(4), 248–260. doi:10.1016/j.phrp.2016.05.005 PMID:27635375

Lim, T.-P., Chye, F. Y., Mohd Rosni, S., Norazah, M. S., & Lee, J.-S. (2016). A structural modeling on food safety knowledge, attitude, and behaviour among Bum Bum Island community of Semporna, Sabah. *Food Control*, *60*, 241–246. doi:10.1016/j.foodcont.2015.07.042

Madaki, M. Y., & Bavorova, M. (2019). Food safety knowledge of food vendors of higher educational institutions in Bauchi state, Nigeria. *Food Control*, *106*, 106703. doi:10.1016/j.foodcont.2019.06.029

Ministry of Health Malaysia. (2014). *5 Guides To Safer Food*. Retrieved from www.myhealth.gov.my/en/5-guides-to-safer-food/%0D

Ministry of Health Malaysia. (2016). *Health Facts 2016 (Reference data for 2015)*. Retrieved from www.moh.gov.my/moh/resources/Penerbitan/Penerbitan Utama/HEALTH FACTS/KKMHEALTH FACTS 2016.pdf%0D

Ministry of Health Malaysia. (2017). *Health Facts 2017 (Reference data for 2016)*. Retrieved from https://myhdw.moh.gov.my/public/documents/20186/150084/HEALTH+FACTS+2017/98041185-ce34-4877-9ea1-4d5341e43187

Mohamad Fithri, A. O., Nurdiyana Syahirah, A. M., Siti Nurnajwa Nadhirah, M. S., Norsuhaida, Z., Nur Farhah Najwa, A., Siti Farah Alwani, M. N., & Zaini, M. Z. (2018). Determination of the Optimum Time for Preparation of Half-Boiled Eggs Free from Salmonella Enterica Serovar Enteritidis. *Journal of Clinical And Health Sciences*, *3*(1), 16–19. doi:10.24191/jchs.v3i1.6152

Mohamad Hapiz, A. R. (2019). A study on causal relationship between food handler's awareness and safe food handling practices in Malaysia. *Politeknik & Kolej Komuniti Journal of Social Sciences and Humanities*, *4*(1), 145–162.

Mohd Firdaus Siau, A., Son, R., Mohhiddin, O., Toh, P. S., & Chai, L. C. (2015). Food court hygiene assessment and food safety knowledge, attitudes and practices of food handlers in Putrajaya. *International Food Research Journal*, *22*(5), 1843–1854.

Mohd Halim, J., Mazni, S., Mohd Shazali, M. S., & Norazmir, M. N. (2016). Hygiene practices and food safety knowledge for biological, chemical and physical hazards. *Social Sciences*, *11*(19), 4633–4637.

Mohd Yusof, A. M., Rahman, N. A. A., & Haque, M. (2018). Knowledge, attitude, and practice toward food poisoning among food handlers and dietetic students in a public university in Malaysia. *Journal of Pharmacy & Bioallied Sciences*, *10*(4), 232–239. doi:10.4103/JPBS.JPBS_141_18 PMID:30568381

Mohd Zain, M., & Naing, N. N. (2002). Sociodemographic characteristics of food handlers and their knowledge, attitude and practice towards food sanitation: A preliminary report. *The Southeast Asian Journal of Tropical Medicine and Public Health*, *33*(2), 410–417. PMID:12236444

Moher, D., Liberati, A., Tetzlaff, J., & Altman, D. G.The Prisma Group. (2009). Preferred reporting items for systematic reviews and meta-analyses: The PRISMA statement. *PLoS Medicine*, *6*(7), e1000097. doi:10.1371/journal.pmed.1000097 PMID:19621072

Mustaffa, N. A., Rahman, R. A., Hassim, M. H., & Ngadi, N. (2017). Evaluation of Knowledge, Attitude and Practices of Food Handlers in Campus Cafeterias. *Chemical Engineering Transactions*, *56*, 1297–1302.

Nasrolahei, M., Mirshafiee, S., Kholdi, S., Salehian, M., & Nasrolahei, M. (2017). Bacterial assessment of food handlers in Sari City, Mazandaran Province, North of Iran. *Journal of Infection and Public Health*, *10*(2), 171–176. doi:10.1016/j.jiph.2016.03.006 PMID:27435639

Nee, S. O., & Abdullah Sani, N. (2011). Assessment of knowledge, attitudes and practices (KAP) among food handlers at residential colleges and canteen regarding food safety. *Sains Malaysiana*, *40*(4), 403–410.

Nesbitt, A., Thomas, M. K., Marshall, B., Snedeker, K., Meleta, K., Watson, B., & Bienefeld, M. (2014). Baseline for consumer food safety knowledge and behaviour in Canada. *Food Control*, *38*, 157–173. doi:10.1016/j.foodcont.2013.10.010

New Straits Times. (2019, February 15). 49 Keramat students struck down with food poisoning; canteen ordered shut. *New Straits Times*.

Norazmir, M. N., Noor Hasyimah, M. A., Siti Shafurah, A., Siti Sabariah, B., Ajau, D., & Norazlanshah, H. (2012). Knowledge and Practices on Food Safety among Secondary School Students in Johor Bahru, Johor, Malaysia. *Pakistan Journal of Nutrition*, *11*(2), 110–115. doi:10.3923/pjn.2012.110.115

Nur Izyan, F. R., Zuraini, M. I., Maria, M. S., Lovelyna, B. J., Maimunah, M., & Saidatul Afzan, A. A. (2019). A preliminary study on food safety knowledge, attitude and practices among home-based food providers in Klang Valley, Malaysia. *Malaysian Applied Biology*, *48*(2), 157–160.

Okour, A. M., Alzein, E., Saadeh, R., & Alfaqih, M. (2020). Food safety knowledge among Jordanians: A national study. *Food Control, 114*, 107216. doi:10.1016/j.foodcont.2020.107216

Parry-Hanson Kunadu, A., Ofosu, D. B., Aboagye, E., & Tano-Debrah, K. (2016). Food safety knowledge, attitudes and self-reported practices of food handlers in institutional foodservice in Accra, Ghana. *Food Control, 69*, 324–330. doi:10.1016/j.foodcont.2016.05.011

Queirós, A., Faria, D., & Almeida, F. (2017). Strengths and limitations of qualitative and quantitative research methods. *European Journal of Education Studies, 3*(9), 369–387.

Rebouças, L. T., Santiago, L. B., Martins, L. S., Menezes, A. C. R., & Araújo, M. (2017). Food safety knowledge and practices of food handlers, head chefs and managers in hotels' restaurants of Salvador, Brazil. *Food Control, 73*, 372–381. doi:10.1016/j.foodcont.2016.08.026

Redmond, E. C., & Griffith, C. J. (2003). Consumer food handling in the home: A review of food safety studies. *Journal of Food Protection, 66*(1), 130–161. doi:10.4315/0362-028X-66.1.130 PMID:12540194

Rodriguez-Martinez, V., Velázquez, G., & Altaif, R. (2020). Deterministic and probabilistic predictive microbiology-based indicator of the listeriosis and microbial spoilage risk of pasteurized milk stored in residential refrigerators. *Lebensmittel-Wissenschaft + Technologie, 117*, 108650. doi:10.1016/j.lwt.2019.108650

Rosnani, A. H., Son, R., Mohhidin, O., Toh, P. S., & Chai, L. C. (2014). Assessment of knowledge, attitude and practices concerning food safety among restaurant workers in Putrajaya, Malaysia. *Food Science and Quality Management, 32*, 20–28.

Ruby, G. E., Ungku Zainal Abidin, U. F., Lihan, S., Jambari, N. N., & Radu, S. (2019). A cross sectional study on food safety knowledge among adult consumers. *Food Control, 99*, 98–105. doi:10.1016/j.foodcont.2018.12.045

Saira Banu, M. R., Humairak, S., Zakiah, M. D., & Siti Adila, Z. A. (2019). Retrospective study on persistent salmonella serotypes in meat samples tested in the veterinary public health section, regional of veterinary laboratory Bukit Tengah, Penang. *Malaysian Journal of Veterinary Research, 10*(1), 34–42.

Samapundo, S., Cam Thanh, T. N., Xhaferi, R., & Devlieghere, F. (2016). Food safety knowledge, attitudes and practices of street food vendors and consumers in Ho Chi Minh city, Vietnam. *Food Control, 70*, 79–89. doi:10.1016/j.foodcont.2016.05.037

Sánchez, M., Neira, C., Laca, A., Laca, A., & Díaz, M. (2019). Survival and development of Staphylococcus in egg products. *Lebensmittel-Wissenschaft + Technologie*, *101*, 685–693. doi:10.1016/j.lwt.2018.11.092

Saw, S. H., Mak, J. L., Tan, M. H., Teo, S. T., Tan, T. Y., Cheow, M. Y. K., Ong, C. A., Chen, S. N., Yeo, S. K., Kuan, C. S., New, C. Y., Radu, S., Phuah, E. T., Thung, T. Y., & Kuan, C. H. (2020). Detection and quantification of Salmonella in fresh vegetables in Perak, Malaysia. *Food Research*, *4*(2), 441–448. doi:10.26656/fr.2017.4(2).316

Siti Nurul Ain, S., Sahilah, A. M., & Razalee, S. (2018). Knowledge, attitude and practice of food utensils hygiene amongst food handlers in Kuala Pilah, Negeri Sembilan, Malaysia. *Sains Malaysiana*, *47*(7), 1527–1533. doi:10.17576/jsm-2018-4707-21

Syahira, S., Huda, B. Z., & Mohd Rafee, B. B. (2019). Factors associated with level of food safety knowledge among form four students In Hulu Langat District, Selangor. *International Journal of Public Health and Clinical Sciences*, *6*(2), 252–265. doi:10.32827/ijphcs.6.2.252

Taha, S., Osaili, T. M., Saddal, N. K., Al-Nabulsi, A. A., Ayyash, M. M., & Obaid, R. S. (2020). Food safety knowledge among food handlers in food service establishments in United Arab Emirates. *Food Control*, *110*, 106968. doi:10.1016/j.foodcont.2019.106968

Tan, S. L., Cheng, P. L., Soon, H. K., Ghazali, H., & Mahyudin, N. A. (2013). A qualitative study on personal hygiene knowledge and practices among food handlers at selected primary schools in Klang valley area, Selangor, Malaysia. *International Food Research Journal*, *20*(1), 71–76.

Tan, S. L., Fatimah, A. B., Muhammad Shahrim, A. K., Lee, H. Y., & Nor Ainy, M. (2013). Hand hygiene knowledge, attitudes and practices among food handlers at primary schools in Hulu Langat district, Selangor (Malaysia). *Food Control*, *34*(2), 428–435. doi:10.1016/j.foodcont.2013.04.045

Tegegne, H. A., & Phyo, H. W. W. (2017). Food safety knowledge, attitude and practices of meat handler in abattoir and retail meat shops of Jigjiga Town, Ethiopia. *Journal of Preventive Medicine and Hygiene*, *58*, 320–327. PMID:29707664

The Malaysian National News Agency (2018, October 13). Three more food poisoning cases reported in Perak. *New Straits Times*.

The Malaysian National News Agency (2019, June 19). 110 college students down with food poisoning. *New Straits Times*.

The Malaysian National News Agency (2020, January 6). "Ayam masak lemak" dish downs 137 kindergarten teachers. *Malaysiakini.*

Tutu, B. O., Hushie, C., Asante, R., & Egyakwa-Amusah, J. A. (2020). Food safety knowledge and self-reported practices among school children in the Ga West Municipality in Ghana. *Food Control, 110*, 107012. doi:10.1016/j.foodcont.2019.107012

Unusan, N. (2007). Consumer food safety knowledge and practices in the home in Turkey. *Food Control, 18*(1), 45–51. doi:10.1016/j.foodcont.2005.08.006

Whited, T., Feng, Y., & Bruhn, C. M. (2019). Evaluation of the high school food safety curriculum using a positive deviance model. *Food Control, 96*, 324–328. doi:10.1016/j.foodcont.2018.09.004

World Health Organization. (2002). Statistical information on food-borne disease in europe microbiological and chemical hazards. *Pan European Conference on Food Safety and Quality*, 60–85.

Young, I., Waddell, L. A., Wilhelm, B. J., & Greig, J. (2020). A systematic review and meta-regression of single group, pre-post studies evaluating food safety education and training interventions for food handlers. *Food Research International, 128*, 108711. doi:10.1016/j.foodres.2019.108711 PMID:31955782

Yu, H., Sirsat, S. A., & Madera, J. M. (2018). Enhancing hospitality students' motivation to learn food safety knowledge using tablet personal computer (TPC)-based simulation game class activity. *Journal of Hospitality, Leisure, Sport and Tourism Education, 23*, 82–94. doi:10.1016/j.jhlste.2018.08.002

ADDITIONAL READING

Bai, L., Wang, M., Yang, Y., & Gong, S. (2019). Food safety in restaurants: The consumer perspective. *International Journal of Hospitality Management, 77*, 139–146. doi:10.1016/j.ijhm.2018.06.023

Dudeja, P., & Singh, A. (2017). Food handlers. In R. K. Gupta, P. Dudeja, & A. S. Minhas (Eds.), *Food Safety in the 21st Century: Public Health Perspective* (pp. 269–280). doi:10.1016/B978-0-12-801773-9.00021-2

Dumitraşcu, L., Ioana, A., Neagu, C., Didier, P., Maître, I., Nguyen-the, C., ... Borda, D. (2020). Time-temperature profiles and *Listeria monocytogenes* presence in refrigerators from households with vulnerable consumers. *Food Control, 111*, 107078. doi:10.1016/j.foodcont.2019.107078

Gallo, M., Ferrara, L., Calogero, A., Montesano, D., & Naviglio, D. (2020). Relationships between food and diseases: What to know to ensure food safety. *Food Research International*, *137*, 109414. doi:10.1016/j.foodres.2020.109414 PMID:33233102

Her, E., Seo, S., Choi, J., Pool, V., & Ilic, S. (2019). Assessment of food safety at university food courts using surveys, observations, and microbial testing. *Food Control*, *103*, 167–174. doi:10.1016/j.foodcont.2019.04.002

Kosa, K. M., Cates, S. C., Godwin, S., & Chabers, I. V. E. (2017). Barriers to using a food thermometer when cooking poultry at home: Results from a national survey. *Food Protection Trends*, *37*(2), 116–125.

Miller, B., & Notermans, S. H. W. (2014). Food Poisoning Outbreaks. In C. A. Batt & M. L. Tortorello (Eds.), Encyclopedia of Food Microbiology (Vol. 1, pp. 954–958). Elsevier. doi:10.1016/B978-0-12-384730-0.00128-2

Murray, K., Tremblay, C., Rghei, A., & Warriner, K. (2018). Challenges and options for enhancing Salmonella control in partially cooked breaded poultry products. *Current Opinion in Food Science*, *20*, 44–50. doi:10.1016/j.cofs.2018.03.003

KEY TERMS AND DEFINITIONS

Adult: Those aged 19 years and above.

Consumer: Individuals involved in food handling at home and did not receive any training related to food handling.

Food Premises: Premises involved in the preparation, serving and handling of food.

Chapter 8
Assessing Consumer Awareness and Opinion Toward Food Safety Practices and Policies in the Maldives

Siti Nurhayati Khairatun
https://orcid.org/0000-0002-5239-8212
Universiti Putra Malaysia, Malaysia

Fathmath Nuzla
Ministry of Health, Maldives

ABSTRACT

The Maldives relies completely on imported food for its staples such as rice, flour, and sugar. It is reported that the Maldives produces less than one-tenth of its overall food requirements. Due to its huge dependence on imported food products, Maldives is exposed to a high risk of contaminated food and foodborne illnesses caused within the supply chain. This chapter aimed to investigate the level of awareness among the public of the Maldives relating to food safety practices in the food industry. An online survey was developed and administered to the consumers in the Maldives. A total number of 369 usable responses were analyzed statistically. The findings highlight that even though there was a high level of awareness of the participants towards food safety in general, participants were poorly aware of the activities and the consumers' roles in the food safety practices. This study serves as a baseline study for future research in this area, particularly in the Maldives.

DOI: 10.4018/978-1-7998-7415-7.ch008

BACKGROUND OF THE STUDY

Food safety is a major concern in all countries around the world. World Health Organization (WHO) estimated 600 million foodborne illnesses and 420,000 deaths in the year 2010 (WHO, 2015). It is reported that the foodborne diseases have caused the total productivity loss associated with foodborne disease in low- and middle-income countries was around US$ 95.2 billion per year, and the annual cost of treating foodborne illnesses is estimated at US$ 15 billion (World Bank, 2018). Food safety practices have become complex with the emerging trends and behaviors in the food supply chain, particularly during the preparation and processing stages.

Human is the primary stakeholder in food consumption framework. For decades, food safety is an aged concern since the day's humans have developed measures to make sure that the food consumed does not harm them in any way. Written records and problem-based solutions have been established to mitigate the contamination and to enhance the safety points of many products. Additionally, numerous public awareness programs have been conducted to increase public knowledge in food safety and to ensure that the food standards set by authorities are effectively implemented. In many countries, the proactive participation of consumer associations allows the consumer representatives involvement in policy reforms and implementation at national and international levels.

While a proper legal framework provides a detailed guideline for the food industry regarding to food safety management within their establishment, consumer awareness is necessary for the prevention and implementation of food safety measures in a community. It is known that a legal framework alone cannot ensure that the food will never pose a risk to any consumer (De Boer & Bast, 2018). Based on the code of ethics for international food trades as outlined in Codex Alimentarius, the consumers have the right to safe and quality food, as well as the right to express their opinions about the food control procedures, including standards set out by the governments and the food industries to ensure safe, sound and wholesomeness of the food (FAO & WHO, 2010).

Numerous studies relating to consumers' level of awareness toward food safety have been done in different demographic settings. Today thc food travels at great distances from different countries, different farms, and factories before being passed on to the consumer. The world's increasing population and the consumers' needs for a variety of food have resulted in a more complex network of the food chain. Nonetheless, a complex food chain can cause contamination at any stage of the food chain. As a result, foodborne illnesses pose a risk to international public health safety and economic development (Uçar, Yilmaz, & Çakiroglu, 2016). While this scenario has happened around the world with no exception to the Maldives. The Maldives has imported more than 95 percent of the food consumed from various parts of the world (Maldives Food and Drug Authority, 2017). Every year there is an estimate of 600 million cases of foodborne illnesses globally (WHO, 2015). Though there is no specific monitoring data available for foodborne diseases in the Maldives, 30442 cases of acute gastrointestinal or diarrheal diseases were reported in 2015 alone (Ministry of Health, 2016). The National Food Safety Policy (NFSP) has highlighted the need for the assessment of foodborne disease surveillance and response through different means such as validation workshops and self-administered questionnaires (Maldives Food Drug Authority, 2017). In an effort to design an effective advocacy framework and to engage positively with the local community, it is crucial to understand their level of awareness and understanding about the food safety practices in the Maldives. Lack of consumers' awareness of policies governing food safety deprives the consumers of means of receiving information on food safety activities done by the regulatory authorities and can cause consumers to lose trust in the food safety system of the country. As for now, there are no published studies in this field in the Maldives.

A conceptual framework is developed within the basis of the health belief model and the theory of planned behavior. The independent variables of this conceptual model are sociodemographic characteristics (gender, age, education level, working situation, food safety training, and marital status). The dependent variables are consumers' awareness of food safety and consumers' awareness of food safety policy. As this study aims to assess the relationship between consumers' opinions and awareness of food safety policy, awareness acts as a mediator variable.

Thus, this paper investigated the consumers' level of awareness toward food safety practices in the Maldives. The significance of this study is to provide a baseline reference for the policymakers in developing effective awareness programs among the consumers in the Maldives particularly.

Literature Review

This literature review is intended to describe the key components of this study with a reference to numerous past studies conducted in various countries.

Consumers' Awareness of Food Safety

The importance of food safety is instrumental in protecting consumers from foodborne illnesses and promoting public health and market safer foods. There are several cases of foodborne diseases that can be prevented by mere education and awareness. While many of the consumers display a certain level of awareness in food safety practices, many studies done suggested that it was limited by the knowledge of food safety acquired (Liu & Niyongira, 2017). Also, a positive relationship is observed in food safety knowledge and education level (Ruby et al., 2019). Studies conduct to measure the level of awareness on personal hygiene practices among consumers of different categories and regions show concerning disparities. A high level of awareness was recorded on some of the personal hygiene questions such as washing hands with soap after using the toilet (Tomaszewska et al., 2018) among consumers from Thailand and Poland. However, a study conducted in Miami concluded that pregnant women and mothers, in general, reported less frequent hand washing and had poor food safety practices (Davila et al., 2009).

Many of the consumers are unaware of food safety practices such as microbiological contamination of food kept that is kept at room temperature for a longer time (Odeyemi et al., 2019). The study conducted in Malaysia shows that less than a quarter of participants had detailed knowledge of temperature effects on bacterial growth (Ruby et al., 2019). Interestingly, a majority of the participants were aware that foodborne bacterial can be fatal though, there was little awareness of specific foodborne pathogens and they were not concerned or aware of the food safety issues while purchasing ready-to-eat street food (Asiegbu, Lebelo, & Tabit, 2016).

In Turkey, public concerns were acknowledged regarding food safety and food handling practices to prevent foodborne illnesses (Ergönül, 2013). Tomaszewska et al. (2018) found that consumers from Poland and Thailand have incorrect knowledge of food preparation practices and the cause of foodborne illnesses. Meanwhile, Spanish consumers have a higher level of awareness of food labels and the majority read the label before purchasing the food (Todt et al., 2009). On the contrary, the practice of checking labels before purchasing food is low in Asia and Africa (Odeyemi et al., 2019).

Consumers' Opinion of Food Safety

Consumers have the right to give their opinion and concerns when food safety and quality are involved (FAO & WHO, 2010). Numerous studies indicated the concerns of consumers of food safety. In Vietnam, consumers were concerned about the safety of food and perceived that there is a high risk in food due to pesticide residues, preservatives, and using hormones in livestock (Ha, Shakur, & Pham Do, 2019). Similarly, another study conducted in Vietnam shows a majority of the consumers are concerned about the use of pesticides and growth chemicals which can potentially lead to residues in vegetables (Ngo et al., 2020). Increased consumer concerns are expressed due to an observable increase in food adulteration and food fraud in recent years (Medina et al., 2019).

Consumers from rural areas are less concerned about food safety as they perceived they have better control of the food through the integrated farming and livestock husbandry. Nonetheless, consumers from urban areas have a higher level of concern as they perceived they have a low level of control over food (vegetables and meat) (Ha et al., 2019). As education is one of the major factors that influence various aspects of life, consumers with a lower level of education are seen to be less concerned in certain areas of food safety while consumers with higher education and higher expenditure are more likely to have high concerns for food safety. A high level of concern for food safety is also seen in women with families and children (Liu & Niyongira, 2017).

History and Development of the National Food Safety Policy in the Maldives

The Maldives is a long-standing member country of WHO since 1965, to assist with and improve food safety activities in the country (WHO, 1998). The WHA53 resolution recognizes food safety as an essential priority for public health and committed WHO and its Member States to series of multidisciplinary and multisectoral actions to promote food safety at international, national, and local levels. Specifically, it focuses on the expansion of WHO responsibility in food safety, and promotes food safety as an essential public health function by utilization of the limited resources available and recommended suitable interventions to improve global food safety (WHO, 2002).

Since the beginning of the 1990s, the WHO regional office South-East Asia has been working simultaneously with the Member States to improve national food safety programs through collaborative activities and by providing technical assistance. In 1993 a review was done on the food safety initiatives in the region which results in the formation of the 10-point regional strategy for food safety in the South-East Asia Region. This blueprint identifies 10 key strategies that would be useful in achieving safe food for all in the region including (i) food safety policy; (ii) food legislation; (iii) food control and inspection; (iv) analytical capability; (v) epidemiological system; (vi) relationships with the food industry and trade; (vii) relationships with food service providers and retailers; (viii) relationships with consumers; (ix) education and training; and (x) research in food safety. Through this 10-point strategy WHO urges all Member States to develop their own national food safety policies (WHO, 2013).

Regional Food Safety Strategy 2013-2017 by WHO states that even though numerous action plans have been developed by the Member States, many of them do not have a specific policy for food safety including Maldives, Indonesia, India, Nepal, Bangladesh, and Thailand. WHO identifies the lack of coordination and cooperation between government agencies involved as the major cause of problems in most Member States. Also, in some Member countries, the responsibilities of the involved government sectors remain unclear with the absence of leadership in the implementation and monitoring of the programs. In support of the WHO call for a food safety policy, the government of the Maldives drafted a bill in 2011 (WHO, 2013).

Under WHO technical assistance, the initial draft of the National Food Safety Policy came into existence in 2015. This policy was developed based on a multisectoral and integrated farm-to-fork approach and was done with the consideration of the traditional, cultural, and religious practices in the context and perspective of the Maldivians. Four policy goals were identified and objectives for each goal were developed. Along with proposed responsible agencies, a strategic approach for each objective was identified in the policy. Also, the action plans for each strategic activity with a time frame and coordinating mechanisms were developed through consultations. The workshop for stakeholders was held in 2016 for concerns and comments on the policy.

Implementation of the policy was planned within 10 years with technical and human resources along with sufficient budget and coordination with the stakeholders. There is an important role to be played by the high-level Inter-ministerial Food Safety Committee and by the National Codex Committee. For the implementation of the NFSP, it will be necessary to enact and enforce food safety-related statutory Acts and Regulations. This can be achieved through the development of adequate infrastructure, capacity building, and collaboration among stakeholders and consumers' education and awareness (Maldives Food and Drug Authority, 2017).

METHODOLOGY

The research design of this study was a quantitative survey. An online survey was developed using a free online survey development cloud-based software. The original survey was in English and a translation was provided in Dhivehi (the native language of Maldives) to attract more native respondents. Prior to data collection, an ethical approval was obtained from UPM Research Management Center (RMC) with reference number UPM/TNCPI/RMC/JKEUPM/1.4.18.2 (JKEUPM). The location selected for this study was in Malé City, Maldives. Malé City has the highest population in the Maldives with a population of 133,412 people (National Burcau of Statistics, 2017). The main reason for this selection was because the co-researcher has access to the local respondents and it was convenient to complete the data collection activities there. The target respondents for this study are all consumers age 18 years and above living in Malé City, Maldives. It must be noted that in the Maldives, the legal age of being an adult starts from 18 years. It is important to recruit respondents of legal age so it will give them the legal right to give consent either or not to participate voluntarily in the study, thus, avoiding any legal issues with the consent granted for the study. Figure 1 illustrates the seven major data collection steps taken in the study which includes, i) develop a research design; ii) identify the location for data collection; iii) identify the population; iv) determine the sample size; v) collect the data via online platforms; vi) analyze data, and vii) report the final results.

Figure 1. Data collection steps in this study

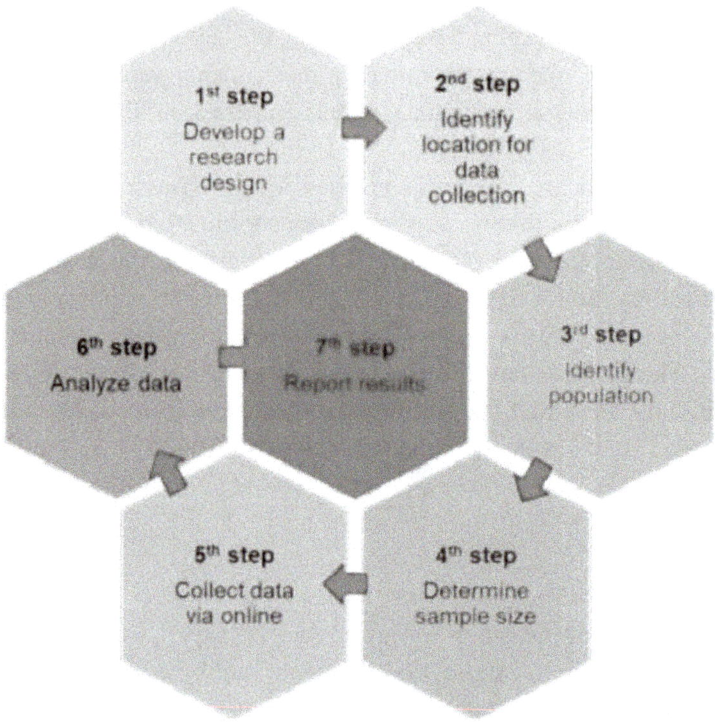

Sample Size

The total population of Maldives is 344,023 and the population total Malé City is 133,412 (National Bureau of Statistics, 2017). However, there is no data published for the population of Malé City over the age of 18 years. Due to this reason the target population was estimated to be 100,000. The sample size was calculated using n = z2 * p * (1 - p) / e2 formula (Daniel & Cross, 2013), utilizing z score and population proportion. Where: z = 1.96 for a confidence level (α) of 95%, p = proportion (expressed as a decimal), N = population size and e = margin of error. The calculated sample size is 383. Below is the sample size calculation used in this study.

$z = 1.96, p = 0.5, N = 100000, e = 0.05$
$n = [1.962 * 0.5 * (1 - 0.5) / 0.052] / [1 + (1.962 * 0.5 * (1 - 0.5) / (0.052 * 100000))]$
$n = 384.16 / 1.0038 = 382.69$
$n \approx 383$

The sample size (with finite population correction) is equal to 383.

Survey Instrument

The survey instrument in this study consisted of 42 questions derived from past literature. The questionnaire consisted of two major sections. Section 1 was designed to capture the socio-demographic profiles of the participants and it consisted of 6 questions. Section 2 was created to investigate the level of awareness and opinion of the participants towards food safety practices. This section was further divided into 4 sub-sections to assess different aspects of food safety as follows:

Sub-section 2.1: This section aimed to evaluate the level of awareness of food hygiene practices among the participants. Questions are designed with a 5-point Likert scale (always, often, sometimes, seldom, never) to measure the frequency of the hygiene practices employed by the participants. This section consists of 6 questions.

Sub-section 2.2: This section had 8 questions, is intended to evaluate the level of awareness on food safety practices by measuring the frequency of these food safety practices employed by the participants and measured using a 5-point Likert scale (always, often, sometimes, seldom, never).

Sub-section 2.3: This section was designed to evaluate the opinions of the participants on food safety in different aspects with 5 points Likert scale to measure how much the participants agree or disagree on the given statements (strongly agree, agree, undecided, disagree, strongly disagree). This section consisted of 9 questions.

Sub-section 2.4: This section was developed to capture the awareness of the participants on the food safety policy and its activities. It was measured with 'yes, no, and unsure' options to identify if the participants were aware of the roles and activities implemented by the government that has been put in place to ensure food safety. There were 13 questions in this section.

Data Collection

Samples were collected through convenience sampling by sharing survey links through social media and messaging platforms such as Facebook, Viber, WhatsApp, and Telegram. The sample collection was completely voluntary and the participants had the right to leave the survey at any point. The sample collection was done over six weeks. A total of 470 responses were received through the online survey, but due to selection criteria and respondents opting out from answering the questions after starting the survey only 369 responses were usable for statistical analysis. The collected samples were generated in a google sheet which later was converted to Microsoft excel file format for data analysis.

Data Analysis

Data were analyzed using Microsoft Excel and Statistical Package for the Social Sciences (SPSS) version 25.0. Descriptive analysis was done using Microsoft Excel and the inferential analysis was done using SPSS 25.0. The raw data was initially collected in google sheets and converted to Microsoft Excel files. The data was further computed and converted to numerical data through SPSS 25.0 for data analysis.

The normality testing for data was done to determine the statistical tests to confirm the relationship or association between different variables. As data distribution was not normal, which was common when using the Likert scale for measurements, non-parametric statistical testing was employed. Mann-Whitney U test was done to test dichotomous variables (gender, working situation, training situation) in the sociodemographic characteristics against the dependent variables (awareness of food safety policy, awareness of food safety practices, and consumers' opinions towards food safety) (see Table 1). Finally, the Kruskal-Wallis test was used to analyze the association between marital status (independent variable) and the dependent variables (awareness of food safety, awareness of food safety training, and consumers' opinions towards food safety) of the study (see Table 2). Spearman's rho correlation was used to analyze the relationship between the level of education, age against awareness on food safety, awareness of food safety policy, and consumers' opinions towards food safety (see Table 3).

The level of awareness score focused mainly on the classification of 'good' (more than 75%) and 'adequate' (50 to 75%) and 'poor' (less than 50%). On the other hand, the food safety opinion section evaluated how much consumers understand food safety practices. The scoring is similar to the score used by Samapundo, Cam Thanh, Xhaferi, & Devlieghere, (2016).

Table 1. Mann-Whitney U test

Analysis of awareness of food safety and gender				
Ranks				
	Gender	**N**	**Mean Rank**	**Sum of Ranks**
AVawareness2.1n2.2	Male	133	140.32	18663.00
	Female	236	210.18	49602.00
	Total	369		
Test Statistics				
	AVawareness2.1n2.2			
Mann-Whitney U	9752.000			
Wilcoxon W	18663.000			
Z	-6.049			
Asymp. Sig. (2-tailed)	.000			
a. Grouping Variable: Gender				
Analysis of awareness of food safety with training situation				
Ranks				
	RCFST	**N**	**Mean Rank**	**Sum of Ranks**
AVawareness2.1n2.2	No	316	176.93	55910.00
	Yes	53	233.11	12355.00
	Total	369		
Test Statistics				
	AVawareness2.1n2.2			
Mann-Whitney U	5824.000			
Wilcoxon W	55910.000			
Z	-3.554			
Asymp. Sig. (2-tailed)	.000			
a. Grouping Variable: RCFST				
Analysis of awareness of food safety with working situation				
Ranks				
	RCwork	**N**	**Mean Rank**	**Sum of Ranks**
AVawareness2.1n2.2	No	281	171.76	48264.00
	Yes	88	227.28	20001.00
	Total	369		

continues on following page

Table 1. Continued

Test Statistics				
	AVawareness2.1n2.2			
Mann-Whitney U	8643.000			
Wilcoxon W	48264.000			
Z	-4.268			
Asymp. Sig. (2-tailed)	.000			
a. Grouping Variable: RCwork				
Analysis of awareness of FSP with gender				
Ranks				
	RCSex	**N**	**Mean Rank**	**Sum of Ranks**
FSPawareness	Male	133	202.33	26910.00
	Female	236	175.23	41355.00
	Total	369		
Test Statistics[a]				
	FSPawareness			
Mann-Whitney U	13389.000			
Wilcoxon W	41355.000			
Z	-2.363			
Asymp. Sig. (2-tailed)	.018			
a. Grouping Variable: RCSex				

Table 2. Kruskal-Wallis test analysis

Analysis of awareness of food safety with marital status			
Ranks			
	RCmarital	**N**	**Mean Rank**
AVawareness2.1n2.2	Single	134	159.63
	Married	218	201.70
	Divorced	17	170.85
	Total	369	

continues on following page

Table 2. Continued

Test Statistics			
	AVawareness2.1n2.2		
Kruskal-Wallis H	13.265		
df	2		
Asymp. Sig.	.001		
a. Kruskal Wallis Test b. Grouping Variable: RCmarital			
Analysis of awareness of FSP with marital status			
Ranks			
	RCmarital	**N**	**Mean Rank**
FSPawareness	Single	134	195.56
	Married	218	181.14
	Divorced	17	151.29
	Total	369	
Test Statistics[a,b]			
	FSPawareness		
Kruskal-Wallis H	3.353		
df	2		
Asymp. Sig.	.187		
a. Kruskal Wallis Test b. Grouping Variable: RCmarital			

Table 3. Spearman's rho correlation

Analysis of awareness of food safety with level of education				
Correlations				
			AVawareness2.1n2.2	Education
Spearman's rho	AVawareness2.1n2.2	Correlation Coefficient	1.000	.017
		Sig. (2-tailed)	.	.745
		N	369	369
	RCeducation	Correlation Coefficient	.017	1.000
		Sig. (2-tailed)	.745	.
		N	369	369

continues on following page

Table 3. Continued

Analysis of awareness of food safety with age				
Correlations				
			AVawareness2.1n2.2	Age categories
Spearman's rho	AVawareness2.1n2.2	Correlation Coefficient	1.000	.146**
		Sig. (2-tailed)	.	.005
		N	369	369
	RCage	Correlation Coefficient	.146**	1.000
		Sig. (2-tailed)	.005	.
		N	369	369
**. Correlation is significant at the 0.01 level (2-tailed).				
Analysis of awareness of FSP with level of education				
Correlations				
			FSPawareness	RCeducation
Spearman's rho	FSPawareness	Correlation Coefficient	1.000	.041
		Sig. (2-tailed)	.	.431
		N	369	369
	RCeducation	Correlation Coefficient	.041	1.000
		Sig. (2-tailed)	.431	.
		N	369	369
Analysis of awareness of FSP with age				
Correlations				
			FSPawareness	RCage
Spearman's rho	FSPawareness	Correlation Coefficient	1.000	.014
		Sig. (2-tailed)	.	.784
		N	369	369
	RCage	Correlation Coefficient	.014	1.000
		Sig. (2-tailed)	.784	.
		N	369	369

continues on following page

Table 3. Continued

Analysis of awareness of FSP with consumers' opinions of food safety				
Correlations				
			FSPawareness	Opinion
Spearman's rho	FSPawareness	Correlation Coefficient	1.000	.205**
		Sig. (2-tailed)	.	.000
		N	369	369
	Opinion	Correlation Coefficient	.205**	1.000
		Sig. (2-tailed)	.000	.
		N	369	369
**. Correlation is significant at the 0.01 level (2-tailed).				

Pilot Testing

The validity of the questionnaire was tested in a pilot study consisting of 20 participants (5% of the study population). This data was not included as data for the final study. After pilot testing, minor changes were done to the structure and sentencing of the questions in section 2, and the inclusion of another parameter for section 1 (sociodemographic characteristics). The validity and the reliability of the different sections of the research instrument were determined by calculating Cronbach's α value (see Table 4). Cronbach's α for different constructs ranges from 0.70 to 0.724 which was within the acceptable limits (Ursachi, Horodnic, & Zait, 2015).

Table 4. Cronbach's alpha reliability test

Reliability Statistics		
Cronbach's Alpha	**Cronbach's Alpha Based on Standardized Items**	**N of Items**
.724	.739	13
Reliability Statistics		
Cronbach's Alpha	**Cronbach's Alpha Based on Standardized Items**	**N of Items**
.700	.702	9

RESULTS AND DISCUSSIONS

Sociodemographic Characteristics

A total of 369 respondents participated in this study. As shown in Table 5 among 369 respondents, 236 were females (64%) and 133 were males (36%). In terms of age categories, the highest percentage of the respondents are of age 23-27 with 26.3%) and the lowest percentage of responses regarding age is 53 years and above (1.1%). Regarding the level of education, 58.3% (215) participants have tertiary and above education, and 22% (82) participants had higher secondary education. Only 8 respondents (2.2%) had only primary education. Among the respondents, 218 (59.1%) are married and 134 (36.2%) of participants are single and the rest 17 (4.6%) are divorced. Out of all the respondents, only 88 (23.8%) respondents deal with food safety in their work, while only 53 (14.4%) have received food safety training.

Table 5. Sociodemographic characteristics of the respondents

Variables	Frequency (N= 369)	Percentage
Gender		
Male	133	36.0
Female	236	64.0
Age categories		
18 – 22	48	13.0
23 – 27	97	26.3
28 – 32	92	24.9
33 – 37	73	19.8
38 – 42	28	7.6
42 – 47	21	5.7
48 – 52	6	1.6
53 and above	4	1.1
Education level		
Primary	8	2.2
Secondary	64	17.3
Higher secondary	82	22.2
Tertiary and above	215	58.3

continues on following page

Table 5. Continued

Variables	Frequency (N= 369)	Percentage
Marital status		
Single	134	36.3
Married	218	59.1
Divorced	17	4.6
Does your work deal with food safety?		
No	281	76.2
Yes	88	23.8
Have you received food safety training?		
No	316	85.6
Yes	53	14.4

Awareness of Consumers on Food Safety

The respondents have a good awareness of hand hygiene, where 286 (77.5%) and 43 (11.7%) wash their hands always and often respectively, before the preparation of food. Similarly, a high percentage of the respondents wash their hands after using the toilet. That is always 79.7 percent and often 13.8 percentage. The highest percentage of awareness is observed in washing hands after handling raw meats and poultry, with 325 (88.1%) participants. Personal hygiene awareness is consistent with other studies, where consumers were found highly knowledgeable on hand hygiene, especially handwashing.

Hand hygiene is perceived as the most critical component of food safety (Ruby et al., 2019) A study conducted in the United Arab Emirates also has a high percentage (62.5%) of good awareness in handwashing (Abushelaibi, Jobe, Al Dhanhani, Al Mansoori, & Al Shamsi, 2016). In comparison, Thailand consumers even though having adequate knowledge about handwashing, are less likely to wash hands before food preparation (Tomaszewska et al., 2018).

It is seen the participants have low awareness when it comes to washing hands for 20 seconds, where only 94 (24.4%) participants always practice this. On the contrary, only 37.9% of adult consumers in Malaysia are unaware of the proper duration of handwashing (Ruby et al., 2019). It is deduced that an inadequate handwashing duration may expose the consumers handling, preparing, or consuming the food to be subjected to foodborne pathogens such as Staphylococcus aureus and Salmonella spp. (Lin, Jensen, & Yen, 2005). Interestingly, (62.3%) 230 participants have good awareness in covering their mouths during coughing and sneezing however only 80

(21.7%) participants wash their hands after sneezing and coughing always. Likewise, Odeyemi et al. (2019) concluded that more than half (58%) of the respondents use hands to cover their mouths during coughing or sneezing.

A survey done in Northern Ireland observes that the most frequently checked information while purchasing food is the "use by" or "best before" information (Food Standard Agency, 2016). In contrast in this study, only 200 (54.2%) of the participants are very much aware of checking the labels for "use by" or "expiry date" before purchasing pre-packaged food. A lower percentage have good awareness when it comes to checking labels for allergens, where only 85 (23.5%) respondents check always. In regards to checking the instruction of packaged food for "condition of use" and "storage", 96 (26.0%) participants have good awareness as they always practice it. Conversely, Odeyemi et al. (2019) concluded that only 48% of respondents check the label for the condition of use and storage condition of the packaged food.

A majority of the participants (79.9%) always wash raw vegetables and raw fruits before eating and cooking. Tomaszewska et al. (2018) noted that comparing between Polish consumers and Thailand consumers, Thais were more aware of the risks of consumption of raw unwashed fruits and vegetables. In regards to storing food in the refrigerator, 216 (58.5%) respondents store raw meats and chicken separately. Similar results are reported by Odeyemi et al. (2019) with 48% of respondents. A benchmark survey done in Hong Kong by the Food and Environment Hygiene Department shows a comparatively high percentage (>90%) of the participants, stored cooked and raw food separately in the refrigerator (Mercado Solutions, 2018).

In the current study, 185 (50.1%) respondents used separate cutting boards always when cutting raw meat and other foods. On the other hand, in a survey conducted in Turkey while assessing the consumers' knowledge on cross-contamination, it was found that fifty-nine percent of the consumers do not think using the same equipment for cooked and raw food can cause any risks (Ergönül, 2013) Similar results by Redmond & Griffith (2003) indicated that 36% and 22% of consumers residing in England did not know the significance of using washed or separate equipment for food preparation. Lastly, 294 (79.7%) respondents always washed the dishes manually with detergent and water.

Limitations and Recommendations

First, since the study proved there is a relationship between awareness of food safety practices and consumers' opinions of food safety, a more comprehensive model about the consumer attitude towards the role of government in food safety could be developed to investigate the awareness of the consumers towards food safety governance. In addition to that since this study used self-reporting questionnaires, further studies can be developed with observations and face-to-face surveys to ensure the validity of the reported level of awareness of hand hygiene and food safety practices in this study. Lastly, a future study can be conducted with a population of the entire Maldives instead of one particular city or island so the data can be generalized.

CONCLUSION

In conclusion, this study has shown a high level of awareness among consumers towards hand safety practices in the Maldives. Regarding consumers' opinions of food safety, it was observed that the majority of participants have an adequate attitude and understanding towards food safety practices. Moreover, statistical analysis confirms that there is a positive correlation between awareness of food safety practices and consumer opinion. It is important to note that most of the consumers who participated in this study have attained education equal to higher secondary education or above. It is necessary to highlight that the biggest challenge in food safety is the absence of a proper legislative framework in governing food safety in the Maldives. Also due to the geographically dispersed of the islands in the Maldives, effective awareness, communications, and foodborne illnesses surveillance are often barriers to an effective advocacy program.

REFERENCES

Abushelaibi, A., Jobe, B., Al Dhanhani, F., Al Mansoori, S., & Al Shamsi, F. (2016). An overview of food safety knowledge and practices in selected schools in the city of Al Ain, United Arab Emirates. *African Journal of Microbiological Research, 10*(15), 511–520. doi:10.5897/AJMR2016.7917

Asiegbu, C. V., Lebelo, S. L., & Tabit, F. T. (2016). The food safety knowledge and microbial hazards awareness of consumers of ready-to-eat street-vended food. *Food Control, 60*, 422–429. doi:10.1016/j.foodcont.2015.08.021

Daniel, W. W., & Cross, C. L. (2013). *Biostatistics: A foundation for analysis in the health sciences. The American Biology Teacher*. John Wiley & Sons, Inc., doi:10.2307/4446617

Davila, E. P., Trepka, M. J., Newman, F. L., Huffman, F. G., & Dixon, Z. (2009). Diarrheal illness among women, infants, and children (WIC) program participants in Miami, Florida: Implications for nutrition education. *Journal of Nutrition Education and Behavior, 41*(6), 420–424. doi:10.1016/j.jneb.2008.11.008 PMID:19879498

De Boer, A., & Bast, A. (2018). Demanding safe foods – Safety testing under the novel food regulation (2015/2283). *Trends in Food Science & Technology*, *72*, 125–133. doi:10.1016/j.tifs.2017.12.013

Ergönül, B. (2013). Consumer awareness and perception to food safety: A consumer analysis. *Food Control*, *32*(2), 461–471. doi:10.1016/j.foodcont.2013.01.018

Food and Agriculture Organization & World Health Organization. (2010). *Protecting health, facilitating trade*. https://www.fao.org/fao-who-codexalimentarius/en/

Food Standard Agency. (2016). *Understanding NI consumer needs around food labelling*. https://www.food.gov.uk/research/research-projects/understanding-northern-ireland-consumer-needs-around-food-labelling

Ha, T. M., Shakur, S., & Pham Do, K. H. (2019). Consumer concern about food safety in Hanoi, Vietnam. *Food Control*, *98*, 238–244. doi:10.1016/j.foodcont.2018.11.031

Lin, C. T. J., Jensen, K. L., & Yen, S. T. (2005). Awareness of foodborne pathogens among US consumers. *Food Quality and Preference*, *16*(5), 401–412. doi:10.1016/j.foodqual.2004.07.001

Liu, A., & Niyongira, R. (2017). Chinese consumers food purchasing behaviors and awareness of food safety. *Food Control*, *79*, 185–191. doi:10.1016/j.foodcont.2017.03.038

Maldives Food and Drug Authority. (2017). *National Food Safety Policy (2017-2026), Pub. L. No. Policy/23-MoH/2017/02, Ministry of Health, Maldives, 40*. http://www.health.gov.mv/Uploads/Downloads//Publications/Publication(29).pdf

Medina, S., Pereira, J. A., Silva, P., Perestrelo, R., & Câmara, J. S. (2019). Food fingerprints – A valuable tool to monitor food authenticity and safety. *Food Chemistry*, *278*, 144–162. doi:10.1016/j.foodchem.2018.11.046 PMID:30583355

Mercado Solutions. (2018). *A report in food safety survey*. https://www.cfs.gov.hk/english/multimedia/multimedia_pub/files/foodsurvey.pdf

Ministry of Health. (2016). *Maldives health profile*. http://www.health.gov.mv/publications/50_Maldives_Health_Profile_2016_D13rdMay.pdf

National Bureau of Statistics. (2017). *Statistical yearbook of Maldives 2017*. Ministry of Finance and Treasury, Malé, Republic of Maldives. http://statisticsmaldives.gov.mv/to-share/yearbook/StatisticalYearbookofMal-dives2017.pdf

Ngo, H. M., Liu, R., Moritaka, M., & Fukuda, S. (2020). Urban consumer trust in safe vegetables in Vietnam: The role of brand trust and the impact of consumer worry about vegetable safety. *Food Control, 108*, 106856. doi:10.1016/j.foodcont.2019.106856

Odeyemi, O. A., Sani, N. A., Obadina, A. O., Saba, C. K. S., Bamidele, F. A., Abughoush, M., Ashgar, A., Dongmo, F. F. D., Macer, D., & Aberoumand, A. (2019). Food safety knowledge, attitudes and practices among consumers in developing countries: An international survey. *Food Research International, 116*, 1386–1390. doi:10.1016/j.foodres.2018.10.030 PMID:30716930

Redmond, E. C., & Griffith, C. J. (2003). Consumer food handling in the home: A review of food safety studies. *Journal of Food Protection, 66*(1), 130–161. doi:10.4315/0362-028X-66.1.130 PMID:12540194

Ruby, G. E., Ungku Zainal Abidin, U. F., Lihan, S., Jambari, N. N., & Radu, S. (2019). A cross sectional study on food safety knowledge among adult consumers. *Food Control, 99*, 98–105. doi:10.1016/j.foodcont.2018.12.045

Samapundo, S., Cam Thanh, T. N., Xhaferi, R., & Devlieghere, F. (2016). Food safety knowledge, attitudes and practices of street food vendors and consumers in Ho Chi Minh City, Vietnam. *Food Control, 70*, 79–89. doi:10.1016/j.foodcont.2016.05.037

Todt, O., Mũoz, E., González, M., Ponce, G., & Estévez, B. (2009). Consumer attitudes and the governance of food safety. *Public Understanding of Science (Bristol, England), 18*(1), 103–114. doi:10.1177/0963662507078019 PMID:19579538

Tomaszewska, M., Trafialek, J., Suebpongsang, P., & Kolanowski, W. (2018). Food hygiene knowledge and practice of consumers in Poland and in Thailand: A survey. *Food Control, 85*, 76–84. doi:10.1016/j.foodcont.2017.09.022

Uçar, A., Yilmaz, M. V., & Çakiroglu, F. P. (2016). Food safety: Problems and solutions. In Significance, Prevention and Control of Food Related Diseases. InTech. doi:10.5772/63176

Ursachi, G., Horodnic, I. A., & Zait, A. (2015). How reliable are measurement scales? External factors with indirect influence on reliability estimators. *Procedia Economics and Finance, 20*, 679–686. doi:10.1016/S2212-5671(15)00123-9

World Bank. (2018*). Food-borne illnesses cost US$ 110 billion per year in low- and middle-income countries.* https://www.worldbank.org/en/news/press-release/2018/10/23/food-borne-illnesses-cost-us-110-billion-per-year-in-low-and-middle-income-countries

World Health Organization. (1998). *Members and associate members of WHO.* https://www.who.int/whr/1998/en/whr98_annex.pdf

World Health Organization. (2002). *WHO global strategy for food safety : Safer food for better health.* https://www.who.int/fsf

World Health Organization. (2013). *Regional food safety strategy 2013 - 2017.* www.searo.who.int

World Health Organization. (2015). *WHO estimates of the global burden of foodborne diseases.* https://apps.who.int/iris/bitstream/handle/10665/199350/9789241565165_eng.pdf?sequence=1

Section 3

Food Adulteration and Defense

This section presents two topics covering the level of awareness of food terrorism activities among consumers and food adulteration.

Chapter 9
Evaluating Level of Awareness of Food Terrorism Activities Among Consumers in Klang Valley, Malaysia

Siti Nurhayati Khairatun
https://orcid.org/0000-0002-5239-8212
Universiti Putra Malaysia, Malaysia

ABSTRACT

Food terrorism issues have gained attention from the food industry globally. Food terrorism is a threat, sabotage, or contamination act to the food supply chain committed intentionally by people with a purpose to harm the public, jeopardize the economy, and disrupt the social and political system. More seriously, injury and death caused by contaminated food are inevitable. In Malaysia, very few studies have delved into food terrorism research. Therefore, this chapter presents a study to evaluate the level of awareness among consumers in the Klang Valley towards food terrorism activities. A quantitative survey was employed to gather data from consumers residing in the Klang Valley. Approximately 384 reusable surveys were analyzed using Pearson correlation and one-way analysis of variance to generate the results. The results indicated that customers in Klang Valley have an average level of awareness of food terrorism activities. This study offers a baseline reference for future research to investigate a wide area of food terrorism in Malaysia.

DOI: 10.4018/978-1-7998-7415-7.ch009

BACKGROUND

Around the globe, the act of terrorism occurred in many ways: hijacking, bombings (mass or suicide), hostage-takings, kidnapping, shooting (random or mass), as well as food contamination in a deliberate manner (Jenkins, 2020; Spink & Moyer, 2011). Terrorism is defined as the unlawful use of violence and intimidation towards civilians at large or a specific group of people (Oxford Reference, 2021). World Health Organization describes food terrorism as an act of intentional contamination using biological, chemical, and physical agents on food supply to cause harm to the public and disrupting the stability of social, economic, or political (Detwiler, 2014). The act of food terrorism has caused million dollars loss to the food-related industry because of the damage caused to the brand image, customers' trust, and reputation of the companies.

In several foodborne disease outbreaks, the contaminated food from intentional sabotage acts was found to have the capability to affect human health such as illness and death. Food terrorism can affect the economy in terms of development in poorly developed areas and worsen the scarcity of food availability (World Health Organization, 2002). The vulnerability of our food system has been highlighted by Tommy G. Thompson, the United States Health and Human Services Secretary, where he said, "I, for the life of me, cannot understand why the terrorists have not attacked our food supply because it is so easy to do." (Cox, 2006). This statement indicates that the most convenient channel to be contaminated deliberately is through our food supply. That being said, some incidents of food terrorism have been reported for the past years in many countries. Table 1 lists out some countries with food terrorism incidents.

Table 1. Countries with food terrorism incidents

Year	Country	Incidents
2020	Indonesia	A fish farmer lost his seven fish ponds due to poisoning by an unknown person
2018	Australia	A coworker was charged for committing a needle scare on packed strawberries resulted from a workplace revenge
2014	Japan	A factory worker was sentenced to life imprisonment after tampering with frozen dumpling with pesticide
2011	USA	A male worker was jailed for contaminating his female coworker's yogurt lunch by injected his semen at the P&G facility
2008	UK	A hypermarket worker was charged for contaminating the store with his urine and feces which had caused £700K loss to the company

On a different page, most younger consumers were concern more about their food safety compared to 10 years ago (Tucker et al., 2006). This is due to active advocacy campaigns relating to food safety awareness by government agencies, non-government organizations, and social scientists. In recent years, research on food terrorism started to gain interest from the research and industry communities. Most research done is on the implementation of food defense systems within the organizations which is related to preventive measures of food terrorism. Table 2 summarizes the current research trend relating to food safety defense from 2017 until 2019 which is available online.

Table 2. Current research trend relating to food safety defense

Author(s)	Title	Summary of the Study
Manning (2019)	Food defense: Refining the taxonomy of food defense threats	Describes strategies of food defense that should be adopted by developing a taxonomy of food defense threats which aims to make the food defense a risk-reduction strategy in the food supply chain
Tumbarski (2019)	Foodborne zoonotic agents and their food bioterrorism potential: A review	Reviews a deliberate action taken by using biological agents (bioterrorism) with the purpose to cause death, disease, fear, and panic among the public to achieve their goals
Sadiku, Musa, & Ashaolu (2019)	Food terrorism	Introduces food terrorism by explaining how food has been used as a weapon with a purpose to harm the public
Meulenbelt (2018)	Assessing chemical, biological, radiological and nuclear threats to the food supply chain	Provides an understanding of the threat of contamination in the food chain through chemical, biological, radiological, or nuclear (CBRN) agents and clarify various parameters such as access to the types of agents that can cause damage and effect of deliberate CBRN
Adelaja, George, Miyahara, & Penar (2018)	Food insecurity and terrorism	Explain the relationship between food insecurity and violence for eight states and views on the consequences of not protecting food from violence in efforts to improve food security globally
Davidson et al. (2017)	From food defense to food supply chain integrity	Elaborate how food defense protects the food supply from any threat. Besides, it also provides a distinction between food security and food safety, as well as its importance to the food supply chain

All current articles above provide a brief explanation about the food terrorism issues. Most previous studies focused on food safety and food security in the food supply chain.

Comparison Between Food Safety and Food Defense System

A food safety system is concerned about contamination hazards that might occur accidentally within the food supply chain. On contrary, the food defense system deals with intentional contamination done to the food supply chain by grudge employees (in most cases) or anyone who has ill-intent to cause public harm or economic loss. According to Food Defense Certification Guideline 2016, food defense is the security to secure the food supplies from being contaminated intentionally which causes public harm or economic disruption. The purpose of this guideline is to protect food from a person or group of people who are intentionally doing something to the food to either cause illness or death on a large scale (FDA, 2019). The key difference between these two systems is the motivation for committing adulteration (see Figure 1).

Figure 1. Classification of accidental and intentional adulteration
Source: Eurofins (2020)

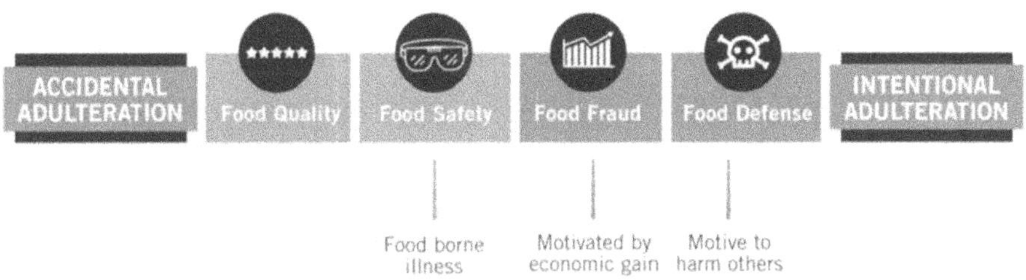

Food quality and safety are exposed to accidental adulteration, while food fraud and defense are vulnerable to intentional adulteration motivated by benefitting for economic gain and harming others, respectively. Sadiku et al. (2019) suggest that effective food defense systems must be the priority of the food industry, regulatory authorities, and consumers. Kinsey et al. (2009) explained that in the federal annual budget, U.S. federal government spent $5.65 billion to protect airline travel; however, this budget should be spent to secure the food system since there are only $93 million allocated to secure the food supply chain. The FDA has taken steps to help food facilities develop their food defense system from deliberate adulteration of the food supply by developing several tools and resources (FDA, 2019).

Karayan (2016) indicated that providing a regulatory framework for food defense, while maintaining it at a practical level, including the need for a requirement of food defense plans, training and record-keeping has enabled these deliberate adulteration regulations to be a huge success. To identify, measure, and assess weaknesses in the food system, the FDA conducts vulnerability assessment (VA) with the identification of the highest concern management measures and mitigation strategies to reduce these weaknesses (FDA, 2018). Besides, Spink (2014) states that the Global Food Safety Initiative (GFSI) has separated Hazard/Food Safety (HACCP), Threat/Food Defense (TACCP), and Vulnerability/Food Fraud (VACCP) in a different column to be addressed individually in the Food Safety Management system (see Figure 2).

Malaysia External Trade Development Corporation (MATRADE) (2016) reported

Figure 2. Food safety management system
Source: Food Fraud and Vulnerability Assessment (VACCP) by Spink (2014)

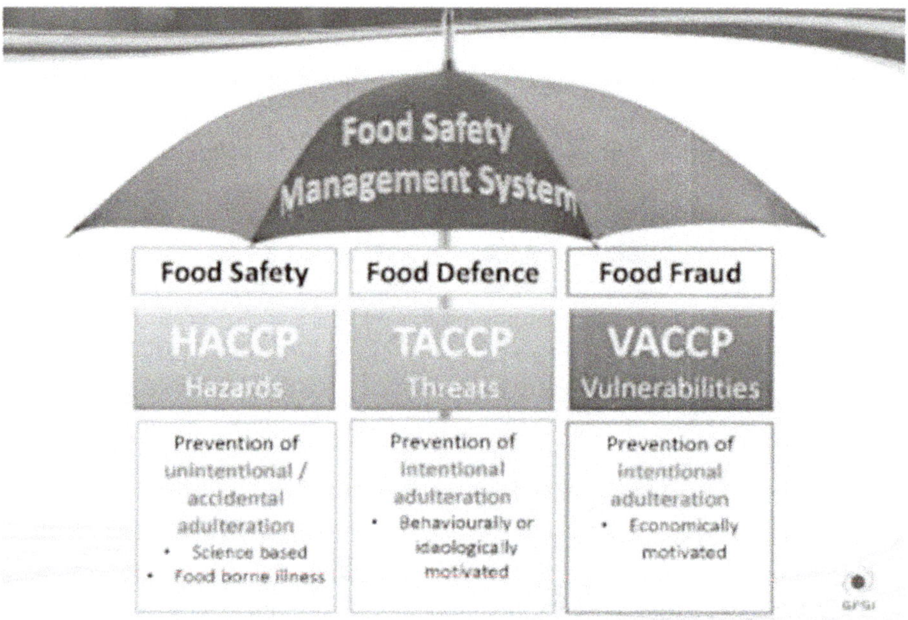

that the United States Food and Drug Administration (USFDA) has introduced a food defense certification system that requires companies in the US to implement a food defense system to prevent food supply from intentionally contaminated. MATRADE also urges that Malaysian companies adapt the food defense system. In 2018, the Ministry of Health Malaysia introduces Food Defense Certification under the Food Safety and Quality Program. The purpose of this certification is to provide written assurance that the food defense plan implemented in the food facilities conform to food defense certification requirements before the food products are being exported to the US. However, there is still a lack of literature available in Malaysia relating to the food terrorism sector. Thus, this study aims to evaluate the level of awareness of food terrorism among consumers in Klang Valley. Three independent variables were used to measure the consumers' awareness are their knowledge, concern, and risk perception about the food terrorism activities (Cassoff et al., 2014). This study offers a baseline reference in food terrorism which could be useful to the policymakers and industry players in drafting food safety management systems and related policy. Figure 1 shows an illustration of a conceptual framework in this study.

Figure 3. Conceptual framework for consumers awareness

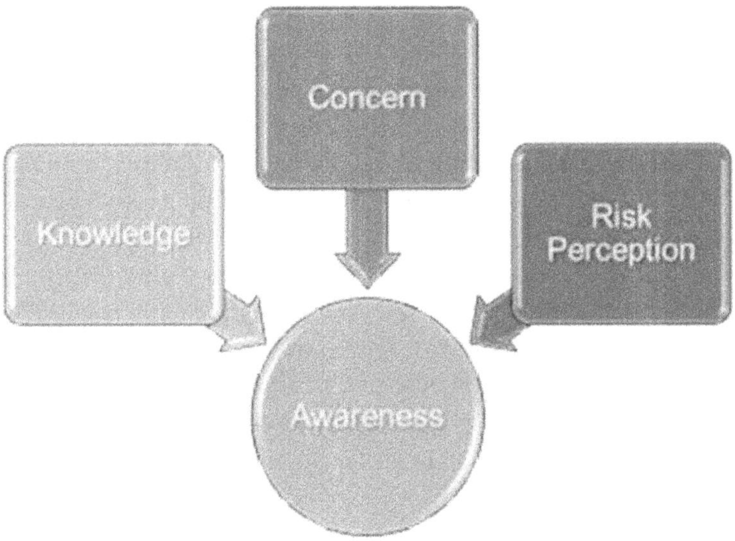

Consumers' Knowledge

Ishak and Zabil (2012) clarify that knowledge influences individual attitudes and beliefs as it been explained by Ajzen's Theory of Planned Behavior (TPB). Eggers et al. (2011) indicated that mostly informants said they had never searched for information related to food terrorism during interview sessions regarding their knowledge about food terrorism. Damage control can be effective if the information relating to food safety or food defense crisis communication is effectively conveyed to the consumers, yet approximately 84% of consumers know how to protect themselves during any crisis that happened when they received the information (Stinson et al., 2007). According to Tucker et al. (2006), due to lack of knowledge, individuals cannot anticipate or avoid the negative effects of unusual threats which of them may be severe and can cause them to feel worried.

WHO (2004) also explains that consumers should be aware and know how to react correctly if food terrorism might have happened, so the information related to food terrorism should be included in consumers' education action plans as a part of preparedness planning. Therefore, knowledge has been associated to conduct preparedness actions and self-reported preparedness attitudes, and knowledge is also antecedent as an action to get ready for future threats (Wood et al., 2012).

Consumers' Concern

Generally, concern is a feeling of worry about something important. Recently, the concerns level of Portuguese consumers is increased to 43.9%, and their level of confidence in food decreased due to the food crisis have big diffusion through the media (Venture-Lucas, 2004). Many surveys on food terrorism have been done in the United States compared to other countries. One large-scale online survey was conducted nationwide relating to consumers' attitudes and concerns about terrorism (Applebaum, 2006. Based on another survey, the consumers prefer anti-terrorism agencies to spend 19% of their budget to secure the supply of food terrorism (Ghosh et al., 2007). They also stated that the overall result of the survey shows the consumers have a higher level of concern towards terrorism on public transportation. Kinsey et al. (2009) also clarified that the National Center for Food Protection and Defense was designed a survey to identify the relationship between consumers' levels of fear and preferences towards protection and secure the food system instead of airlines and other potential targets by allocating some funds. A higher level of awareness of food problems in emergency response is also needed to ensure rapid and effective prevention of the effects of food violence (World Health Organization, 2002). Sadiku et al. (2019) discovered that to increase the awareness among the government organizations, industry agents, and consumers, FDA has disseminated information relating to food terrorism threats. As explained by World Health Organization (2004), consumers need to be more aware of the threats and play their roles by preventing any possible threats. Through the global safety net, about 600 million people are affected every year in food fraud cases, adulterated products, and ingredients containing harmful chemicals or pathogens as safety regulations are tightened; however, not all of the food and drinks we consume are as secure as we think.

Consumers' Perceived Risk

Risk perception is explained as people's subjective judgments about the feasibility of negative occurrences such as injury, illness, disease, and death. People need to have risk perception so they know how to handle the situation based on their judgment and what hazards to focus on to mitigate the risk. Risk perception has two primary domains: the cognitive and emotional domains, which relate to how much people know about and understand risks, and which relate to how they feel about them, respectively (Paek & Hove, 2017). In food terrorism, consumers' risk perception is significant to determine which possible behavior can be amounting to food sabotaging and further cause harm to the people. Thus, consumers can take precautionary steps to prevent themselves from being injured or killed in consuming the contaminated food.

Figure 4. Data collection process chart

METHODOLOGY

There are seven steps involved in the data collection process in this study. Figure 4 lists out each step of the data collection including determine the research design, identify the population, specify the sample size, develop a questionnaire, administer the survey, gather the data, and analyze the data.

At the beginning of this study, it is important to determine the research design to provide answers to the research question. According to Sileyew (2019), a research design is aiming to anticipate a suitable framework for research. Quantitative research design is considered more suitable for this study as it involves a mass community of customers. Moreover, the primary objective of this study is to evaluate their level of awareness relating to food terrorism activities, which could be quantified scientifically. In the second step, the population was identified to frame and manage the respondents' demographic, particularly their residency area. The purpose of this step is to generalize the results. For this study, the respondents must reside in Klang Valley (see Figure 5). Klang Valley, covering the Federal Territory of Kuala Lumpur and Selangor, had the highest population density compared to other states in Malaysia (Department of Statistics Malaysia, 2020). Due to lack of budget and time constraints, this study focused on the population in Klang Valley only, not the entire Malaysia. Convenient sampling was carried out to reach out to the entire Klang Valley population.

The next step was to specify the sample size derived from the population for this study. Krejcie and Morgan's sample size calculation was referred to below to calculate the sample size (Krejcie & Morgan, 1970).

Formula:
S = required sample size (n)
X2 = the table value of chi-square for 1 degree of freedom at the desired confidence level (3.841)
N = the population size
P = the population proportion (assumed to be 0.50 since this would provide the maximum sample size)
d = the degree of accuracy expressed as a proportion (0.05)

After the calculation has been made by using the formula, the sample size needed for this study from the selected population was 385 respondents.

Figure 5. Map of Klang Valley, Malaysia
Source: Google map

Calculation:
S = X2NP(1-P) ÷ d2(N-1) +X2P(1-P)
Required sample size (n) = (3.841) (6530000) (0.50) (1-0.50)
(0.05)2 (6530000-1) + (3.841) (0.50) (1-0.50)
= 6270432.50
16325.95775
= 385

In the stage of developing a questionnaire survey, it is crucial to provide inclusion statements to filter the respondents. The inclusion statement includes the respondent's age must be above 18, currently residing in the Klang Valley area, and must be a

Malaysian. The questionnaire is adapted from Stinson et al. (2011); Aldoory et al. (2010), which consisted of four sections: demographic, knowledge, concern, and perceived risk about food terrorism activities. To achieve a valid and reliable questionnaire, a pilot test was conducted before the actual administration of the survey. Generally, pilot testing means an attempt to try the survey out first on a few individuals to find out if the survey, key informant interview guide, or perception frame will work in the "real world" (TCEC, 2011). There were 10 respondents recruited as representatives for the pilot test. After running the pilot test, there were no changes made to the questionnaire and it was found that each respondent would need approximately 7-10 minutes to complete the survey.

The survey was created in Google Form and consisted of four sections: Section A is about demographics, Section B is about consumers' knowledge, Section C is about consumers' concern, and Section D is about consumers' perceived risk relating to food terrorism. The Google form link was administered via an online platform, primarily using social media including Facebook and Instagram to gather data. The social media platform is convenient, fast, and inexpensive, as an effective tool to collect data. The possible limitations of using this media are the respondents could be non-Malaysians and living outside the Klang Valley area. The survey was launched in September 2019 for a month. All raw data were saved in cloud storage. Data cleaning was done after the survey link was closed in early October 2020. A total of 385 usable surveys were analyzed.

In the last step, data analysis was completed to generate results by inspecting, rearranging, modifying, and transforming data making (Chapman, 2018). For this study, descriptive analysis was used to summarize the data: mode, mean, median, standard deviation, and variance. Further data analysis employed Statistical Package for the Social Sciences (SPSS) software to assess the level of consumers' awareness of food terrorism activities through comparison of the mean of the variables among different ages of the consumers. Next, the Pearson Correlation analysis was run to know whether consumer knowledge can be a factor that influences the level of consumer awareness of food terrorism activities. The one-way analysis of variance (ANOVA) was used to determine whether there are any statistically significant differences between the means of two or more independent groups.

RESULTS AND DISCUSSIONS

Validity and Reliability Test

The questionnaire constructs must be valid and reliable. The reliability and validity tests were conducted by using the data gathered from the pilot test. For the reliability test, Cronbach's Alpha value was 0.771. It was shown that the value was above 0.70, then the items in the questionnaire were considered acceptable and reliable. Pearson correlation analysis was completed and all items were valid since the *r*-value was 0.386 and the sig-*r* value was 0.000. It showed that there was a positive and significant relationship. Thus, the questionnaire was considered valid.

Section A: Demographic

The descriptive analysis was used to describe and understand the demographic of respondents. In general, the consumer comes from a variety of demographic backgrounds in terms of their gender, age, marital status, occupation, and level of education. Table 3 shows a summary of descriptive analysis for the demographic profile of respondents.

Table 3. Demographic data of respondents

Variables	Category	Frequency(N)	Percentage (%)
Gender	Female	275	71.4
	Male	110	28.6
Age	21-30	351	91.2
	31-40	19	4.9
	41-50	11	2.9
	51-60	4	1.0
Marital status	Single	336	87.3
	Married	49	12.7
Occupation	Student	261	67.8
	Employed	109	28.3
	Unemployed	15	3.9
Education Level	Certificate	36	9.4
	Tertiary	349	90.6

N= 385

There were 275 females and 110 males were involved as respondents to answer the questionnaire for this study. The percentage represented female respondents was 71.4%, while the percentage of male respondents was 28.6%. In this study, the majority of the respondents reported were age between 21 to 30 years old with 351 respondents (91.2%) followed by those aged between 31 to 40 years old with 19 respondents (4.9%) and 41 to 50 years old with 11 respondents (2.9%). Only 4 respondents (1%) were aged between 51 to 60 years old. Besides, 336 respondents were single (87.3%) and the rest, 49 respondents were married (12.7%). Overall, 67.8% (261 respondents) were students, 28.3% (109 respondents) were employed and 3.9% (15 respondents) were unemployed. Of 385 respondents, as many as 36 respondents which 9.4% have certificate level, while 349 respondents (90.6%) have tertiary education.

Section B: Consumers' Knowledge

In this section, respondents were asked the definition of food terrorism and to identify which of the following incidents given was an action of food terrorism that happened in food chain supply. This is to measure whether the consumers know what is food terrorism.

Figure 6. Consumers' opinions on the definition of food terrorism

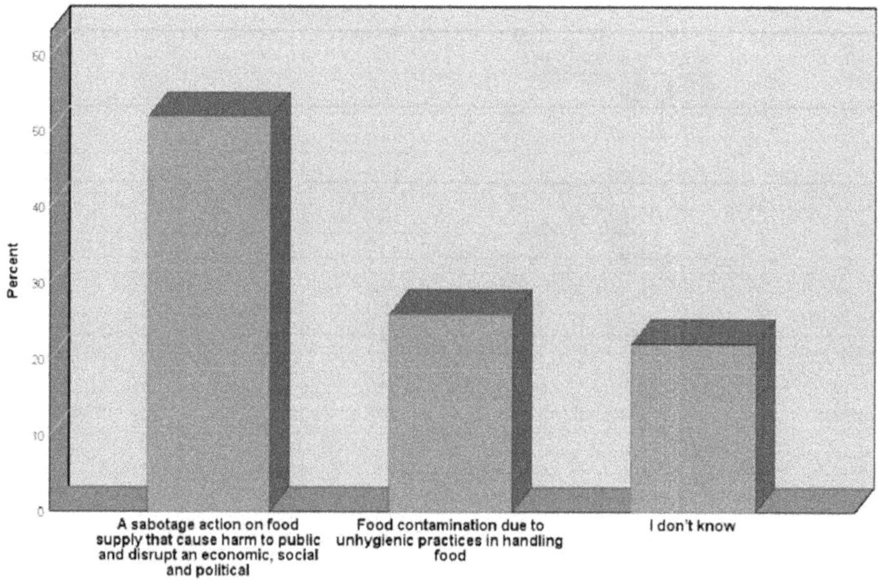

As shown in Figure 6, about 51.6% of respondents knew food terrorism was a sabotage action on the food supply that could affect public health and disrupts the environment, culture, and politics. While 26% of respondents have thought that food terrorism is food contamination due to unhygienic practices in handling food and the rest as much 22.1% respondents did not know what is food terrorism.

Figure 7. Consumers' correlate incidents as food terrorism

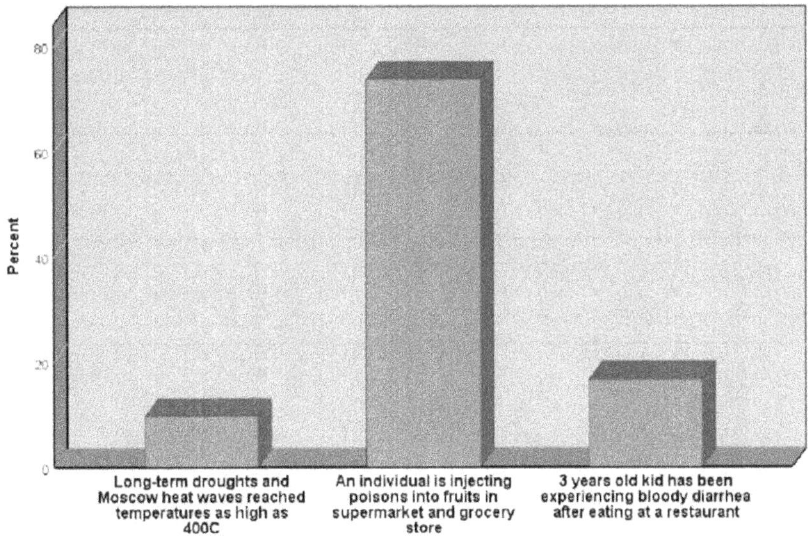

Figure 7 shows the percent of respondents who correlated the incidents as food terrorism. The majority of respondents (73.8%) were able to correlate a particular incident such as, "An individual is injecting poisons into fruits in the supermarket and grocery stores" is an action of food terrorism that happened in the food supply chain. However, 16.6% of respondents were unable to correlate the incident as food terrorism answered "3-year-old kid has been experiencing bloody diarrhea after eating at a restaurant" and 9.6% of respondents answered, "Long-term droughts and Moscow heatwaves reached temperatures as high as 400°C and wheat production almost paralyzed in Russia".

For the next section, the question was whether or not the consumers have heard the news about food terrorism and what were the news sources about food terrorism. A descriptive analysis was conducted.

Figure 8 above shows the numbers of respondents in percentage whether or not they have heard about food terrorism. The majority of respondents in Klang Valley (68.8%) answered that they had never heard the news about food terrorism, while 31.2% of respondents said that they had heard the news of food terrorism. Eggers et al. (2011) pointed out in their research that the use of social media mediums such as Twitter could be an effective news channel as a food safety initiative to raise awareness and direct people to appropriate sources. This result is supported as the majority of consumers in Klang Valley have heard about food terrorism from social media. In Malaysia, the deliberate contamination incidents were reported as food sabotage rather than as food terrorism. Therefore, the consumers gained little exposure to the issue of food terrorism that happened in another country through television and social media.

Figure 8. Responses relating to question if they have heard any news about food terrorism

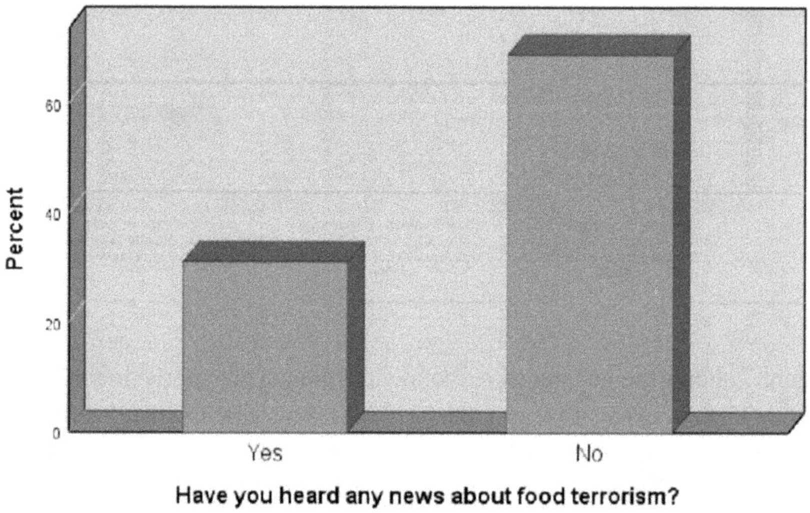

Figure 9. Source of news about food terrorism

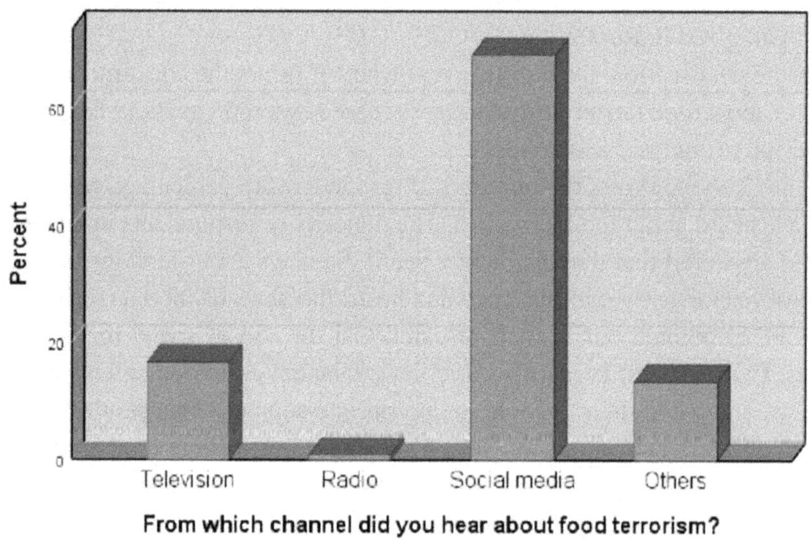

Based on the percentage in above Figure 9, on average, most consumers have heard about food terrorism from social media (69.2%), and the rest heard about the threat from television (16.7%), other channels (13.3%), and radio (0.08%). It can be concluded that most consumers still did not hear about food terrorism from the common news channels.

Section C: Consumers' Concern

Next, respondents were given news clippings related to food terrorism issues and they were asked to give their answers by using a six-point Likert scale indicating 1 as strongly disagree to 6 as strongly agree. The means and standard deviations for the items on how consumers concern about food terrorism issues through the news given were summarized (refer to Table 4).

Table 4. Consumers concern about food terrorism

Items	Mean	SD
I would like to know more about the food sabotage addressed in the situation above	5.03	0.928
I am very curious about the issue described above	5.10	0.890
I believe the case in the situation above could involve me or someone close to me	4.86	1.125
I believe the issue described in the situation could affect me personally	4.77	1.103
Do you pay attention to the situation above after reading the first sentence?	4.94	0.943
Do you search for more information about the issue presented in the situation above after reading it?	4.61	1.077

Note. SD = standard deviation, shown in three decimal points

Section D: Consumers' Risk Perception

One-way ANOVA analysis was employed to measure perceived risk items towards the news message with the consumers' age. Table 5 shows the output of the ANOVA analysis for consumers' perceptions about food terrorism practices through the news clippings has given between different ages of the consumer in Klang Valley. This ANOVA analysis was to measure whether there is a statistically significant difference between the group means. The output showed that at $\alpha = 0.05$, the significance mean is $0.63 > 0.05$, so the variance is assumed. Since the value of ($F(3, 382) = 1.969$, $p = 0.118$) which is the p-value above 0.05 ($p > 0.05$); therefore, there was no significant difference in the mean on how consumers risk perception on food terrorism issues through the news given between different age of consumer in Klang Valley.

Table 5. ANOVA analysis on consumers perceived risks

	Sum of squares	df	Mean square	F	Sig.
Between groups	3.348	3	1.116	1.969	0.118
Within groups	215.968	382	0.567		
Total	219.316	385			

Note. df = degree of freedom associated with the sources of variance, F = the ratio of two means square and Sig. = significance value

Based on the Tukey post hoc test, it also indicated that the mean score for 21-30 years old group (4.88 $\pm$ 0.748, $p = 0.76$) did not significantly differ with 31-40 years old group (5.06 $\pm$ 0.790, $p = 0.71$), 41-50 years old group (4.75 $\pm$ 0.700, $p = 0.41$) and 51-60 years old group (4.08 $\pm$ 1.142, $p = 0.14$). This shows that the source of information contained in the news message about food terrorism cases was not a major factor that could attract their attention towards this threat even though their age is different.

The level of consumers' awareness of food terrorism was presented in Figure 10. It is based on their perception regarding the news issue given about food terrorism. The level of consumer awareness on food terrorism threat is at an average level (46%). About 41% of respondents have a low awareness level on food terrorism issues. Only 13% of respondents have a high level of awareness of this food terrorism threat. It can be concluded, majority of consumers in Klang Valley were unaware and not concerned about the food terrorism threat. Therefore, it was likely that consumers would be more aware of the threat if they read more on media information about food threats being widely disseminated across various channels. In line with this, it can help to increase the level of consumers' awareness of the food terrorism threat.

However, the result showed some respondents were aware that food sabotage is also considered food terrorism but they were unable to correlate any incidents presented in the survey as a real food terrorism activity. It could be possibly because that respondents only answered this survey with little knowledge relating to food terrorism. Respondents were also asked about the products that were likely to be targeted for food sabotage. The types of food products included were fresh produce, dairy, meat, seafood, baked goods, canned goods, boxed goods, bottled water, and delivered foods. The assessment was used the 6-point rating scale to reflect the respondents' belief in the product with 1 indicating that the product was not at all likely to be attacked by food sabotage action and 6 indicating that the product was extremely likely to be targeted for food sabotage. There are differences in respondents' perceived risk of which food was most likely to be the target of a food terrorism threat.

Figure 10. Level of consumers' awareness of food terrorism threat

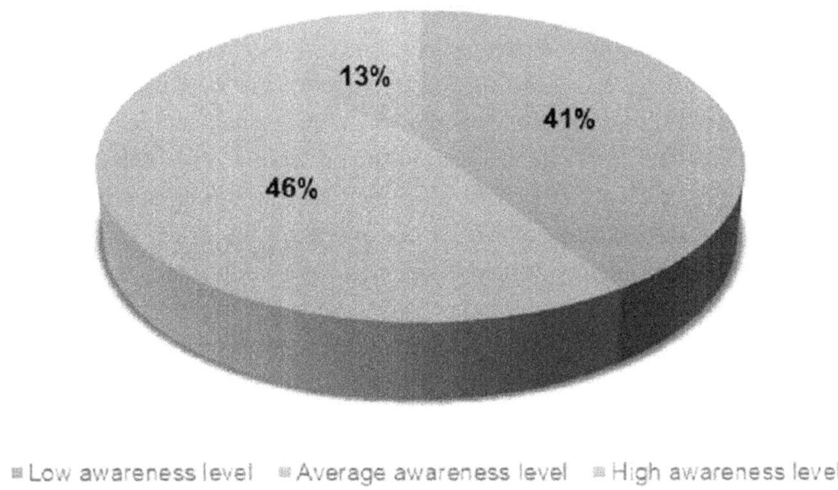

Table 6. Products that are likely to be targeted for food terrorism

Products	Mean	SD	Variance
Bottled water	4.10	1.374	1.888
Bakery product	4.31	1.201	1.443
Seafood	4.46	1.264	1.598
Delivered foods	4.55	1.264	1.597
Boxed goods	4.65	1.143	1.306
Fresh produce	4.82	1.165	1.357
Meat	4.85	1.085	1.177
Canned foods	4.87	1.115	1.244
Dairy	5.00	0.941	0.885

Based on respondents' opinions as shown in Table 6, the dairy product was voted as very likely to be targeted for food sabotage with the mean of 5.00 ± 0.941. Respondents believed that dairy product was more at risk to terrorist attack. Then, canned goods (4.87 ± 1.115), meat (4.85 ± 1.085), fresh produce (4.82 ± 1.165), boxed goods (4.65 ± 1.143), and delivered foods (4.55 ± 1.264) were among the food that the respondents believed can be likely targeted by the terrorist. While, products that were not likely to be sabotaged were seafood (4.46 ± 1.264), bakery products (4.31 ± 1.201), and bottled water (4.10 ± 1.374). This result was contrary to the study done by Stinson et al. (2011) which found that most respondents from the United States, Italy, UK, Germany, Spain, and Japan chose bottled water as the most likely target for intentional terrorist contamination.

The question asked was, "Who is responsible for food safety and food defense" (See Figure 11). The purpose of this question was to assess consumers' opinions on which sector whether farmers, processors, distributors, retailers, consumers, government, or others that are responsible for protecting our food supply chain. Respondents can select more than one sector that they thought are responsible for food safety and food defense. When respondents were asked to identify who is responsible for food safety and food defense, the result shows that processors were thought to be most responsible for food safety and food defense since it was selected by 84.2% of respondents. The second-highest sector that was selected as responsible by 68.1% of respondents was the government. Distributors also have a higher percent selected by respondents as much as 62.1%, followed by 53.0% of respondents were selected farmers and 47.5% of respondents were selected consumer as the sector that responsible for food safety and food defense. Retailers were ranked the lowest as only 45.2% of respondents were selected. About 10.4% of respondents also thought that there are other sectors instead of farmers, processors, distributors, retailers, consumers, and government that are responsible for protecting our food supply chain. As mentioned by Stinson et al. (2011), there is no solid basis for determining which part of the food sector is considered most responsible for food safety and food defense; thus, farmers, processors, distributors, retailers, consumers, and government all bear the responsibility to protect food supplies from accidental and accidental contamination.

Figure 11. Sectors that are most responsible for determining food safety and defense

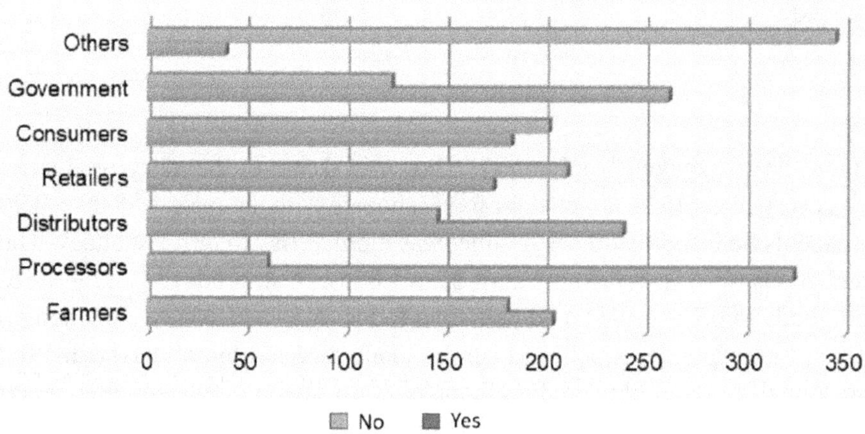

Consumers Knowledge Influences their Level of Awareness

Table 7 shows that the Pearson correlations analysis was conducted to determine whether consumers' knowledge influences their level of awareness of food terrorism. The *r*-value is 0.384 and the significant value of 0.000. There was a weak and positive relationship between consumers' knowledge and the level of consumers' awareness of food terrorism. Since the significant value, $0.000 < 0.05$, there was a significant relationship between consumers' knowledge and the level of consumers' awareness of food terrorism at 0.000 level of significance. Therefore, this study shows that knowledge can influence the level of awareness of food terrorism. By having the relevant knowledge, the consumers will be more prepared and aware of the food terrorism threats.

Table 7. Correlation between consumers' knowledge and awareness

		Mean Awareness	Mean Knowledge
Mean Awareness	Pearson Correlation Sig. (2-tailed) N	1 385	0.384" 0.000 385
Mean Knowledge	Pearson Correlation Sig. (2-tailed) N	0.384" 0.000 385	1 385

Note. ** Correlation is significant at the 0.01 level (2-tailed)

Consumers' Concern about Food Terrorism

Respondents were asked how concerned are they about protecting the food supply and how confident are they that the food supply was safe from deliberate contamination. A 6-point rating scale was used to describe consumers' concerns and confidence feeling on our food supply chain. To determine the consumers' concern in protecting the food supply, the result indicated 1 as not concerned at all and 6 as extremely concerned. The next question asked was about how confident the consumers were if the food supply is safe or not, the result showed 1 as not confident at all and 6 as extremely confident.

Table 8 shows the level of concern and confidence of respondents towards food supply which most respondents (72.2%) were extremely concerned about protecting the food supply. It is about 26.2% of respondents were just slightly concerned and only 1.5% of respondents were not at all concerned about protecting the food supply. Also, the result showed 61.5% of respondents have slight confidence in how safe the food supply was protected from deliberate contamination. Only 32.7% of respondents were extremely confident that the food supply was protected from food sabotage, and the rest of respondents (5.7%) were not confident at all about the food supply protected from food sabotage. Based on the average, consumers have a moderate concern and confidence level towards the safety of food supplies.

Table 8. Consumers level of concern and confidence towards food terrorism

Concern rating (N=385)						
	1	**2**	**3**	**4**	**5**	**6**
How concerned are you about protecting the food supply?	0.5%	1.0%	3.9%	22.3%	39.2%	**33.0%**
Mean level of concern= 4.98 ± 0.942	2	4	15	86	151	**127**
Confidence rating (N=385)						
	1	**2**	**3**	**4**	**5**	**6**
How confident are you that the food supply is protected from contaminants?	1.3%	4.4%	19.7%	41.8%	24.9%	**7.8%**
Mean level of concern= 4.08 ± 1.024	5	17	76	161	96	**30**

Respondents were also asked how safe was the food that they eat in their daily life. As shown in Figure 12, the majority of respondents (37.7%) indicated that the food they eat is safe, but they had a minor concern about food safety in terms of protecting the food supply from deliberate contamination. About 30.4% of respondents felt that overall, the food they ate was very safe, but they only had a minor concern about the food they ate. Some of the respondents who are about 19.7% feel that the food they eat is safe, and they had a moderate concern for food safety. Other than that, 7.8% of respondents feel their food is very safe, yet they did not have any concern about the food that they were eating. While 2.9% of respondents felt that the food they eat is not safe and they were much concerned about how safe the food that they consumed in their daily life. However, there were only 1.6% of respondents did not care whether the food they eat was safe or not.

Further, respondents were asked to determine how much should the food protection budget be allocated to food safety or food defense system implementation.

These results were generated from the data within different respondents' ages ranging from 21-30 years old, 31-40 years old, 41-50 years old, and 51-60 years old to choose the choices provided (see Table 8 above). On average, 20.15% of respondents thought that the food protection budget must be allocated to food safety activities. Only 2.73% of respondents thought the budget provided must be allocated to the food defense system implementation. While 2.15% of respondents did not know whether the food protection budget to secure the food supply chain should be allocated to our food safety activities or food defense system implementation. This result showed that most respondents did not know that the food safety activities and food defense implementation were crucial to protect our food supply from being contaminated intentionally.

Figure 12. Concern rate for food safety

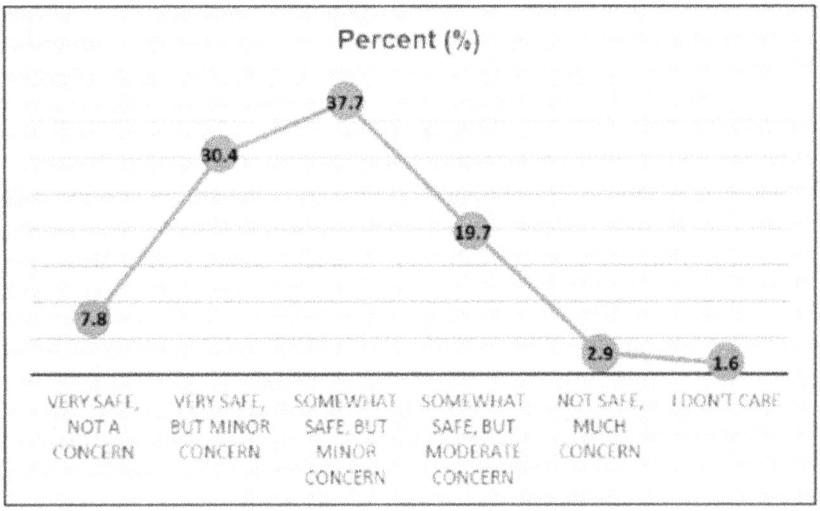

Table 8. Allocation budget for food safety and defense protection

	21-30	31-40	41-50	51-60	Average
All responses (N=385)					
Food safety	73.0	4.2	2.9	0.5	20.15
Food defense	9.9	0.5	0.0	0.5	2.73
I don't know	8.3	0.3	0.0	0.0	2.15

LIMITATIONS AND RECOMMENDATIONS

Some limitations have been identified in this study and some recommendations are proposed to improve the future study. The current study used a quantitative approach using a survey questionnaire to provide answers to the research objectives. The data collected was quantified by measuring scale and this has its drawback. Future studies could be using a qualitative approach to obtain in-depth data from the interview.

A thematic analysis could be used to generate meaningful themes which are more robust in highlighting the real issues. Next, the current study was completed in Malaysia only. Realizing that food terrorism activities and related issues are gaining attention lately, information on food terrorism must be handy and well-presented.

FUTURE RESEARCH DIRECTIONS

Some future research directions could be considered for future study. The lack of literature on food terrorism-related issues offers opportunities to explore food defense and terrorism domains in Malaysia. Based on the results generated from the study, the following recommendations are made: First, this current study was completed on population in the Klang Valley area only. Future research should focus on wider demographics including different population areas and occupations, to assess the level of consumers' awareness of food terrorism. For instance, all major cities in Malaysia such as Penang, Johor Bahru, Kuching, and Kota Kinabalu. The different populations could provide meaningful data to the study. Next, future research could focus on food industry players and how the food defense management system is being implemented in their facilities. A future study may look at challenges and opportunities faced by the industries in adapting the system into their manufacturing lines. Another future research direction is to obtain in-depth outputs from food safety experts regarding the importance and implication of adapting the food defense management system by the food industry. The experts' opinions could be used as a baseline reference to a better understanding of the food terrorism threats in the food industry.

CONCLUSION

In conclusion, this study was accomplished to evaluate the consumers' awareness regarding the food terrorism activities which could occur in the community. Based on the results, it is apparent that the consumers, particularly in Klang Valley, have an average level of awareness about food terrorism in general. Consumers should be exposed to food terrorism news or information through all media available in Malaysia. The government should develop an advocacy action plan to educate the community and the industry players about food terrorism.

ACKNOWLEDGMENT

This research received no specific grant from any funding agency in the public, commercial, or not-for-profit sectors.

REFERENCES

Adelaja, A., George, J., Miyahara, T., & Penar, E. (2019). Food Insecurity and Terrorism. *Applied Economic Perspectives and Policy*, *41*(3), 475–497. doi:10.1093/aepp/ppy021

Aldoory, L., Kim, J. N., & Tindall, N. (2010). The influence of perceived shared risk in crisis communication: Elaborating the situational theory of publics. *Public Relations Review*, *36*(2), 134–140. doi:10.1016/j.pubrev.2009.12.002

Applebaum, R. S. (2006). Terrorism and the nation's food supply perspectives of the food industry: Where we are, what we have, and what we need. *Journal of Food Science*, *69*(2), crh48–crh50. Advance online publication. doi:10.1111/j.1365-2621.2004.tb15493.x

Cassoff, J., Gruber, R., Sadikaj, G., Rushani, F., & Knäuper, B. (2014). What motivational and awareness variables are associated with adolescents' intentions to go to bed earlier? *Current Psychology (New Brunswick, N.J.)*, *33*(2), 113–129. doi:10.100712144-013-9201-6

Chapman, C. (2018). Interpretive methodological expertise and editorial board composition. *Critical Perspectives on Accounting*, *51*, 47–51. doi:10.1016/j.cpa.2017.10.007

Cox, S. (2006, August 22). *US food supply 'vulnerable to attack'*. BBC News. http://news.bbc.co.uk/2/hi/americas/5274022.stm

Davidson, R. K., Antunes, W., Madslien, E. H., Belenguer, J., Gerevini, M., Perez, T. T., & Prugger, R. (2017). From food defence to food supply chain integrity. *British Food Journal*, *119*(1), 52–66. doi:10.1108/BFJ-04-2016-0138

Department of Statistics Malaysia. (2020). *Current population estimates Malaysia 2021*. https://www.dosm.gov.my/

Detwiler, D. (2014, March 18). *The reality of food terrorism*. FoodSafety magazine. https://www.food-safety.com/articles/4020-the-reality-of-food-terrorism

Eggers, S., Verril, L., Bryant, C. M., & Thorne, S. L. (2011). Developing consumer focused risk communication strategies related to food terrorism. *International Journal of Food Safety, Nutrition and Public Health*, *4*(1), 45. doi:10.1504/IJFSNPH.2011.042574

Eurofins. (2020). *Why you need a food defense plan*. https://www.eurofinsus.com/food-testing/resources/why-you-need-a-food-defense-plan/

FDA. (2018). *Vulnerability assessments and key activity types*. https://www.fda.gov/food/food-defense/food-defense-programs

FDA. (2019). *Protecting the food supply from intentional adulteration, such as acts of terrorism*. https://www.fda.gov/food/conversations-expertsfood-topics/protecting-food-supply-intentional-adulteration-such-actsterrorism

Ghosh, K., Degeneffe, D., Kinsey, J., & Stinson, T. (2007). How would Americans allocate anti-terrorism spending? Findings from a national survey of attitudes about terrorism. *Homeland Security Affairs*, *3*. https://www.hsaj.org/articles/148

Ishak, S., & Zabil, N. F. M. (2012). Impact of consumer awareness and knowledge to consumer effective behavior. *Journal of Asian Social Science*, *8*(13). Advance online publication. doi:10.5539/ass.v8n13p108

Jenkins, J. P. (2020). Terrorism. *Encyclopedia Britannica*. https://www.britannica.com/topic/terrorism

Karayan, H. (2016). *Food defense: Protecting food against intentional adulteration*. https://www.sgs.com/en/news/2016/12/fsma-food-defenseprotecting-food-against-intentional-adulteration

Kinsey, J. D., Stinson, T. F., Degeneffe, D. J., Ghosh, K., & Busta, F. F. (2009). *Consumers response to a new food safety issue: Food terrorism*. https://iufost.edpsciences.org/articles/iufost /pdf/2006/01/iufost06000666.pdf

Krejcie, R. V., & Morgan, D. W. (1970). Determining sample size for research activities. *Educational and Psychological Measurement*, *30*(3), 607–610. doi:10.1177/001316447003000308

Manning, L. (2019). Food defence: Refining the taxonomy of food defence threats. *Trends in Food Science & Technology*, *85*, 107–115. doi:10.1016/j.tifs.2019.01.008

MATRADE. (2016). *FDA issues new regulation to protect food safety*. http://www.matrade.gov.my /en/about-matrade/media/news-clippings/146 press-releases/press-releases-2016/3526-fda-issues-new-regulation-toprotect-food-safety

Meulenbelt, S. (2018). Assessing chemical, biological, radiological and nuclear threats to the food supply chain. *Global Security: Health. Science and Policy*, *3*(1), 14–27.

Oxford Reference. (2021). *Terrorism*. https://www.oxfordreference.com/view/10.1093/oi /authority.20110803103209420

Paek, H., & Hove, T. (2017). *Risk perceptions and risk characteristics*. Oxford Research Encyclopaedia of Communication. doi:10.1093/acrefore/9780190228613.013.283

Sadiku, M. N. O., Musa, S. M., & Ashaolu, T. J. (2019). Food terrorism. *International Journal of Trend in Scientific Research and Development*, *3*(4), 134–135. doi:10.31142/ijtsrd23642

Sileyew, K. J. (2019). *Research design and methodology*. https://www.intechopen.com/online-first/research-design-and-methodology

Spink, J. (2014). *GFSI direction on food fraud and vulnerability assessment (VACCP)*. http://foodfraud.msu.edu/2014/05/08/gfsidirection-on-food-fraud-and-vulnerability-as

Spink, J., & Moyer, D. C. (2011). Defining the public health threat of food fraud. *Journal of Food Science*, *76*(9), R157–R163. doi:10.1111/j.1750-3841.2011.02417.x PMID:22416717

Stinson, T. F., Albisu, L. M., Canavari, M., Larson, R., & Gracia, A. (2011). Differences in household attitudes on food defence and food safety: An international comparison. *International Journal of Food Safety, Nutrition and Public Health*, *4*(1), 29–44. doi:10.1504/IJFSNPH.2011.042573

TCEC. (2011). *Pilot testing data collection instruments*. https://tcec.sf.ucdavis.edu/pilot-testing

Tucker, M., Whaley, S. R., & Sharp, S. S. (2006). Consumer perceptions of food-related risks. *International Journal of Food Science & Technology*, *41*(2), 135–146. doi:10.1111/j.1365-2621.2005.01010.x

Tumbarski, Y. D. (2020). Foodborne zoonotic agents and their food bioterrorism potential: A review. *Bulgarian Journal of Veterinary Medicine*, *23*(2), 147–159. doi:10.15547//bjvm.2232

Venture-Lucas, M. R. (2004). *Consumer perceptions and attitudes towards food safety in Portugal*. The 84th EAAE Seminar 'Food Safety in a Dynamic World'. 10.22004/ag.econ.24986

Wood, M. M., Mileto, D. S., Kano, M., Kelley, M. M., Regan, R., & Bourque, L. B. (2012). Communicating actionable risk for terrorism and other hazards. *An International Journal: Risk Analysis*, *32*(4), 601–615. doi:10.1111/j.1539-6924.2011.01645.x PMID:21689127

World Health Organization. (2002). *Potential effects of food terrorism. Terrorist threats to food: Guidance for establishing and strengthening prevention and response systems.* https://apps.who.int/iris/bitstream/handle/10665/42619/9241545844.pdf?sequence=1&isAllowed=y

World Health Organization. (2004). *Annex 5: Precautions against the sabotage of drinking-water, food, and other products. Public health response to biological and chemical weapons: WHO guidance.* https://www.who.int/csr/delibepidemics/annex5.pdf?ua

Chapter 10
Food Adulteration:
A Challenge for Safer Food

Murlidhar Meghwal
Independent Researcher, India

Mahalakshmi M.
National Institute of Food Technology Entrepreneurship and Management, Kundli, India

Mahalakshmi R.
Indira Gandhi National Open University Regional Centre, Karnal, India

Simran Rani
Indira Gandhi National Open University Regional Centre, Karnal, India

Carolina Krebs de Souza
https://orcid.org/0000-0003-1340-5085
Regional University of Blumenau, Brazil

Sonam
Indira Gandhi National Open University Regional Centre, Karnal, India

Simmi Jain
M. O. P. Vaishnav College for Women, Chennai, India

Ankur Ojha
National Institute of Food Technology Entrepreneurship and Management, Kundli, India

Nitin Kumar
National Institute of Food Technology Entrepreneurship and Management, Kundli, India

Lekhraj Katariya
Export Inspection Council, Ministry of Commerce and Industry, Government of India, India

Kiran Meghwal
Mohanlal Sukhadia University, Udaipur, India

Mahalakshmi S.
M. O. P. Vaishnav College for Women, Chennai, India

Tuany Gabriela Hoffmann
Universidade Regional de Blumenau, Brazil

DOI: 10.4018/978-1-7998-7415-7.ch010

ABSTRACT

Food adulteration is a prevalent issue in the food industry. It affects safety and quality of food and causes harm to the health of the consumer. To reduce incidence of adulteration in food, it is necessary to study adulterants being added to food. Food safety concerns arise when adulteration occurs for a variety of reasons. The act includes use of expired ingredients, deceptive adulterated food labels, addition of harmful compounds to food, and more. Tests and techniques for suspected food items that are commonly adulterated with certain food or non-food products are studied and described here. The detection of adulteration in food sectors can be done qualitatively and quantitatively. Various methods like chemical analysis, spectroscopic, and chromatographic techniques used to detect adulteration are reviewed. Various laws and regulations are in effect around the world in order to prevent adulteration and ensure food safety to protect the consumers. Regulatory agencies play an important role in putting a check to food adulteration by monitoring the quality of food and penalizing defaulters.

INTRODUCTION

Food is one of the basic needs of every living being. Humans make food available in a processed or semi-processed form and consume it to support various physiological processes. Food commodities have always been vulnerable to fraudulent admixture or adulteration with cheaper inferior materials. Food fraud or adulteration is common, and such food adulteration degrades food quality, makes it dangerous to consume, and causes health problems in humans. Food fraud is defined as the intentional substitution, addition, tampering, or misrepresentation of food, food ingredients, or food packaging, as well as the making of false or misleading assertions about a product for financial advantage (Banti, 2020).

According to the Prevention of Food Adulteration Act, 1954, a food article is considered adulterated:

1. if the article sold by a vendor is not of the nature, substance, or quality demanded by the purchaser and is to his prejudice, or is not of the nature, substance, or quality which it purports or represented to be;
2. if the article contains any other substance which affects, or if the article is so processed as to affect, injuriously the nature, substance or quality thereof;
3. if any inferior or cheaper substance has been substituted wholly or in part for the article to affect injuriously the nature, substance or quality thereof;

4. if any constituent of the article has been wholly or in part abstracted to affect injuriously the nature, substance or quality thereof;
5. if the article had been prepared, packed, or kept under insanitary conditions whereby it has become contaminated or injurious to health;
6. if the article consists wholly or in part of any filthy, putrid, rotten, decomposed, or diseased animal or vegetable substance or is insect-infested or is otherwise unfit for human consumption;
7. if the article is obtained from a diseased animal;
8. if the article contains any poisonous or other ingredients which renders it injurious to health;
9. if the container of the article is composed, whether wholly or in part, of any poisonous or deleterious substance which renders its contents injurious to health;
10. if any coloring matter other than that prescribed in respect thereof is present in the article, or if the amounts of the prescribed coloring matter which is present in the article are not within the prescribed limits of variability;
11. if the article contains any prohibited preservative or permitted preservative above the prescribed limits;
12. if the quality or purity of the article falls below the prescribed standard or its constituents are present in quantities not within the prescribed limits of variability, which render it injurious to health;
13. if the quality or purity of the article falls below the prescribed standard or its constituents are present in quantities not within the prescribed limits of variability but which does not render it injurious to health;

With regard to food, the term "quality" means the products meet the requirements of an entire complex of criteria, properties, and peculiarities, which characterize the product's degree of suitability based on its assessment and consumption. "Food safety" is a condition that ensures food will not cause harm to the consumer when prepared and/or eaten according to its intended use. It entails the handling, preparation, and storage of food in ways that prevent foodborne illnesses. Quality and safety remain major challenges in the production of high-quality foods.

In order to ensure the quality of food, it is necessary to understand the foods that are regularly adulterated, the adulterants that are commonly employed in them, and the detrimental effects that these adulterants impose. This chapter examines many facets of adulteration, the health risks imposed by these adulterants, detection techniques, and legal responses around the world.

HAZARDS IN FOOD

The National Advisory Committee on Microbiological Criteria for Foods (NACMCF) defines a hazard as any biological, chemical, or physical attribute that poses an unacceptable health risk to consumers (see Figure 1).

Chemical Hazards

Any substance that is employed in or obtained through a chemical process or processes is referred to as a chemical. Chemicals are included in all food products, and any chemical can be harmful at a certain dosage level. However, some chemicals are not permitted in food, while others have set permissible levels. Pesticide residues, drug residues, unlawful additives, contamination, and heavy metals are just a few of the chemical risks that can be found in food. Good agricultural and manufacturing practices are essential to avoid these chemical substances from entering the food supply. Nutrients, sweetening agents, and flavor imparting agents are examples of food additives that can be introduced intentionally or unintentionally. These additives can react with foods and cause chemical changes in them. Heavy metals can be found in a variety of foods, including tea, seafood, and fruits such as apples, watermelon, and pineapples, as well as beverages.

Figure 1. Classification of chemical hazards

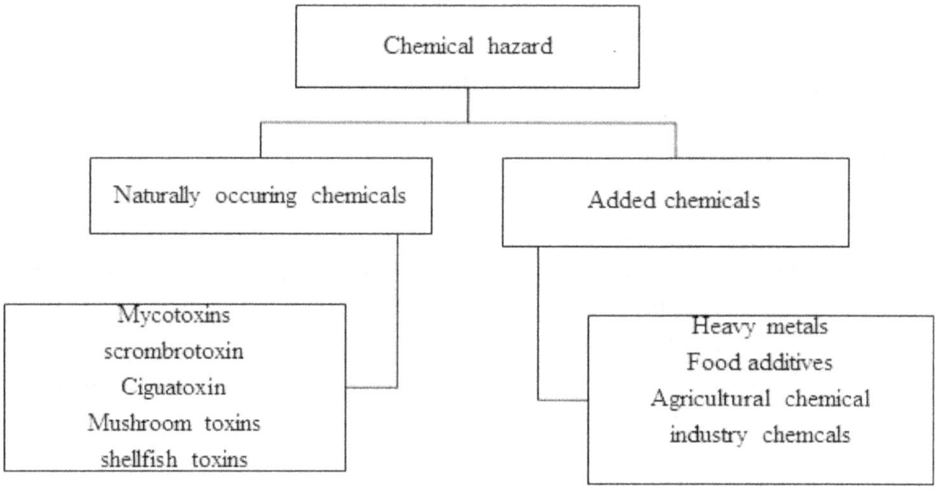

Biological Hazards

Bacteria, yeast, molds, parasites, and viruses are examples of biological risks. Bacterial dangers can lead to foodborne illnesses or poisoning. Ingesting a sufficient amount of pathogenic microorganisms may lead to illness, their multiplication, and the production of toxins, which causes foodborne infection. Ingestion of preformed toxins generated and secreted by some bacteria when they grow in foods causes foodborne poisoning.

Physical Hazards

Physical risks are foreign materials that may be present in food as a result of improper handling of food throughout the production process, from farm to fork. Physical hazards are dangerous because of their hardness, sharpness, and size, as well as the fact that they can inflict wounds or choke on humans. It could be avoidable or unavoidable. Foreign elements such as dirt in tuber crops and stalks in fruits are prevalent. Various safety strategies can be used to prevent physical risks such as glass, wood, stones, plastic, metal, insulation, bone, and plastic.

Allergens

Food allergens are food safety hazards that damage health by producing allergic reactions to certain food components, and they are vital to managing the food safety management system (Cianferoni & Spergel, 2009). Allergens can be transferred to food that is not supposed to contain allergens, during production, storage, shipment, or later processing. Product manufacturers make sure that consumers are aware of such allergic ingredients in their food. The list of food items that trigger allergic responses is as follows:

1. Cereals containing gluten;
2. Crustaceans and products of these;
3. Eggs and egg products;
4. Fish and fish products;
5. Peanuts, soybeans, and products of these;
6. Milk and milk products; cow's milk contains whey proteins and casein which shows allergic reactions; fermentation of milk by Lactobacillus causes proteolysis of β-lactoglobulin causes allergic reactions (Ehn et al., 2005)
7. Tree nuts and nut products

TYPES OF ADULTERATION

Intentional Adulteration

Economically motivated adulteration (EMA) is defined by the Food and Drug Administration (FDA) of the United States as "the fraudulent, intentional substitution or addition of a substance in a product for the purpose of increasing the apparent value of the product or lowering the cost of its production." When a food item is intentionally altered, it is referred to as intentional adulteration. It involves the addition of substandard substances with attributes that are comparable to those of the foods to which they are added. The adulterant could be either physical or chemical. Some examples of intentional adulteration include the addition of water to liquid milk, extraneous matter to ground spices, or the removal or substitution of milk solids from the natural product, etc.

The motivation for intentional contamination can range from economics to bioterrorism, the latter being an increasing concern. Food bioterrorism is "an act or threat of deliberate contamination of food for human consumption with chemical, biological or radio nuclear agents to cause injury or death to civilian populations and/or disrupting social, economic or political stability". The melamine scandal in China in 2008 is an example of EMA how its intentional addition to dairy products had huge repercussions for both consumers and producers (Changbai et al., 2010).

Incidental/ Accidental Adulteration

Incidental adulteration is caused by a lack of suitable hygienic conditions for food and beverages from the production plant to the consumption table. Pesticide residues, rodent droppings, and larvae in meals are examples of unintentional adulterants. Arsenic, lead, and mercury contamination of metals can also happen by mistake. Pests such as rats and insects, which trespass on food to a high degree and cause impurity in the form of excreta, body fluids, and deterioration through microorganisms, are also considered incidental adulterants.

DETECTION OF ADULTERANTS

Simple Screening Tests

Table 1. Simple screening test for adulterants in some food

Food item	Adulterants	Method for detection
Paneer, condensed milk, koya	Starch	Take a small portion of the product in a test tube, add water, and boil. Cool to room temperature. Add 1-2 drops of Iodine solution. Blue color indicates the presence of starch.
Milk	Water	Put a drop of milk on a polished vertical surface. The drop of pure milk either stops or flows slowly leaving a white trail behind it, whereas, milk adulterated with water will flow immediately without leaving a mark
	Hydrogen peroxide	Addition of 5 drops of 2% solution of paraphenylenediamine into 5 ml of milk sample which is suspected and appearance of blue color indicates the hydrogen peroxide added in milk.
	Formalin	Addition of 5 ml of conc. Sulfuric acid with ferric chloride in 10 ml of milk sample. Appearance of violet or blue color at the junction of two liquid layers indicates the presence of formalin.
Pure Ghee or Butter	Vanaspati	Take one teaspoonful of melted ghee or butter with an equal quantity of Conc. hydrochloric acid in a test tube. Add to it a pinch of cane sugar. Shake well for one minute and let it stand for five minutes. Crimson red color in the lower layer shows the presence of vanaspati.
Ice cream	Washing Powder	Put some lemon juice, bubbles are observed if washing powder is present.
Honey	Water	A cotton wick dipped in pure honey burns when ignited with a matchstick. Presence of water will not allow the honey to burn, and if it does, it will produce a cracking sound.
	Sugar	Take a transparent glass of water. Add a drop of honey to the glass. Pure honey will not disperse in water. The presence of additional sugar is indicated if a drop of honey disperses in water.
Sugar powder, salt	Chalk	Dissolve sugar/salt in a glass of water, chalk will settle down.
Sugar	Washing Soda	Add water and dip red litmus paper. Add concentrated Hcl to the test tube. If red litmus paper turns in blue color indicates the presence of washing soda. Take 5 g of Bura in the test tube. Add concentrated HCL in the test tube. The presence of washing soda in Bura produces an effervescent effect.
Silver foil	Aluminum foil	On ignition, silver foil burns away completely leaving glistening white spherical ball of the same mass while aluminum foil reduces to black grey color ash
Coffee	Chicory	Gently sprinkle the coffee powder on surface of water in a glass. The coffee floats over the water but chicory begins to sink within few seconds. Also, the falling chicory powder particles leave behind them a trail of color, due to a large amount of caramel contained.

Table 2. Common adulterants in milk and milk products

Milk and milk products	Adulterants
Milk	Water, soya milk, starch, groundnut milk, and wheat flour
Butter	Vegetable oil, anatta, banana, oleomargarine
Ice cream	Starch, rice powder, or wheat flour
Ghee	Vanaspati, anatta, and oleomargarine

Chromatographic Methods

Chromatography is a technique used for both quantitative and qualitative analyses for the separation, identification, purification of small components such as carbohydrates, amino acids, fatty acids in a mixture. For example, HPLC is used in the separation of amino acids, carbohydrates, and steroids. Wang et al., (2015) studied the HPLC valid adulteration detection method of honey for the added starch syrup of both C3 and C4 types. Gas chromatography is used in the separation of ester, alcohol, and lipids. Jabeur et al., (2014) differentiated olive oil adulterated with vegetable oils by determining the composition of fatty acid methyl esters.

Spectroscopy Methods

Extra virgin olive oil adulterated with various fats and oils such as soybean oil, peanut oil, canola oil screened by Fourier transform near-infrared spectroscopy observed by Azizian et al., (2015). In the study by Oroian et al., (2017). Raman spectroscopy with PLS-LDA was successful for the detection of honey adulterated with malt wort rather than glucose and fructose. The study by Khan et al., (2013) investigated the limit of urea in milk less than the recommended cut-off limit and quantification for urea could be a promising technique for finding out adulteration in milk.

Physical Methods

Physical approaches include microscopic or macroscopic detection, as well as analyzing food's physical qualities. In the case of fungal microbiological analysis, micro or macroscopic ways of visually analyzing are useful (Mangal et al., 2014). Cane syrup in honey can be detected using optical microscopy by identifying sclerous rings of cane sugar (Louveaux et al., 1978).

ADULTERATION IN DIFFERENT FOOD

Milk and Milk Products

From farm to table, milk and its products are subjected to adulteration at various phases of production (see Table 2). When milk is extracted from the udder, it is pure. But it is contaminated by bacteria, yeast, fungi, different chemicals, dust particles, and adulterants owing to unsanitary activities (Singuluri & Sukumaran, 2014). Common adulterants in milk are soy, pea proteins, vegetable protein, rennet whey, acid whey, cheese whey, non-dairy fats, vegetable fats, sodium bicarbonate, sodium carbonate, urea, tetracycline, and formalin.

From the point of production, milk is transported to the consumers through intermediaries during which it is adulterated in many ways. Suppliers of milk commonly use these methods to increase the value of milk by dilution, extraction of valuable components like milk fat, or a combination of the above methods along with the addition of cheap bulking additives to bring the total solids to an acceptable level. These are commonly seen in developing countries. (Faraz et al., 2013)

Water was found to be the most common adulterant in milk followed by the detergent in a survey conducted by FSSAI in 2011 (Nirwal et al., 2013). Milk production during summer is greatly reduced due to factors like heat stress and scarcity of fodder etc. (Faraz et al., 2013). Water is added to increase the volume of milk (Poonia et al., 2016).

Detergents when added emulsify and dissolve the oil in water and give a frothy solution, the characteristic white color of milk or it can be used to mask the fat value of milk. Detergents may also enter the milk due to the low maintenance of milk tanks during processing (Singuluri et al., 2014). Starch is added to milk to increase solid-not-fat content. Hydrogen peroxide is added along the supply chain to increase its freshness.

Nitrogen-rich compounds like melamine and urea are nitrogen-rich and are added to milk to increase their protein content. The addition of urea to milk improves consistency, provides whiteness, and helps to standardize SNF to the desired level and to give false positive measurements of protein content in milk. (Poonia et al., 2016, Singuluri et al., 2014). Milk can be tainted with urea in two ways: by adding urea to it intentionally, or by adding nonspecific synthetic milk to natural milk.

To extend the shelf life of milk, preservatives such as formalin and antibiotics are added. Boric acid, Formalin, Sodium carbonate (Na2CO3), Sodium bicarbonate (NaHCO3), Salicylic acid, Benzoic acid, Sodium azides can preserve the milk for long time (Reddy et al., 2017). Hydrogen peroxide is also added to prolong its freshness (Singuluri et al., 2014).

Adulteration of milk can be detected by qualitative and quantitative techniques. Mostly edible adulterants such as soy, pea proteins, starch, glucose, buffalo milk are detected by color-based chemical reactions. Quantitative methods such as near-infrared Raman spectroscopy. Figure 2 shows test results of adulteration in milk.

Figure 2. Test results showing (a) starch, (b) urea, adulteration in milk

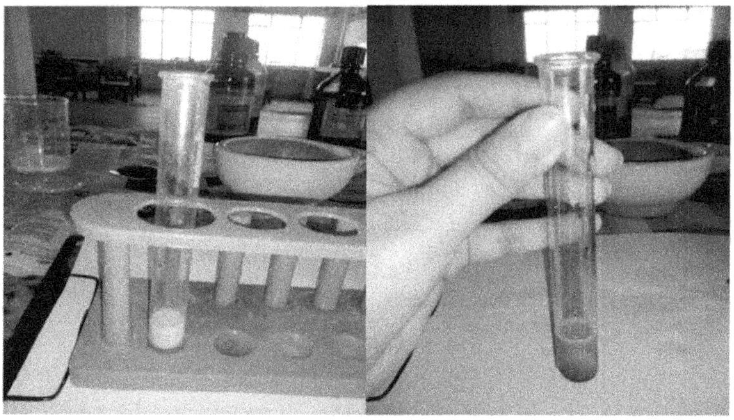

(a) Pure milk Adulterated milk with starch

(b) Pure milk Adulterated milk with urea

Spices and Condiments

Food fraud is becoming more common as the market for ground spices and herbs grows. Spices stand among the top five adulterated food types. The long complex supply chains and increase in ground herbs and spices are drivers of EMA. Other factors that influence adulteration in spices are crop seasonality and availability, weather, cultural and geopolitical events, food safety laws, the prevalence of corruption, and advances in technology to mask fraud (Galvin-King et al., 2018).

Adulteration in spices can be done with non-functional parts of the plant (e.g. intentional addition of sticks and stems in ground black pepper) during grinding. Extraneous matter from parts of a different plant that is similar to the herb are added at the cutting/grinding/ blending stage (e.g. intentional addition of foreign plant material in oregano). Exhausted material which is also referred as spent, defatted and depleted material is added without being declared. (e.g. by-product of essential oil, oleoresin and extrusion extraction at grinding/ blending stage). Color enhancement is done by addition of non-permitted or undeclared color at grinding/ blending stage to improve acceptability (e.g. Sudan dyes in ground turmeric). Sudan-I is a carcinogenic substance found in red chili powder. Sudan-I is not allowed to be used in food. Artificial colors and flavorings are increasingly being added to spices and herbs that are not required to be labeled as ingredients by law (see Table 3). Some adulterants added to spices, such as papaya seeds and gum added to black pepper and asafoetida, are unlikely to be harmful. Figure 3 depicts various adulteration of spices.

Table 3. Common adulterants in spices

Spices	Adulterants
Black pepper	Papaya seeds
Turmeric powder	Metanil yellow (dye)
Oregano	Sumac, myrtle leaves, olive leaves
Cinnamon, nutmeg	Coffee husk
Cumin	Fennel seeds, peanut shell
Paprika, chili powder	White pepper, curcuma, barium sulphate, brick powder
Saffron	Marigold to red stigma, beetroot

Food Grains

Adding sand or broken stones to increase the weight, as well as mixing plastic beads that appear like grains of the same size and color, are typical ways food grain adulteration (see Table 4). When pigeon pea is contaminated with kesari dal, it has negative consequences in people such as lathyrism. To attract customers, coloring chemicals are added to green gram and pigeon pea, causing allergic responses. Pesticides used to protect crops from insects and pests in the field and during storage can also be classified as pollutants that can harm humans.

Figure 3. Test results showing adulteration in (a) turmeric powder, (b, c) coriander powder, (d) black pepper

(a) Pure turmeric powder Adulterated turmeric powder with *metanil* yellow

(b) Pure coriander powder Adulterated coriander powder with salt

Table 4. Common adulterants in grains and their products

Food products	Adulterants
Pigeon pea or chick pea	Grass pea
Wheat/millet	Buckwheat
Durum wheat	Common wheat
Wheat based products	Melamine
Wheat flour	Peanut flour
Wheat flour, corn flour, rice flour	Melamine
Basmati rice	Non-Basmati

Figure 4. Test results showing sugar adulteration in honey

(c) Pure coriander powder | Adulterated coriander powder with dung

(d) Pure black pepper | Adulterated black pepper with papaya seed

Sugar and Confectionary

When high fructose corn syrup was introduced in the food industry, honey adulteration became a global problem. Honey has a higher market price than other sweeteners because of its nutritional value. Honey is commonly tainted with other sweeteners such sugar syrups, cane, beet, and maple syrup, and synthetic honey has become more popular in recent years.

Adulteration of honey began in the 1970's after high fructose corn syrup was introduced. Since the major component of honey is sugar with monosaccharides like fructose and glucose being a major part it is susceptible to adulteration easily with cheaper sweeteners for economic gain (see Figure 4). Some of these cheaper alternatives that have been detected in honey are sugar syrups, molasses subjected to inversion, HFCS, beet sugar and syrups of natural origin like maple syrup (Choudhary et al., 2020). Adulteration in sugar occurs using washing powder and other substances (see Figure 5).

Figure 5. Test results showing adulteration in sugar using (a) washing powder, (b) chalk powder

Pure honey Adulterated honey

Pure Honey Adulterated Honey

Oils and Fats

Adulteration of oils and fats is simple and easy but detection is difficult. Ghee is frequently adulterated by mixing with hydrogenated oils and animal fats (see Figure 6). Synthetic colors and flavors are added to other fats to adjust and make them similar in appearance to ghee (Choudhary, 2020). Adulteration of edible oils can have negative health consequences for consumers. Groundnut, mustard oil with palmolein, palm, and argemone are the most common adulterants (see Table 5). Mustard oil is mixed with the essence of allyl isocyanate to give it pungency and a yellow color to make it look like mustard oil, which is hazardous.

Table 5. Common adulterants in oils

Ingredients	Adulterants
Edible oils	BHA or BHT
Mustard oil	Papaya seed
Vanaspati	Argemone
Groundnut oil	Palmolein, palm oil

Coffee and Tea

Chicory, cereals, caramel, additional parchment, starch, malt, and figs are all added to instant coffee. The common laboratories used methods for the ground coffee adulteration includes electron microscopy and optical microscopy (see Figure 7). Separation methods such as liquid and gas chromatography, capillary electrophoresis was used for the detection of various coffee adulteration like maltodextrin, maize, starch, barley. Tea is one of the most popular beverages, although it is sometimes tainted with sawdust, sand, iron fillings, and dyes, and is sold in loose form.

HARMFUL EFFECTS OF FOOD ADULTERATION

One's health is one of the most important parts of life. Adulteration can cause severe foodborne diseases. Recognizing foodborne diseases as an outbreak is frequently overlooked. There are some dangerous food adulterants that are purposely added to food to extend its shelf life, disguise food rotting, and boost the value of the product's commercial features or traits (Mathur, 2005). As a result, the quality, and nutritive value of the food is hampered, which is harmful to one's health and can lead to food poisoning if consumed.

Figure 6. Test results showing adulteration in ghee using (a) vanaspati, (b) starch

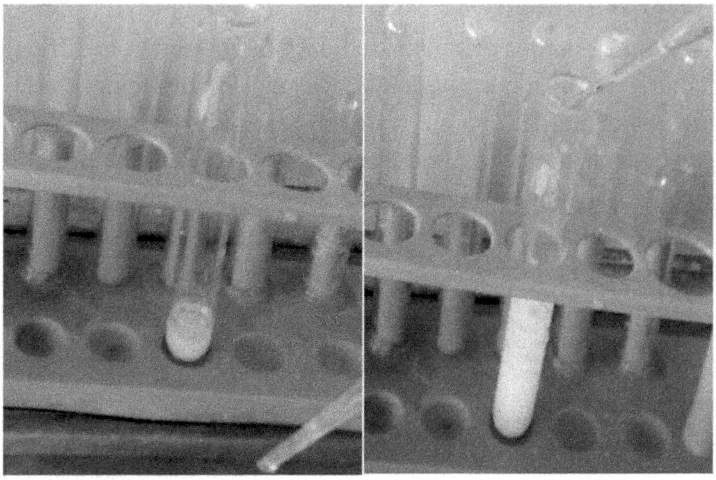

(a) Pure bura sugar Adulterated bura sugar with washing powder

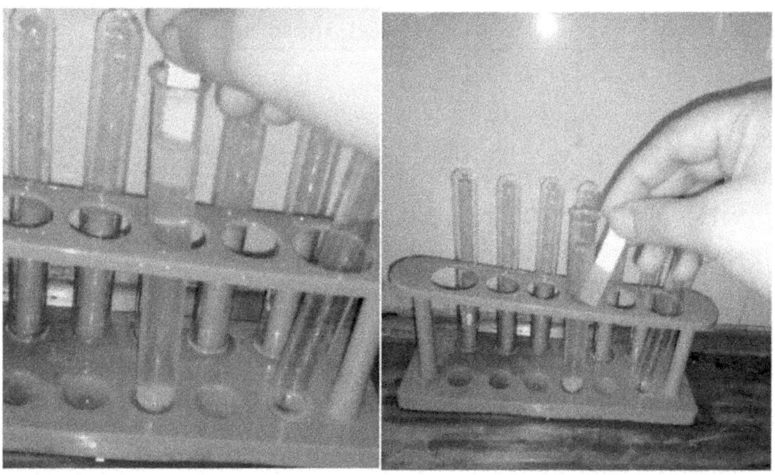

(b) Pure bura sugar Adulterated bura sugar with chalk powder

Many large-scale food poisoning incidents have been reported, such as Ergot Bajra, which causes nausea, vomiting, and giddiness, or faba beans and kesari dal (*Lathyrus sativus*), which cause favism (RBC breakdown) and lathyrism (paralytic illness), respectively (Mathur, 2005). People who are aware of naturally occurring food poisonings avoid them, but if any other unwanted substances are intentionally added or subtracted from the food product, it becomes difficult to separate the adulterant from the food product, resulting in a slew of negative consequences, including critical health hazards. Food adulterants can be classified based on their impact on the consumer:

Critical: Adulterants that can cause death or seriously endanger the health of consumers.

Major: Adulterants affecting a significant percentage of the individual product units due to the level of adulterant present. Long-term consumption can have an impact on the health of the consumer.

Intermediate: Adulterants that impact the product's quality without affecting the health of the consumer.

Minor: Adulterants that typically have only an economic impact e.g., varietal or geographic misrepresentation.

Figure 7. Test result showing iron filling adulteration in tea leaves

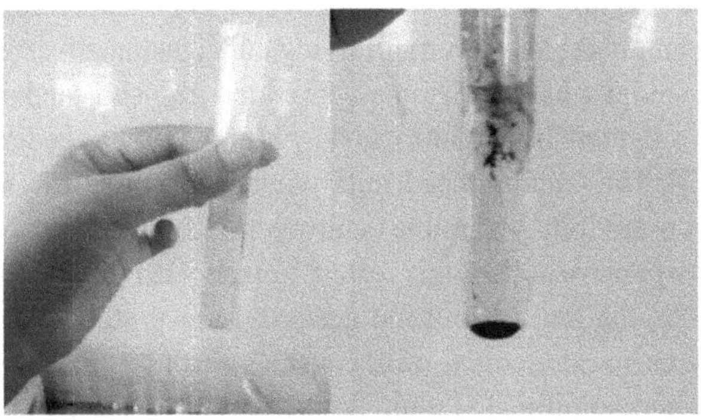

(a) Pure ghee Adulterated ghee with vanaspati

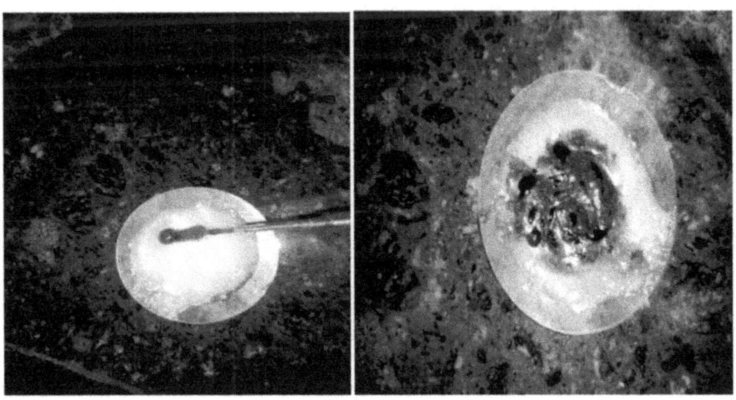

(b) Pure ghee Adulterated ghee with starch

Adulteration in food is done for various reasons which include financial gain, to increase the quantity or shelf life of food, and to improve its appearance. Carelessness and lack of proper hygienic conditions of processing, storing, transportation and selling leads to unintentional adulteration. Whether purposefully or unintentionally, but as a result, toxic or hazardous compounds end up in the food. Food adulteration is a significant problem in the food supply chain, lowering the quality and safety of food that reaches consumers (Ayelew et al., 2013). Milk and milk products, edible oil, spices, food grains, cereals, sweetening agents, and other foods are frequently falsified. The following are some of the dietary adulterants and their hazardous effects:

Milk and Milk Products

Because milk is largely water by nature, it is perishable and susceptible to microbial deterioration if not kept chilled. During summer, milk must be kept refrigerated to avoid curdling. If not, lactic acid forms, altering the nature and qualities of milk, resulting in curd formation. Many times, milk vendors need to deliver milk to remote locations, thus they add adulterants or chemicals to neutralize the acidity in the milk and prevent bacteria from multiplying and causing adverse health effects. The melamine scandal where melamine was added into substandard milk or milk diluted with water to increase the nitrogen levels and thereby the protein content in nutritional testing. The contaminated milk powder left around 300,000 poisoned and killed at least 6 children. Melamine (a nitrogen-rich resin) is added to milk since it is relatively inexpensive. When this resin is mixed with diluted milk, the protein level appears to be higher, causing indigestion, kidney stones, and renal failure in youngsters. Urea in adulterated milk causes gastrointestinal irritation, lack of appetite, nausea, and vomiting (Mathur, 2005).

Oil

Oil, commonly known as liquid gold, is a costly commodity that is essential for human survival. A deadly pandemic that erupted in Spain in 1982, paralyzed people who ate rapeseed oil laced with aniline and eventually losing their hands and feet and finally necrosis and death. Oil, like milk, is contaminated to a higher level, as listed below:

1. The addition of argemone oil to cooking oil causes epidemic dropsy, which is characterized by oedema around the ankles, gastrointestinal abnormalities, blood vessel modifications, changes in the eye, and cardiac insufficiencies.
2. Toxic oil syndrome is caused by the addition of pentachlorophenol and aniline.

3. Mineral oil, commonly known as liquid paraffin, causes anal seepage and irritation, as well as interfering with the absorption of fat-soluble vitamins (A,D,E,K).
4. When castor oil is combined with cooking oil, it causes symptoms such as nausea, vomiting, and colic pain, as well as having a strong laxative effect if swallowed.
5. Paralysis of the hands and feet is caused by tricresyl phosphate (Mathur,2005)
6. Certain forbidden colors, neem oil, and even karanja oil are added to edible oil or vegetable oil, causing liver damage such as hepatitis oedema, eyesight loss, and heart dilation.
7. In vanaspati, the main adulterant can be any cheaper oil, such as groundnut, cottonseed, or linseed, resulting in an economic loss as well as visual and cardiac illnesses, which are automatically allowed to wreck one's health at the worst (Pardeshi, 2019).

Spices and Condiments

Spices are commonly utilized in Indian cuisines. Indians have a variety of spices to suit the varying climates throughout the year. And spices are utilized on a daily basis to make the cuisine delectable, mouthwatering, and healthful. To achieve the color of the ingredients, adulterants in spices such as sawdust or brick powders are used. The following are the spices that are most likely to be adulterated:

1. Chili powder- contains sawdust, brick powder, non-permitted colors such as Sudan dye, salt, and talc powder, which causes toxic stomach disorders, carcinogenic (particularly blood and lung cancer) (Srivastava, 2015), vision loss, and respiratory ailments.
2. Turmeric powder- contains lead chromate, foreign starch, common salt, husks, earthy matter, metanil yellow, and other aniline dyes (non-permitted colors), which can cause carcinogenic consequences, brain damage, anemia, and miscarriages in pregnant women.
3. Whole spices - dirt, dust, straw, insect, damaged seeds, other seeds, rodent hair and excreta which is a source of microbial contamination and an economic loss.
4. Black pepper - Papaya seeds, light black pepper, coating with mineral oil. The addition of pinheads to black pepper has an adverse effect on the liver and stomach.
5. Coriander powder - starch, cow dung, horse dung, talcum powder or sometimes sawdust is added too which leads to severe microbiological intoxication (Pardeshi,2019).

Beverages

1. Coffee - Roasted powder of wheat, gram, date seed, chicory and tamarind husk, scorched persimmon stone powder are all the adulterants which are added to the coffee powder resulting in Diarrhea, stomach disorders, giddiness and severe joint pains, economic loss.
2. Tea – Addition of artificial color, tea wastes, gram husk, coffee husk, cashew nut endosperm, by product of leather industry, tamarind seed powder, sawdust, exhausted tea, chicory powder to tea are commonly seen. Iron filling, in particular, which is frequently found in large amounts in tea powder, can cause cancer, tetanus, appendicitis, and small intestinal problems, as well as the likelihood of tetanus infection (Food Adulteration: The Demonic Onslaught on Health and Wellness, 2015).
3. Soft drink alcoholic and other beverages, fruit products - Non permitted colors, artificial sweeteners such as saccharin, dulcin are the major adulterants found in the soft drinks and alcohol-based drink, as well as other beverages in the market, can be extremely poisonous, cause cancer, and cause severe health problems. Methanol is a chemical that, when mixed with alcohol, causes blurred vision, blindness, and death. As it interacts, it becomes poisonous, resulting in death. Brominated vegetable oils are added to soft drinks, anemia and heart hypertrophy. Diethylene glycol is a substance added to wine that causes toxicity in the body.

Cereals

1. Maida/suji (rawa) - Adulterants like cheaper flour, boric acid, sand, soil, insects, iron fillings, rodent hair and excreta, excess bran, chalk powder are added which can cause severe gastrointestinal pain.
2. Grains (millet, wheat, oats, rye, rice etc.) - Toxins from *Fusarium sporotrichioides*, in the grains, cause Alimentary toxic aleukia (ATA). Alimentary toxic aleukia, is a mycotoxin-induced condition characterized by nausea, vomiting, diarrhea, leukopenia (aleukia), hemorrhaging, skin inflammation, and sometimes death.

Pulses

1. Dal whole and split pulses - Dust, pebble, stone, straw, weed seeds, damaged grain, weevilled grain, hidden insects, rodent hair & excreta, kernel bunt, ergot (bajra), khesari dal, clay, gravels, webs, non -permitted colors are most common adulterants found in pulses, causing toxic, irreversible paralysis, tumor and cancer, anemia, epilepsy, and neurotoxicity.

Fruits and Vegetables

1. Apple - Lead arsenate is sprayed on fruits such as apples. Dizziness, chills, cramps, paralysis, and death are all symptoms of arsenic poisoning.
2. Raw vegetables and fruits - *Entamoeba histolytica* l, a biological contamination, is added to raw vegetables and fruits, causing amoebic dysentery.

Food Colors

The following are some of the food colors that are added to the primary food ingredients, as well as the negative consequences they have:

1. Auramine - Inhibits growth, dysfunction of the liver and kidney.
2. Rhodamine B - Growth retardation, hemolysis of RBCs, degenerative changes in the liver and kidney, adversely affects the immune system.
3. Sudan dye - sudan dye is toxic to the liver and produce kidney lesions.
4. Malachite green - decreased food intake, growth rate and fertility rate, damage to organs like liver, kidney, heart and spleen as well as lesions of skin, eyes, lungs and bones.
5. Orange ll - growth retardation, increased mortality and haematological changes.
6. Metanil yellow - degenerative changes in the stomach, ileum, rectum, liver, kidney, ovary, and testes, giddiness, weakness, vomiting, and cyanosis.
7. Lead chromate - stomach ache, nausea, constipation, and anemia.
8. Tartrazine - Allergic reactions and worsening of asthma attack in case of asthamatic people.

LEGAL INTERVENTIONS, PREVENTION, PROHIBITION, AND RESTRICTIONS ON FOOD FRAUD AND FOOD ADULTERATION

Indian Laws

Prior to 2006, different food laws existed in India that pertained to different commodities. The following laws were in effect at the time:

Prevention of Food Adulteration Act, 1954

The Prevention of Food Adulteration Act was passed in 1954 with the goal of preventing food adulteration. Adulterated and misbranded foods, as well as any food that did not match the law's standards, were prohibited from being sold (The

Prevention of Food Adulteration Act & Rules, 2004). The law was enacted to ensure that customers have access to pure and wholesome food, as well as to avoid deception and fraud, resulting in fair trading practices.

Fruit Product Order (FPO), 1955

The FPO was established in 1955 under Section 3 of the Essential Commodities Act. The FPO's goal was to ensure that fruit and vegetable products were produced in sanitary and hygienic conditions on the grounds. It required all manufacturers of fruit and vegetable products, as well as some non-fruit items like syrup, non-fruit vinegar, and sweetened aerated water, to receive certification (Jha, 2016).

Milk & Milk Products Order, 1992

The Ministry of Agriculture's Department of Animal Husbandry executed the Milk and Milk Products Order, 1992, under Section 3 of the Essential Commodities Act, 1955 (Jha, 2016). The purpose of this directive was to regulate the production, supply, and distribution of milk in order to assure the supply of high-quality milk and milk products to consumers, as well as to ensure that sanitary conditions in processing machinery and facilities were maintained (Ram, 2011).

Meat Food Products Order, 1973

The Ministry of Food Processing Industries issued the Meat Food Products Order in 1973 under the terms of the Essential Commodities Act of 1955 (Ram, 2011). It establishes guidelines for the manufacturing of safe meat foods. All meat, fish, and poultry products are subjected to stringent quality control at all phases of manufacturing.

Edible Oils Packaging Order, 1998

The Central Government issued the Edible Oils Packaging (Regulation) Order, 1998 on September 17, 1998, in response to the Delhi dropsy tragedy, which claimed several lives owing to the ingestion of contaminated mustard oil (Ram, 2011). The order was passed to ensure that customers have access to safe and high-quality edible oils in pre-packaged form at pre-determined pricing (Jha, 2016).

Vegetable Oil Products Order, 1998

The "Vegetable Oil Products (Regulation) Order, 1998" was enacted to regulate the manufacturing, distribution, and sale of all vegetable oil products. This directive governed the vegetable oil products business through the Ministry of Consumer Affairs and Food and Public Distribution (Jha, 2016).

Solvent-Extracted Oil, De-Oiled Meal and Edible Flour (Control) Order, 1967

This order was also promulgated under Section 3 of the Essential Commodities Act, 1955. This order was issued to regulate and govern the manufacture and distribution of vegetable oils, de-oiled meal, and edible flour, as well as to ensure the quality of these products acquired through the solvent extraction process (Ram, 2011).

FSS Act – Food Safety and Standards Act, 2006

The above-mentioned food laws caused a great deal of misunderstanding in their application. As a result, under the Food Safety and Standards Act of 2006, these rules were merged under a single umbrella, eliminating multi-level and multi-departmental control and combining them into a single chain of command. As a result, the FSS act provides a single point of reference (see Figure 8).

The FSS Act is 'an act to consolidate the laws relating to food and to establish the Food Safety and Standards Authority of India for laying down science-based standards for articles of food and to regulate their manufacture, storage, distribution, sale and import, to ensure availability of safe and wholesome food for human consumption and for matters connected therewith or incidental thereto' (Food Safety and Standards Act, 2006).

FSSAI

Under Section 4, FSS Act, 2006, the Food Safety and Standard Authority of India (FSSAI) was established in the year 2011. The Ministry of Health & Family Welfare, Government of India, is the administrative ministry responsible for the FSSAI. The FSSAI is the authority that regulates food safety and standards.

Figure 8. An overview of the culmination of various food laws (prior to 2006) to FSS Act, 2006

Pure tea leaves Adulterated tea leaves with iron

Laws Regarding Adulteration in Food

A food must not include any pollutant, naturally occurring hazardous chemicals, poisons, hormones, or heavy metals above the limits set forth in Chapter IV of the FSS Act of 2006. Insecticides or pesticide residues, veterinary medication residues, antibiotic residues, solvent residues, pharmacologically active compounds, and microbiological counts should not exceed the tolerance limit specified by the act (Food Safety and Standards Act, 2006).

Standards and Permissible Limits: Physical Adulterants and Chemical Residues

Residues are the remnants of various inputs used during the production process like plant protection products and veterinary drugs. It is possible to regulate the conditions of usage and the levels of these substances in food. Contaminants are pollutants collected from the environment, food, and packaging. Contaminants are undesired and frequently unavoidable and are not caused by the inputs utilized during manufacturing (Benford, 2013)

Metal contaminants have become more diverse as a result of the industrial revolution and globalization. Many of these contaminants come from the soil. Chemical factories, mining, textile industries, e-waste processing and run-off, erosion, and irrigation with dirty water are all possible causes (Rai et al., 2019). The FSS Act establishes specified thresholds at which a residue or adulterant causes consumer harm. Copper, arsenic, tin, zinc, cadmium, mercury, methyl mercury, chromium, nickel etc. are some of the metal pollutants detected in food (Food Safety and Standards (Contaminants, Toxins and Residues) Regulations, 2011). The Appendices include the limitations in parts per million by weight that metal contamination levels shall not exceed.

Insecticides may not be directly applied to food, and fumigants may only be used in compliance with the Insecticide Act of 1968 regulations, according to the FSS Act. Endosulfan, fenitrothion, aldrin, dieldrin, carbaryl, chlordane, and other pesticides have been detected in foods. (Food Safety and Standards (Contaminants, Toxins, and Residues) Regulations, 2011). The Appendices include the safe limits for these.

Natural poisons such as mycotoxins, ochratoxin, and patulin exist. Mycotoxins are poisonous chemicals produced by moulds in their natural state (fungi). Mycotoxins are often produced by *Aspergillus flavus* and *Aspergillus parasiticus*. Patulin is a mycotoxin produced by Aspergillus, Penicillium, and Byssochlamys. All these, when consumed, have adverse effects on health (World Health Organization, 2018). The limits beyond which crop contaminants, naturally occurring toxic substances, antibiotics and pharmacologically active substances should not be present in food are mentioned in Table 6, 7 and 8, respectively.

Table 6. Limits for crop contaminants in food

No.	Contaminant	Article of Food	Limit µg/kg
1.	Aflatoxin	All articles of food	30
2.	Aflatoxin M_1	Milk	0.5
3.	Patulin	Apple juice & Apple juice ingredients in other beverages	50
4.	Ochratoxin A	Wheat, barley & rye	20

Source: *Food Safety and Standards (Contaminants, Toxins and Residues) Regulations, 2011.*

Table 7. Limits for naturally occurring toxic substances in food

No.	Name of the Substance	Maximum limit
1.	Agaric Acid	100 ppm
2.	Hydrocyanic Acid	5 ppm
3.	Hypericine	1 ppm
4.	Saffrole	10 ppm

Source: *Food Safety and Standards (Contaminants, Toxins and Residues) Regulations, 2011.*

Table 8. Limits for antibiotics and other pharmacologically active substances in food

No.	Antibiotics	Tolerance Limit mg/ kg (ppm)
1	Tetracycline	0.1
2	Oxytetracycline	0.1
3	Trimethoprim	0.05
4	Oxolinic acid	0.3

Source: Food Safety and Standards (Contaminants, Toxins and Residues) Regulations, 2011.

Bureau of Indian Standards (BIS)

The Bureau of Indian Standards (BIS) is empowered by the Bureau of Indian Standards Act, 2016. It establishes standards for product specifications, testing techniques, and licensing for businesses in virtually every industry, from agriculture to textiles to electronics. BIS is overseen by the Ministry of Consumer Affairs, Food, and Public Distribution. The ISI mark is a quality label issued by BIS that has pushed manufacturers in India and adjacent markets to make high-quality products for over 55 years (Food Regulatory Portal-One Nation, One Food Law).

Standards have been laid for spices and condiments, fruit and vegetable products, animal products and processed foods. Despite the fact that this standard is voluntary, the Indian government has imposed mandatory certification on a variety of products through various rules in public interest (Food Regulatory Portal-One Nation, One Food Law).

AGMARK

Agmark operates under the Directorate of Marketing and Inspection under the Ministry of Agriculture. Under the Agricultural Produce Grading and Marking Act and Rules, 1937, it is a voluntary method for certification of raw and processed agricultural goods. The Act specifies requirements for a variety of agricultural products, including some processed foods. Each commodity is assigned one of two to three grades based on its quality. Grade criteria for 222 agricultural commodities have been announced so far (Agmark Certification Scheme).

It has grading & marking rules for pulses, cereals, fruits & vegetables and vegetable oils, William pears, vermicelli, macaroni, spaghetti, wheat atta, wheat porridge, cereals, pulses, oilseeds, ghee, spices, honey, creamery butter, etc. (Directorate of Marketing & Inspection, 2020). The AGMARK Standards define the maximum amount of foreign pollutants and other extraneous matter that can be present in certain products. Refer to the appendices for AGMARK standards for grade-wise permissible foreign physical pollutants in wheat (percentage by weight) and AGMARK Standards for pulses (whole grain)- maximum limits of tolerance, percent by weight.

Codex Alimentarius

The Latin term for "food code" is Codex Alimentarius (Van der Meulen, 2010). The CODEX Alimentarius Commission was created in 1963 as an intergovernmental authority. The main goal was to establish norms and standards to assure consumer protection and safety, as well as fair trade practices in the global food trade (Stankovic, 2016). CODEX Alimentarius strives to ensure that everyone—everywhere—has access to safe, nutritious food. International food standards, guidelines, and codes of practice developed by CODEX Alimentarius help to ensure food safety and quality in international trade (Indian Food Safety and Standards Authority, 2018). The Codex Alimentarius includes provisions with respect to methods of analysis and sampling, food hygiene, contaminants, food additives, residues of pesticides and veterinary drugs, labeling and presentation, and import and export inspection and certification. The CODEX standards mention the level of physical adulterants, extraneous matter, damaged food material and organic and inorganic extraneous matter (Stankovic, 2016). For Codex Alimentarius Commission Standard for wheat and durum wheat refer to Appendices.

International Food Regulatory Organizations

World Trade Organization (WTO)

The WTO was established on January 1, 1995 as the result of the Uruguay round of trade negotiations and signed in Marrakesh on 15 April 1994 (WTO Agreement). The WTO is the institutional continuation of the General Agreement on Tariffs and Trade 1947 (GATT) (Van der Meulen, 2014).

The objective of the WTO is to remove barriers to international trade. The basic treaty addressing trade in goods is the General Agreement on Tariffs and Trade (GATT). To address concerns about food safety, human health, animal and plant health two WTO treaties were concluded: the Agreement on Technical Barriers to Trade (the TBT Agreement) and the Agreement on the Application of Sanitary and Phytosanitary Measures (the SPS Agreement). The Sanitary and Phytosanitary Measures Agreement gives procedures for the assessment of risk and the determination of appropriate levels of sanitary or phytosanitary protection (Van der Meulen, 2010).

ISO

ISO is an independent, multinational, non-governmental organization whose mission is to promote the development of standards around the world in order to enable international trade and collaboration. It is made up of 165 national standards organizations. It aids in the harmonization of food standards around the world, as well as the protection of customers and users, as well as the revision of quality, safety, and dependability standards (ISO). The ISO brings together professionals through its members to share information and establish voluntary, market-relevant, consensus-based International Standards that aid in the provision of global solutions (ISO).

USA: FDA

The Food and Drug Administration is the oldest consumer protection agency in the US federal government. The Food and Drug Administration is responsible for safeguarding the safety and security of human and veterinary pharmaceuticals, biological products, and medical devices in the United States. It also protects the nation's food supply (US Food and Drug Administration, 2018).

The following acts and the causing thereof are prohibited:

Introducing any adulterated or misbranded food into interstate commerce, as well as any act of adulteration and misbranding, is illegal under the Federal Food, Drug, and Cosmetic Act, 21 USC 331. The refusal to allow admission or inspection, as well as the change, mutilation, destruction, obliteration, or removal of all or part of the labeling, or any other conduct involving food, are all prohibited under the act (Federal Food Drug and Cosmetic Act, 21 U.S.C. § 331).

UK: FSA

Food Safety Act, 1990establishes rules for food safety and consumer protection in the United Kingdom. The act's major goal is to ensure that food fits the demands of customers and is of the nature, quality, and substance that is expected. It assures that no material is added to or withdrawn from food, and that food is not processed in a way that harms customers' health. (Food Standards Agency, 2009).

The Food Standards Act of 1999 created the Food Standards Agency and established criteria for its operation. The legislation allows for provisions to be made for the notification of foodborne disease tests and other related matters (Food Standards Act, 1999). The Food Standards Agency (FSA) is responsible for food safety and food hygiene in England, Wales and Northern Ireland. Food safety regulations are enforced by the FSA in collaboration with local authorities (GOV UK). It's also in charge of labeling policy in Wales and Northern Ireland, as well as nutrition policy in the latter (GOV UK). Consumers can trust the food they buy and eat because the agency utilizes its expertise and clout to ensure that it is safe (Food Standards Agency).

Food Standards Scotland is a non-ministerial government department of the Scottish Government, established by the Food Act of 2015 of Scotland. It is responsible for food safety and standards, nutrition, food labeling and meat inspection in Scotland. The responsibilities of the UK-wide organization, the Food Standards Agency, has been taken over by this department in Scotland. It assures accurate labeling and that safe food is available throughout the food supply chain (Food Standards Scotland).

European Union: European Food Safety Authority (EFSA)

EFSA is a European agency set up in 2002 after a multitude of food crises in the 1990s. It is funded by the European Union. It operates independently of the European legislative and executive institutions and EU Member States (European Food Safety Authority). It strives to provide scientific guidance using the knowledge of its experts and workers to protect European consumers and manage public health issues. Its mission is to provide scientific advice and highlight food-related concerns. The European Food Safety Authority (EFSA) is continually developing and adapting data to ensure that the food safety system is at the forefront of scientific and administrative thinking and practice (European Union).

CONCLUSION

Activities to anticipate the risk of adulteration have an inherent flaw because risk quantification is usually based on past data that may or may not be available, or may or may not reflect the actual danger now or in the future. Food fraud or adulteration is common, and such food adulteration degrades food quality, makes it dangerous to consume, and causes health problems in humans. The goal of this chapter was to discuss the acts of adulteration in various food sectors, health implications of the same and look into the current strategies for monitoring and detecting potential adulterants. It may be concluded from the preceding examination and debate that adulteration added to food items can have a significant impact on health without our understanding. Though visual examination cannot guarantee the absence of adulterants because harmful pollutants are present at extremely low levels, it can ensure the absence of insects, fungus, and other foreign elements prior to purchase.

Food safety concerns arise when adulteration occurs for a variety of reasons, including the use of expired ingredients, deceptive adulterated food labels, or the addition of harmful compounds to food (Nasreen & Ahmed, 2014). Authenticity of food items is one of the most important factors to be maintained in the food industry, at all stages from manufacturing to the end of the supply chain, all over the world (Dong et al., 2016). Furthermore, food fraud and adulteration force consumers to change their product preferences and seek out more authentic and safe foods, resulting in financial losses for manufacturers. The most effective way to avoid food adulteration and fraud is to implement a robust risk management system (Spink et al., 2016). Food safety programs and practices such as Six Sigma, HACCP, Good Manufacturing Practices, and Good Hygienic Practices throughout the food supply chain can be used in the food industry to ensure the creation of quality products. Implementation of regulating mechanisms, sampling and monitoring food items, raising awareness among food makers across the supply chain, and establishing various detection tools will pave the path for combating food fraud and adulteration (Tibola et al., 2018).

REFERENCES

Azizian, H., Mossoba, M. M., Fardin-Kia, A. R., Delmonte, P., Karunathilaka, S. R., & Kramer, J. K. (2015). Novel, rapid identification, and quantification of adulterants in extra virgin olive oil using near-infrared spectroscopy and chemometrics. *Lipids*, *50*(7), 705–718. doi:10.100711745-015-4038-4 PMID:26050093

Banti, M. (2020). Food adulteration and some methods of detection [review]. *International Journal Nutrition Food Science*, *9*(3), 86–94. doi:10.11648/j.ijnfs.20200903.13

Benford, D. J. (2013). Risk assessment of chemical contaminants and residues in foods. *Persistent Organic Pollutants and Toxic Metals in Foods*, 173–187.

Changbai Xiu, C., & Klein, K. K. (2010). Melamine in milk products in China: Examining the factors that led to deliberate use of the contaminant. *Food Policy*, *35*(5), 463–470. doi:10.1016/j.foodpol.2010.05.001

Choudhary, A., Gupta, N., Hameed, F., & Choton, S. (2020). An overview of food adulteration: Concept, sources, impact, challenges and detection. *International Journal of Chemical Studies*, *8*(1), 2564–2573. doi:10.22271/chemi.2020.v8.i1am.8655

Cianferoni, A., & Spergel, J. M. (2009). Food allergy: Review, classification and diagnosis. *Allergology International*, *58*(4), 457–466. doi:10.2332/allergolint.09-RAI-0138 PMID:19847094

Codex Alimentarius. (2021). *About Codex Alimentarius.* http://www.fao.org/fao-who-codexalimentarius/ about-codex/en/

Directorate of Marketing and Inspection. (2020). *Grades & Standard: AGMARK Standards*. https://dmi.gov.in/GradesStandard.aspx

Dong, H., Luo, D., & Luo, H. (2016). Effective use of food traceability in cereal grain food supply chains. In *Advances in Food Traceability Techniques and Technologies* (pp. 303–319). Woodhead Publishing. doi:10.1016/B978-0-08-100310-7.00016-8

Ehn, B. M., Allmere, T., Telemo, E., Bengtsoon, U., & Ekstrand, B. (2005). Modification of IgE binding to b-lactoglobulin by fermentation and proteolysis of cow's milk. *Journal of Agricultural and Food Chemistry*, *53*(9), 3743–3746. doi:10.1021/jf048121w PMID:15853429

European Food Safety Authority (EFSA). (n.d.). *About us*. https://www.efsa.europa.eu/en/aboutefsa

Faraz, A., Lateef, M., Mustafa, M. I., Akhtar, P., & Yaqoob, M. (2013). Detection of adulteration, chemical composition and hygienic status of milk supplied to various canteens of educational institutes and public places in Faisalabad. *Journal Animal Plant Science*, *23*, 119–124.

Food Safety and Standards Authority of India. (2018). *About Codex*. https://archive.fssai.gov.in/home/fss-legislation/about-codex.html

Food Safety and Standards Authority of India. (2020). *Food law: Codex Alimentarius Commission*. https://fssai.gov.in/cms/codex.php

Food Standards Act. (1999). https://www.legislation.gov.uk/ukpga/1999/28#reference-c1312451

Food Standards Agency. (n.d.). *Homepage*. https://www.food.gov.uk/

Food Standards Scotland. (n.d.). https://www.foodstandards.gov.scot

Galvin-King, P., Haughey, S. A., & Elliott, C. T. (2018). Herb and spice fraud; the drivers, challenges and detection. *Food Control*, *88*, 85–97. doi:10.1016/j.foodcont.2017.12.031

GOV.UK. (n.d.). *Food Standards Agency*. https://www.gov.uk/government/organisations/food-standards-agency

ISO. (2021). *About us*. https://www.iso.org/about-us.html

Jabeur, H., Zribi, A., Makni, J., Rebai, A., Abdelhedi, R., & Bouaziz, M. (2014). Detection of Chemlali extra-virgin olive oil adulteration mixed with soybean oil, corn oil, and sunflower oil by using GC and HPLC. *Journal of Agricultural and Food Chemistry*, *62*(21), 4893–4904. doi:10.1021/jf500571n PMID:24811341

Jha, S. N. (2016). Food standards and permissible limits. *Rapid detection of food adulterants and contaminants*, 63–106.

Khan, K. M., Krishna, H., Majumder, S. K., & Gupta, P. K. (2015). Detection of urea adulteration in milk using near-infrared Raman spectroscopy. *Food Analytical Methods*, *8*(1), 93–102. doi:10.100712161-014-9873-z

Mangal, M., Bansal, S., & Sharma, M. (2014). Macro and micromorphological characterization of different *Aspergillus* isolates. *Legume Research*, *37*(4), 372–378. doi:10.5958/0976-0571.2014.00646.8

Mathur, P. (2005). *Food microbiology and safety*. Indira Gandhi National Open University.

Nasreen, S., & Ahmed, T. (2014). Food adulteration and consumer awareness in Dhaka City, 1995-2011. *Journal of Health, Population and Nutrition*, *32*(3), 452. PMID:25395908

Oroian, M., Ropciuc, S., & Paduret, S. (2018). Honey adulteration detection using Raman spectroscopy. *Food Analytical Methods*, *11*(4), 959–968. doi:10.100712161-017-1072-2

Pardeshi, S. (2019). Food adulteration: Injurious adulterants and contaminants in foods and their health effects and its safety measures in India. *International Journal of Scientific Development and Research*, *4*(6), 231–232.

Poonia, A., Jha, A., Sharma, R., Singh, H. B., Rai, A. K., & Sharma, N. (2016). Detection of adulteration in milk: A review. *International Journal of Dairy Technology*, *70*(1), 23–42. doi:10.1111/1471-0307.12274

Rai, P. K., Lee, S. S., Zhang, M., Tsang, Y. F., & Kim, K. H. (2019). Heavy metals in food crops: Health risks, fate, mechanisms, and management. *Environment International*, *125*, 365–385. doi:10.1016/j.envint.2019.01.067 PMID:30743144

Ram, A. (2011). *Indian food laws in global context a critique*. http://hdl.handle.net/10603/87550

Reddy, D., Venkatesh, K., & Reddy, C. (2017). Adulteration of milk and its detection: A review. *International Journal of Chemical Studies*, *5*(4), 613–617.

Singuluri, H., & Sukumaran, M. K. (2014). Milk adulteration in Hyderabad, India–A comparative study on the levels of different adulterants present in milk. *Journal of Chromatography & Separation Techniques*, *5*(1), 212. doi:10.4172/2157-7064.1000212

Spink, J., Moyer, D. C., & Whelan, P. (2016). The role of the public private partnership in food fraud prevention—Includes implementing the strategy. *Current Opinion in Food Science, 10*, 68–75. doi:10.1016/j.cofs.2016.10.002

Srivastava, S. (2015). Food adulteration affects the nutrition and health of human beings. *Journal of Biological Sciences and Medicine, 1*(1), 68.

Stankovic, I. (2016). Codex Alimentarius. Encyclopedia of Food and Health, 191–196.

Tibola, C. S., da Silva, S. A., Dossa, A. A., & Patrício, D. I. (2018). Economically motivated food fraud and adulteration in Brazil: Incidents and alternatives to minimize occurrence. *Journal of Food Science, 83*(8), 2028–2038. doi:10.1111/1750-3841.14279 PMID:30020548

Van der Meulen, B. (2010). *Development of food legislation around the world. In Ensuring Global Food Safety Exploring Global Harmonization.* Academic Press. doi:10.1016/B978-0-12-374845-4.00002-3

Van der Meulen, B. (2014). Food law. Encyclopedia of Agriculture and Food Systems, 186–195.

Wang, S., Guo, Q., Wang, L., Lin, L., Shi, H., Cao, H., & Cao, B. (2015). Detection of honey adulteration with starch syrup by high performance liquid chromatography. *Food Chemistry, 172*, 669–674. doi:10.1016/j.foodchem.2014.09.044 PMID:25442605

World Health Organization. (2018, May 9). *Mycotoxins.* https://www.who.int/news-room/fact-sheets/detail/mycotoxins

KEY TERMS AND DEFINITIONS

Direct Food Safety Risks: Consumer is put at immediate risk, i.e., addition of melamine to milk powder results in acutely toxic exposure; hiding of substances resulting in undeclared allergens.

Economically Motivated Adulteration (EMA): The deliberate adulteration of food, motivated by greed or a political agenda. In case of economically motivated adulteration (EMA), aims is not to harm people but to inflate profits by fraudulent means.

Food Crimes: Food crime as serious fraud and related criminality in food supply chains. This definition also includes activity impacting on drink and animal feed. It can be seriously harmful to consumers, food businesses and the wider food industry.

Food Defense: An effort to prevent acts of adulteration that are intended to cause harm to a food business or to consumers, such as acts of terrorism or attempted extortion.

Food Security: Defined by when all people at all times have access to sufficient, safe, nutritious food to maintain a healthy and active life.

Horizon Scanning: A term that has been co-opted to the language of food fraud. Horizon scanning is the act of looking for and analyzing threats and opportunities that will emerge in the medium to long term. Within the food industry, horizon scanning refers to the act of collecting information about current trends in food production and predicted incidences that could increase the likelihood of food fraud for a particular food material.

Indirect Food Safety Risks: Consumer is put at risk through long-term exposure i.e. high levels of heavy metals in food supplements causing harm or lack of benefit-over longer period of time.

Intentional Adulteration: Although food fraud activities often involve the intentional adulteration of food with unauthorized substances, within the food safety industry, has recently been given a more specific meaning. And it is not related to food fraud at all. It is related to food defense, and more specifically to activities intended to cause wide scale harm to consumers.

TACCP: Threat Assessment Critical Control Point (TACCP) helpful for prevention of malicious threats to food.

Technical Food Fraud Risk: There is no direct or indirect food safety risk, this indicates that material traceability may have been compromised and companies are no longer able to guarantee the safety of their food products.

Vulnerability: It is a term of art and a basis for assessment methods in several contexts, including climate impact analysis, disaster management, and food security analysis. 'Vulnerability' is used because food fraud 'risks' do not exactly fit with the accepted definition of risk as something that has occurred frequently, will occur again and for which there is enough data to make quantitative assessments.

Compilation of References

Abdellah, A., Noordin, M. I., & Ismail, W. A. W. (2015). Importance and globalization status of good manufacturing practice (GMP) requirements for pharmaceutical excipients. *Saudi Pharmaceutical Journal, 23*(1), 9–13. doi:10.1016/j.jsps.2013.06.003 PMID:25685037

Abdul Nazer, A., Angeline, F. W., Sunil Kumar, P., & Nazer Zulfika, A. (2018). A KAP Study on Food Safety and Hygiene Among Private University Students in Kedah State, Malaysia. *Journal of Natural Remedies, 18*(3), 113–121. doi:10.18311/jnr/2018/22289

Abdul Rahman, N. (2019, May 15). 36 tahfiz students down with food poisoning. *New Straits Times*.

Abdulmumeen, H. A., Risikat, A. N., & Sururah, A. R. (2012). Food: Its preservatives, additives and applications. *International Journal of Chemical and Biochemical Science, 1*, 36–47.

Abdul-Mutalib, N.-A., Abdul-Rashid, M.-F., Mustafa, S., Amin-Nordin, S., Awang Hamat, R., & Osman, M. (2012). Knowledge, attitude and practices regarding food hygiene and sanitation of food handlers in Kuala Pilah, Malaysia. *Food Control, 27*(2), 289–293. doi:10.1016/j.foodcont.2012.04.001

Abushelaibi, A., Jobe, B., Al Dhanhani, F., Al Mansoori, S., & Al Shamsi, F. (2016). An overview of food safety knowledge and practices in selected schools in the city of Al Ain, United Arab Emirates. *African Journal of Microbiological Research, 10*(15), 511–520. doi:10.5897/AJMR2016.7917

Adelaja, A., George, J., Miyahara, T., & Penar, E. (2019). Food Insecurity and Terrorism. *Applied Economic Perspectives and Policy, 41*(3), 475–497. doi:10.1093/aepp/ppy021

Ahmad, S. (2016). Penggunaan polisterin diharamkan di WP mulai tahun depan. *Berita Harian Online*. http://www.Bharian.Com.My/Node/201159

Ahmad, M. H., Nache, M., Hinrichs, J., & Hitzmann, B. (2016b). Estimation of the nutritional parameters of various types of wheat flours using fluorescence spectroscopy and chemometrics. *International Journal of Food Science & Technology, 51*(5), 1186–1194. doi:10.1111/ijfs.13080

Ahmad, M. H., Nache, M., Waffenschmidt, S., & Hitzmann, B. (2016a). A fluorescence spectroscopic approach to predict analytical, rheological and baking parameters of wheat flours using chemometrics. *Journal of Food Engineering, 182*, 65–71. doi:10.1016/j.jfoodeng.2016.03.006

Ai, K., Liu, Y., & Lu, L. (2009). Hydrogen-bonding recognition-induced color change of gold nanoparticles for visual detection of melamine in raw milk and infant formula. *Journal of the American Chemical Society*, *131*(27), 9496–9497. doi:10.1021/ja9037017 PMID:19537721

Alamprese, C., Casale, M., Sinelli, N., Lanteri, S., & Casiraghi, E. (2013). Detection of minced beef adulteration with turkey meat by UV–vis, NIR and MIR spectroscopy. *Lebensmittel-Wissenschaft + Technologie*, *53*(1), 225–232. doi:10.1016/j.lwt.2013.01.027

Alander, J. T., Bochko, V., Martinkauppi, B., Saranwong, S., & Mantere, T. (2013). A review of optical nondestructive visual and near-infrared methods for food quality and safety. *International Journal of Spectroscopy*, *2013*, 341402. doi:10.1155/2013/341402

Alava, J. M., Sahi, S. S., Garcia-Alvarez, J., Tur, A. O., Chavez, J. A., & Garcıa, M. J. (2007). Use of ultrasound for the determination of flour quality. *Ultrasonics*, *46*(3), 270–276. doi:10.1016/j.ultras.2007.03.002 PMID:17462688

Aldoory, L., Kim, J. N., & Tindall, N. (2010). The influence of perceived shared risk in crisis communication: Elaborating the situational theory of publics. *Public Relations Review*, *36*(2), 134–140. doi:10.1016/j.pubrev.2009.12.002

Al-Ghazali, M., Al-Bulushi, I., Al-Subhi, L., Rahman, M. S., & Al-Rawahi, A. (2020). Food safety knowledge and hygienic practices among different groups of restaurants in Muscat, Oman. *International Journal of Food Sciences*, *2020*, 1–8. Advance online publication. doi:10.1155/2020/8872981 PMID:33415137

Ali, A. E., Msarah, M. J., & Sahilah, A. M. (2017). Environment contaminant of *Bacillus cereus* isolated from ready to eat meat curry collected at various locations in Malaysia. *International Food Research Journal*, *24*(6), 2640–2644.

Alimentarius, C. (2020). *Code of practice on food allergen management for food business operators.* Codex Alimentarius International Food Standards. https://www.fao.org/fao-who-codexalimentarius/shproxy/en/?lnk=1&url=https%253A%252F%252Fworkspace.fao.org%252Fsites%252Fcodex%252FStandards%252FCXC%2B80-2020%252FCXC_080e.pdf

Al-Kandari, D., Al-abdeen, J., & Sidhu, J. (2019). Food safety knowledge, attitudes and practices of food handlers in restaurants in Kuwait. *Food Control*, *103*, 103–110. doi:10.1016/j.foodcont.2019.03.040

Almeida, M. R., Oliveira, K. D. S., Stephani, R., & de Oliveira, L. F. C. (2011). Fourier-transform Raman analysis of milk powder: A potential method for rapid quality screening. *Journal of Raman Spectroscopy: JRS*, *42*(7), 1548–1552. doi:10.1002/jrs.2893

Al-Mohaithef, M., Abidi, S. T., Javed, N. B., Alruwaili, M., & Abdelwahed, A. Y. (2021). Knowledge of safe food temperature among restaurant supervisors in Dammam, Saudi Arabia. *Journal of Food Quality*, *2021*, 1–8. Advance online publication. doi:10.1155/2021/2231371

Alomar, D., Gallo, C., Castaneda, M., & Fuchslocher, R. (2003). Chemical and discriminant analysis of bovine meat by near infrared reflectance spectroscopy (NIRS). *Meat Science*, *63*(4), 441–450. doi:10.1016/S0309-1740(02)00101-8 PMID:22062513

Alston, P., & Quinn, G. (2017). The nature and scope of states parties' obligations under the international covenant on economic, social and cultural rights. In *Economic, Social and Cultural Rights* (pp. 3–76). Routledge. doi:10.4324/9781315257044-2

Alvarado, U., Zamora, A., Liu, J., Saldo, J., & Castillo, M. (2020). Rapid Quantification of Riboflavin in Milk by Front-Face Fluorescence Spectroscopy: A Preliminary Study. *Foods*, *9*(1), 6. doi:10.3390/foods9010006 PMID:31861817

Alvarez, P. A., & Boye, J. I. (2012). Food production and processing considerations of allergenic food ingredients: A review. *The Journal of Allergy*. PMID:22187573

Alvensleben, R. V. (2002). Verbraucherbild—Verbraucherverantwortung— Verbrauchererziehung. Ziele und Fakten. *ZLR-Zeitschrift fu¨r das Gesamte Lebensmittelrecht, 2*, 139–150.

Amenta, V., Aschberger, K., Arena, M., Bouwmeester, H., Moniz, F. B., Brandhoff, P., Gottardo, S., Marvin, J. P. H., Mech, A., Pesudo, L. Q., Rauscher, H., Schoojans, R., Vettori, M. V., Weigel, S., & Peters, R. J. (2015). Regulatory aspects of nanotechnology in the agri/feed/food sector in EU and non-EU countries. *Regulatory Toxicology and Pharmacology*, *73*(1), 463–476. doi:10.1016/j.yrtph.2015.06.016 PMID:26169479

Andersen, C. M., Vishart, M., & Holm, V. K. (2005). Application of fluorescence spectroscopy in the evaluation of light-induced oxidation in cheese. *Journal of Agricultural and Food Chemistry*, *53*(26), 9985–9992. doi:10.1021/jf051143d PMID:16366684

Andress, E. L., & Harisson, J. A. (2011). *Food storage for safety and quality*. University of Georgia. https://nchfp.uga.edu/how/store/UGA_foodstorage_2011.pdf

Angelo, K. M., Nisler, A. L., Hall, A. J., Brown, L. G., & Gould, L. H. (2017). Epidemiology of restaurant-associated foodborne disease outbreaks, United States, 1998-2013. *Epidemiology and Infection*, *145*(3), 523–534. doi:10.1017/S0950268816002314 PMID:27751201

Angulo, F. J., Jones, T. F., & Angulo, F. J. (2006). Eating in restaurants: A risk factor for foodborne disease? *Clinical Infectious Diseases*, *43*(10), 1324–1328. doi:10.1086/508540 PMID:17051501

Ankita, J. L. (2018). *Development of rapid detection methods for acidic food spoilage and calcium carbide based artificial ripening of fruits* (Doctoral dissertation). Department of Food Science and Technology.

APE Cooperation (APEC) & Food Safety Cooperation Forum's Partnership Training Institute Network (FSCF PTIN). (2018). *Improving Food Safety through Hand Washing and Drying Capacity Building*. http://fscf-ptin.apec.org/docs/training/Handwash_FN-June-2018.pdf

Applebaum, R. S. (2006). Terrorism and the nation's food supply perspectives of the food industry: Where we are, what we have, and what we need. *Journal of Food Science*, *69*(2), crh48–crh50. Advance online publication. doi:10.1111/j.1365-2621.2004.tb15493.x

Ardic, M., & Kahve, H. H. (2015). Chemical migration in food technology. *Academic Journal of Science*, *4*, 163–168.

Argyri, A. A., Panagou, E. Z., & Nychas, G. J. (2014). Monitoring microbial spoilage of foods by vibrational spectroscopy (FT-IR & Raman). *Novel Food Preservation and Microbial Assessment Techniques, 386.*

Arslan, S., Eyi, A., & Küçüksari, R. (2014). Toxigenic genes, spoilage potential, and antimicrobial resistance of *Bacillus cereus* group strains from ice cream. *Anaerobe*, *25*, 42–46. doi:10.1016/j.anaerobe.2013.11.006 PMID:24309214

Asiegbu, C. V., Lebelo, S. L., & Tabit, F. T. (2016). The food safety knowledge and microbial hazards awareness of consumers of ready-to-eat street-vended food. *Food Control*, *60*, 422–429. doi:10.1016/j.foodcont.2015.08.021

Asim, H. S., Elnemr, I., Goktepe, I., Feng, H., Park, H. K., Alzeyara, S., AlHajri, M., & Kushad, M. (2019). Assessing safe food handling knowledge and practices of food service managers in Doha, Qatar. *Food Science & Technology International*, *25*(5), 440–448. doi:10.1177/1082013219830843 PMID:30791700

Asmawi, U. M. M., Norehan, A. A., Salikin, K., Rosdi, N. A. S., Munir, N. A. T. A., Basri, N. B. M., ... Nor, N. M. (2018). An Assessment of Knowledge, Attitudes and Practices in Food Safety Among Food Handlers Engaged in Food Courts. *Current Research in Nutrition and Food Science*, *06*(2), 346–353. doi:10.12944/CRNFSJ.6.2.09

Aygun-Fatma. (2016). D., Aygun-Fatih., & Cam, H. (2016). Successful treatment of *Bacillus cereus* bacteremia in a patient with propionic acidemia. *Case Reports in Pediatrics*, 1–2.

Ayvaz, H., Sierra-Cadavid, A., Aykas, D. P., Mulqueeney, B., Sullivan, S., & Rodriguez-Saona, L. E. (2016). Monitoring multicomponent quality traits in tomato juice using portable mid-infrared (MIR) spectroscopy and multivariate analysis. *Food Control*, *66*, 79–86. doi:10.1016/j.foodcont.2016.01.031

Azaman, N. N. M., Kamarulzaman, N. H., Mad Nasir Shamsudin, M. N., & Selamat, J. (2006). Stakeholders' knowledge, attitude and practice (KAP) towards aflatoxins contamination in peanut-based products. *Food Control*, *70*, 249–256. doi:10.1016/j.foodcont.2016.05.058

Azanza, M. P., Gatchalian, C. F., & Ortega, M. P. (2000). Food safety knowledge and practices of street food vendors in a Philippines university campus. *International Journal of Food Sciences and Nutrition*, *51*, 235–246. doi:10.1080/09637480050077121 PMID:11027035

Azizian, H., Mossoba, M. M., Fardin-Kia, A. R., Delmonte, P., Karunathilaka, S. R., & Kramer, J. K. (2015). Novel, rapid identification, and quantification of adulterants in extra virgin olive oil using near-infrared spectroscopy and chemometrics. *Lipids*, *50*(7), 705–718. doi:10.100711745-015-4038-4 PMID:26050093

Babu, K. S., & Amamcharla, J. K. (2018). Application of front-face fluorescence spectroscopy as a tool for monitoring changes in milk protein concentrate powders during storage. *Journal of Dairy Science*, *101*(12), 10844–10859. doi:10.3168/jds.2018-14885 PMID:30316594

Baert, K., Devlieghere, F., Jacxsens, L., & Debevere, J. (2005). Quality management systems. In *The food industry in safety in the agri-food chain* (pp. 877–879). Wageningen Academic Publishers.

Baietto, M., & Wilson, A. D. (2015). Electronic-nose applications for fruit identification, ripeness and quality grading. *Sensors (Basel)*, *15*(1), 899–931. doi:10.3390150100899 PMID:25569761

Balaganesan, S., & Siti Nur Afifah, J. (2019). A Survey On Food Handling Knowledge And Practices Among Housewives In Selayang, Selangor. *MAEH Journal of Environmental Health*, *1*(1), 38–43.

Balali, G. I., Yar, D. D., Afua Dela, V. G., & Adjei-Kusi, P. (2020). Microbial contamination, an increasing threat to the consumption of fresh fruits and vegetables in today's world. *International Journal of Microbiology*, *2020*, 1–13. Advance online publication. doi:10.1155/2020/3029295 PMID:32565813

Baldwin, E. A., Bai, J., Plotto, A., & Dea, S. (2011). Electronic noses and tongues: Applications for the food and pharmaceutical industries. *Sensors (Basel)*, *11*(5), 4744–4766. doi:10.3390110504744 PMID:22163873

Ballin, N. Z. (2010). Authentication of meat and meat products. *Meat Science*, *86*(3), 577–587. doi:10.1016/j.meatsci.2010.06.001 PMID:20685045

Ballin, N. Z., Vogensen, F. K., & Karlsson, A. H. (2009). Species determination–Can we detect and quantify meat adulteration? *Meat Science*, *83*(2), 165–174. doi:10.1016/j.meatsci.2009.06.003 PMID:20416768

Bansal, S., Singh, A., Mangal, M., Mangal, A. K., & Kumar, S. (2017). Food adulteration: Sources, health risks, and detection methods. *Critical Reviews in Food Science and Nutrition*, *57*(6), 1174–1189. doi:10.1080/10408398.2014.967834 PMID:26054861

Banti, M. (2020). Food adulteration and some methods of detection [review]. *International Journal Nutrition Food Science*, *9*(3), 86–94. doi:10.11648/j.ijnfs.20200903.13

Barbosa-Canovas, G.V., & Swanson, B. (2019). Nonthermal inactivation of endoproteases by pulsed electric field technology. *Pulsed Electric Fields in Food Processing: Fundamental Aspects and Applications*, 135.

Baser, F., Ture, H., Abubakirova, A., Sanlier, N., & Cil, B. (2017). Structural modeling of the relationship among food safety knowledge, attitude and behavior of hotel staff in Turkey. *Food Control*, *73*, 438–444. doi:10.1016/j.foodcont.2016.08.032

Bavani, M. (2016). *Polystyrene not the real culprit*. http://www.thestar.com.my/metro/community/2016/07/11/polystyrene-not-the-real-culprit-experts-say-ban-of-product-will-not-solve-waste-issues

Bell, R. (2012). *Introductory Fourier transform spectroscopy*. Elsevier. Academic Press.

Benford, D. J. (2013). Risk assessment of chemical contaminants and residues in foods. *Persistent Organic Pollutants and Toxic Metals in Foods*, 173–187.

Berzaghi, P., Segato, S., Cozzi, G., & Andrighetto, I. (2006). Mid and near infrared spectroscopy to identify illegal treatments in beef cattle. *Veterinary Research Communications*, *30*(1), 109–112. doi:10.100711259-006-0022-z

Bidwell, M., Briscoe, F., Fernandez-Mateo, I., & Sterling, A. (2013). The employment relationship and inequality: How and why changes in employment practices are reshaping rewards in organizations. *The Academy of Management Annals*, *7*(1), 61–121. doi:10.5465/19416520.2013.761403

Bilung, L. M., Jaraee, J., & Vincent, M. (2018). Detection, genetic diversity and antibiotic resistance profiles of Bacillus cereus isolated from sago processing plants in Malaysia. *Malaysian Journal of Microbiology*, *14*, 320–324.

Bilung, L. M., Tesfamariam, F., & Andriesse, R. (2018). Presence of *Bacillus cereus* from local unhusked (rough) rice samples in Sarawak, Malaysia. *Journal of Sustainability Science and Management*, *13*(1), 181–187.

Biokar Diagnostic. (2019). *COMPASS Bacillus.: 1-2. Detection and enumeration of Bacillus cereus group in human and animal food products*. http://www.biokar-diagnostics.com

Blanchfield, J. R. (2005). Good manufacturing practice (GMP) in the food industry. In Handbook of Hygiene Control in the Food Industry (pp. 324-347). Woodhead Publishing.

Bogdanov, S., Jurendic, T., Sieber, R., & Gallmann, P. (2008). Honey for nutrition and health: A review. *Journal of the American College of Nutrition*, *27*(6), 677–689. doi:10.1080/07315724.2008.10719745 PMID:19155427

Bolton, E. (2009). Guidelines for Assessing the Microbiological Safety of Ready-to-Eat Foods Placed on the Market. Health Protection Agency.

Bonora, S., Francioso, O., Tugnoli, V., Prodi, A., Di Foggia, M., Righi, V., Nipoti, P., Filippini, G., & Pisi, A. (2009). Structural characteristics of 'Hayward' kiwifruits from elephantiasis-affected plants studied by DRIFT, FT-Raman, NMR, and SEM techniques. *Journal of Agricultural and Food Chemistry*, *57*(11), 4827–4832. doi:10.1021/jf9002957 PMID:19413311

Boyce, J., Charles, C., & Binkley, B. M. (2008). Consumer perspectives: Take-out packaging and food safety. *British Food Journal*, *110*(8), 819–828. doi:10.1108/00070700810893340

Brown, A., Lamb, C., Brown, M., & Neal, A. (2000). Review of the developmental and reproductive toxicity of styrene. *Regulatory Toxicology and Pharmacology*, *3*(3), 228–247. doi:10.1006/rtph.2000.1406 PMID:11162717

Brown, L. G., Ripley, D., Blade, H., Reimann, D., Everstine, K., Nicholas, D., Egan, J., Koktavy, N., & Quilliam, D. N. (2012). Restaurant food cooling practices. *Journal of Food Protection*, *75*(12), 2172–2178. doi:10.4315/0362-028X.JFP-12-256 PMID:23212014

Bultman, M. W., Fisher, F. S., & Pappagianis, D. (2013). The ecology of soil-borne human pathogens. In *Essentials of Medical Geology* (pp. 477–504). Springer. doi:10.1007/978-94-007-4375-5_20

Burgess, C. M., Arroyo, C., Bolton, D. J., Danaher, M., O'Connor, L., O'Mahony, P. J., & Tlustos, C. (2019). Food safety: A public health issue of growing importance. *Introduction to Human Nutrition*, 388.

Buzby, J. C., & Hyman, J. (2012). Total and per capita value of food loss in the United States. *Food Policy*, *37*(5), 561–570. doi:10.1016/j.foodpol.2012.06.002

California Department of Education (CDE). (2000). *Proper storage temperatures for USDA commodities*. https://www.cde.ca.gov/ls/nu/fd/mb00404.asp

Caramês, E. T., Alamar, P. D., Poppi, R. J., & Pallone, J. A. L. (2017). Rapid assessment of total phenolic and anthocyanin contents in grape juice using infrared spectroscopy and multivariate calibration. *Food Analytical Methods*, *10*(5), 1609–1615. doi:10.100712161-016-0721-1

Carlin, F., Brillard, J., Broussolle, V., Clavel, T., Duport, C., Jobin, M., Guinebretière, M. H., Auger, S., Sorokine, A., & Nguyen-Thé, C. (2010). Adaptation of Bacillus cereus, an ubiquitous worldwide-distributed foodborne pathogen, to a changing environment. *Food Research International*, *43*(7), 1885–1894. doi:10.1016/j.foodres.2009.10.024

Cassoff, J., Gruber, R., Sadikaj, G., Rushani, F., & Knäuper, B. (2014). What motivational and awareness variables are associated with adolescents' intentions to go to bed earlier? *Current Psychology (New Brunswick, N.J.)*, *33*(2), 113–129. doi:10.100712144-013-9201-6

Castle, L. (2006). Chemical migration into food: An overview. In Chemical migration and food contact materials. Woodhead Publishing.

Center for Food Safety (CFS). (2017). *HACCP system: Seven principles of HACCP system*. https://www.cfs.gov.hk/english/programme/programme_haccp/programme_haccp_7requirement.html

Centers for Disease Control and Prevention (CDC). (2018). *Surveillance for Foodborne Disease Outbreaks, United States, 2016, Annual Report*. U.S. Department of Health and Human Services, CDC.

Centers for Disease Control and Prevention (CDC). (2020a). *Food safety for power outages*. CDC.

Centers for Disease Control and Prevention (CDC). (2020b). *Four steps to food safety: Clean, separate, cook, chill*. https://www.cdc.gov/foodsafety/keep-food-safe.html

Centers for Disease Control and Prevention (CDC). (2020c). *Food cooling practice improvements: Key takeaways from 3 food safety reports*. https://www.cdc.gov/nceh/ehs/ehsnet/plain_language/food-cooling-improvements.html

Centers for Disease Control and Prevention (CDC). (2021). *When and how to wash your hands*. Handwashing: Clean hands save lives. https://www.cdc.gov/handwashing/when-how-handwashing.html

Chandrapala, J., Oliver, C., Kentish, S., & Ashokkumar, M. (2012). Ultrasonics in food processing – Food quality assurance and food safety. *Trends in Food Science & Technology*, *26*(2), 88–98. doi:10.1016/j.tifs.2012.01.010

Changbai Xiu, C., & Klein, K. K. (2010). Melamine in milk products in China: Examining the factors that led to deliberate use of the contaminant. *Food Policy*, *35*(5), 463–470. doi:10.1016/j.foodpol.2010.05.001

Chapman, C. (2018). Interpretive methodological expertise and editorial board composition. *Critical Perspectives on Accounting*, *51*, 47–51. doi:10.1016/j.cpa.2017.10.007

Cheftel, J. C. (1995). Review: High pressure, microbial inactivation and food preservation. *Food Science & Technology International*, *1*(2-3), 75–90. doi:10.1177/108201329500100203

Chegeni, B. (n.d.). How to be audit ready. *Global Food Safety Resource*. https://globalfoodsafetyresource.com/how-to-be-audit-ready/

Cheng, J. H., Dai, Q., Sun, D. W., Zeng, X. A., Liu, D., & Pu, H. B. (2013). Applications of non-destructive spectroscopic techniques for fish quality and safety evaluation and inspection. *Trends in Food Science & Technology*, *34*(1), 18–31. doi:10.1016/j.tifs.2013.08.005

Cheng, Y., Dong, Y., Wu, J., Yang, X., Bai, H., Zheng, H., Ren, D., Zou, Y., & Li, M. (2010). Screening melamine adulterant in milk powder with laser Raman spectrometry. *Journal of Food Composition and Analysis*, *23*(2), 199–202. doi:10.1016/j.jfca.2009.08.006

Child, J. (2015). *Organization: Contemporary principles and practice*. John Wiley and Sons. doi:10.1002/9781119176862

Choi, S., Kim, H., Kim, Y., Kim, B., Beuchat, L. R., & Ryu, J. H. (2014). Fate of bacillus cereus and naturally occurring microbiota on milled rice as affected by temperature and relative humidity. *Food Microbiology*, *38*, 122–127. doi:10.1016/j.fm.2013.08.016 PMID:24290634

Choudhary, A., Gupta, N., Hameed, F., & Choton, S. (2020). An overview of food adulteration: Concept, sources, impact, challenges and detection. *International Journal of Chemical Studies*, *8*(1), 2564–2573. doi:10.22271/chemi.2020.v8.i1am.8655

Christiansson, A. (2002). Bacillus Cereus. Encyclopedia of Dairy Sciences, 123–128. doi:10.1016/B0-12-227235-8/00032-8

Chua, A. L., Aziah, I., Balaram, P., Bhuvanendran, S., Anthony, A. A., Mohmad, S. N., Nasir, N. M., Hassan, H., Naim, R., Meran, L. P., Hussin, H. M., & Ismail, A. (2015). Identification of Carriers Among Individuals Recruited in the Typhoid Registry in Malaysia Using Stool Culture, Polymerase Chain Reaction, and Dot Enzyme Immunoassay as Detection Tools. *Asia-Pacific Journal of Public Health*, *27*(2), NP2740–NP2748. doi:10.1177/1010539512458521 PMID:23000800

Chua, B. L., Karim, S., Lee, S., & Han, H. (2020). Customer restaurant choice: An empirical analysis of restaurant types and eating-out occasions. *International Journal of Environmental Research and Public Health*, *17*(17), 1–23. doi:10.3390/ijerph17176276 PMID:32872267

Cianferoni, A., & Spergel, J. M. (2009). Food allergy: Review, classification and diagnosis. *Allergology International*, *58*(4), 457–466. doi:10.2332/allergolint.09-RAI-0138 PMID:19847094

CIES. (2008). *The food business forum*. http://www.ciesnet.com

Cockerill, F., Patel J., Alder J., Bradford P., Dudley M., Eliopoulos G., … Zimmer, B. (2013). Performance Standards for Antimicrobial Susceptibility Testing; Twenty-Third Informational Supplement (M100-S23-2013). *Clinical and Laboratory Standards Institute, 49*(9).

Codex Alimentarius. (2012). *About Codex*. http://www.codexalimentarius.org/about-codex/en/

Codex Alimentarius. (2021). *About Codex Alimentarius*. http://www.fao.org/fao-who-codexalimentarius/ about-codex/en/

Cohen, M. X. (2014). *Analyzing neural time series data: Theory and practice*. MIT Press. doi:10.7551/mitpress/9609.001.0001

Coleman, E., Delea, K., Everstine, K., Reimann, D., & Ripley, D. (2013). Handling practices of fresh leafy greens in restaurants: Receiving and training. *Journal of Food Protection*, *76*(12), 2126–2131. doi:10.4315/0362-028X.JFP-13-127 PMID:24290691

Considine, K. M., Kelly, A. L., Fitzgerald, G. F., Hill, C., & Sleator, R. D. (2008). High-pressure processing–effects on microbial food safety and food quality. *FEMS Microbiology Letters*, *281*(1), 1–9. doi:10.1111/j.1574-6968.2008.01084.x PMID:18279335

Cook, D. J., Mulrow, C. D., & Haynes, R. B. (1997). Systematic Reviews: Synthesis of Best Evidence for Clinical Decisions. *Annals of Internal Medicine*, *126*(5), 376–380. doi:10.7326/0003-4819-126-5-199703010-00006 PMID:9054282

Coorey, R., Ng, D. S. H., Jayamanne, V. S., Buys, E. M., Munyard, S., Mousley, C. J., Njage, P. M. K., & Dykes, G. A. (2018). The impact of cooling rate on the safety of food products as affected by food containers. *Comprehensive Reviews in Food Science and Food Safety*, *17*(4), 827–840. doi:10.1111/1541-4337.12357 PMID:33350119

Cox, S. (2006, August 22). *US food supply 'vulnerable to attack'*. BBC News. http://news.bbc.co.uk/2/hi/americas/5274022.stm

Cozzolino, D., & Murray, I. (2004). Identification of animal meat muscles by visible and near infrared reflectance spectroscopy. *Lebensmittel-Wissenschaft + Technologie*, *37*(4), 447–452. doi:10.1016/j.lwt.2003.10.013

Cui, B., Li, S. Y., Wang, L. D. L., Chen, X., Ke, J., & Tian, Y. (2021). Hand hygiene knowledge and self-reported hand washing behaviors among restaurant kitchen chefs in Jiangsu Province, China. *International Journal of Environmental Research and Public Health*, *18*(4), 1–14. doi:10.3390/ijerph18042149 PMID:33671843

Dales, A. (2018). *A shelf life guide*. Los Angeles Regional Food Bank. https://www.cerritos.edu/basic-needs/_includes/docs/ShelfLifeGuide.pdf

Daniel, W. W., & Cross, C. L. (2013). *Biostatistics: A foundation for analysis in the health sciences. The American Biology Teacher*. John Wiley & Sons, Inc., doi:10.2307/4446617

Dankowska, A., Małecka, M., & Kowalewski, W. (2015). Detection of plant oil addition to cheese by synchronous fluorescence spectroscopy. *Dairy Science & Technology*, *95*(4), 413–424. doi:10.100713594-015-0218-5 PMID:26097644

Dark, G., McLean, D., & Weatherhead, S. (2015). *Kitchen operations*. Pearson Higher Education AU.

Darko, S., Mills-Robertson, F. C., & Wireko-Manu, F. D. (2015). Evaluation of some hotel kitchen staff on their knowledge on food safety and kitchen hygiene in the Kumasi Metropolis. *International Food Research Journal*, *22*(6), 2664–2669.

Das, A. K., Nanda, P. K., Das, A., & Biswas, S. (2019). Hazards and safety issues of meat and meat products. In *Food safety and human health* (pp. 145–168). Academic Press. doi:10.1016/B978-0-12-816333-7.00006-0

Das, A. K., Sultana, Z., Kabir, A., & Kabir, M. S. (2019). Effect of washing on reducing bacterial loads in common vegetables sold in Dhaka City. *Bangladesh Journal of Microbiology*, *35*(2), 96–101. doi:10.3329/bjm.v35i2.42637

Davidson, R. K., Antunes, W., Madslien, E. H., Belenguer, J., Gerevini, M., Perez, T. T., & Prugger, R. (2017). From food defence to food supply chain integrity. *British Food Journal*, *119*(1), 52–66. doi:10.1108/BFJ-04-2016-0138

Davila, E. P., Trepka, M. J., Newman, F. L., Huffman, F. G., & Dixon, Z. (2009). Diarrheal illness among women, infants, and children (WIC) program participants in Miami, Florida: Implications for nutrition education. *Journal of Nutrition Education and Behavior*, *41*(6), 420–424. doi:10.1016/j.jneb.2008.11.008 PMID:19879498

Davis, R., & Mauer, L. J. (2010). Fourier transform infrared (FT-IR) spectroscopy: A rapid tool for detection and analysis of foodborne pathogenic bacteria. Current *Research. Technology and Education Topics in Applied Microbiology and Microbial Biotechnology*, *2*, 1582–1594.

De Boer, A., & Bast, A. (2018). Demanding safe foods – Safety testing under the novel food regulation (2015/2283). *Trends in Food Science & Technology*, *72*, 125–133. doi:10.1016/j.tifs.2017.12.013

De Oliveira, C. A. F., Da Cruz, A. G., Tavolaro, P., & Corassin, C. H. (2016). Food safety: Good Manufacturing Practices (GMP), Sanitation Standard Operating Procedures (SSOP), Hazard Analysis and Critical Control Point (HACCP). In Antimicrobial food packaging (pp. 129-139). Academic Press.

de Souza, L. M., de Santana, F. B., Gontijo, L. C., Mazivila, S. J., & Neto, W. B. (2015). Quantification of adulterations in extra virgin flaxseed oil using MIR and PLS. *Food Chemistry*, *182*, 35–40. doi:10.1016/j.foodchem.2015.02.081 PMID:25842305

Department of Standard Malayisia (DOSM). (2020). *Conpendium of Environment Statistics Malaysia 2020, released 27 November 2020.* https://www.dosm.gov.my

Department of Statistics Malaysia. (2020). *Current population estimates Malaysia 2021.* https://www.dosm.gov.my/

Detwiler, D. (2014, March 18). *The reality of food terrorism.* FoodSafety magazine. https://www.food-safety.com/articles/4020-the-reality-of-food-terrorism

Dewanti-Hariyadi, R., & Gitapratiwi, D. (2014). Prevalence of Foodborne Diseases in South East and Central Asia. In Encyclopedia of Food Safety (Vol. 1, pp. 287–294). Academic Press.

Dewey-Mattia, D., Manikonda, K., Hall, A. J., Wise, M. E., & Crowe, S. J. (2018). Surveillance for foodborne disease outbreaks - United States, 2009-2015. *MMWR. Surveillance Summaries, 67*(10), 1–11. Advance online publication. doi:10.15585/mmwr.ss6710a1 PMID:30048426

Di Egidio, V., Sinelli, N., Limbo, S., Torri, L., Franzetti, L., & Casiraghi, E. (2009). Evaluation of shelf-life of fresh-cut pineapple using FT-NIR and FT-IR spectroscopy. *Postharvest Biology and Technology, 54*(2), 87–92. doi:10.1016/j.postharvbio.2009.06.006

Di Rosa, A. R., Bressan, F., Leone, F., Falqui, L., & Chiofalo, V. (2019). Radio frequency heating on food of animal origin: A review. *European Food Research and Technology, 245*(9), 1787–1797. doi:10.100700217-019-03319-8

Dillman, D. A. (2007). *Mail and internet surveys: The tailored design method* (2nd ed.). John Willey and Son.

Directorate of Marketing and Inspection. (2020). *Grades & Standard: AGMARK Standards.* https://dmi.gov.in/GradesStandard.aspx

Dong, H., Luo, D., & Luo, H. (2016). Effective use of food traceability in cereal grain food supply chains. In *Advances in Food Traceability Techniques and Technologies* (pp. 303–319). Woodhead Publishing. doi:10.1016/B978-0-08-100310-7.00016-8

Dora-Liyana, A. L., Mahyudin, N. A., Ismail-Fitry, M. R., Ahmad-Zaki, A., & Rasiyuddin, H. (2018). Food Safety and Hygiene Knowledge, Attitude and Practices among Food Handlers at Boarding Schools in the Northern Region of Malaysia. *International Journal of Academic Research in Business & Social Sciences, 8*(17), 238–266.

Drabas, E., & Wojciechowski, A. (2006). IFS and BRC – International standards of food products quality assurance. *Kalejd. Miesn, 2*, 86–94.

Dumitraşcu, L., Ioana, A., Neagu, C., Didier, P., Maître, I., Nguyen-the, C., ... Borda, D. (2020). Time-temperature profiles and Listeria monocytogenes presence in refrigerators from households with vulnerable consumers. *Food Control, 111*, 107078. doi:10.1016/j.foodcont.2019.107078

Durek, J., Fröhling, A., Bolling, J., Thomasius, R., Durek, P., & Schlüter, O. K. (2016). Non-destructive mobile monitoring of microbial contaminations on meat surfaces using porphyrin fluorescence intensities. *Meat Science*, *115*, 1–8. doi:10.1016/j.meatsci.2015.12.022 PMID:26773794

Ecolab. (2011). Food safety audits 101: A brief history and preparation essentials. *Food Safety Magazine.* https://www.food-safety.com/articles/3855-food-safety-audits-101-a-brief-history-and-preparation-essentials

Edwards, M. C., & Stringer, M. F. (2007). Observations on patterns in foreign material investigations. *Food Control*, *18*(7), 773–782. doi:10.1016/j.foodcont.2006.01.007

EFSA Panel on Biological Hazards (BIOHAZ). (2013). Scientific Opinion on the maintenance of the list of QPS biological agents intentionally added to food and feed (2013 update). *EFSA Journal*, *11*(11), 3449.

Eggers, S., Verril, L., Bryant, C. M., & Thorne, S. L. (2011). Developing consumer focused risk communication strategies related to food terrorism. *International Journal of Food Safety, Nutrition and Public Health*, *4*(1), 45. doi:10.1504/IJFSNPH.2011.042574

Ehn, B. M., Allmere, T., Telemo, E., Bengtsoon, U., & Ekstrand, B. (2005). Modification of IgE binding to b-lactoglobulin by fermentation and proteolysis of cow's milk. *Journal of Agricultural and Food Chemistry*, *53*(9), 3743–3746. doi:10.1021/jf048121w PMID:15853429

Ellis, D. I., & Goodacre, R. (2001). Rapid and quantitative detection of the microbial spoilage of muscle foods: Current status and future trends. *Trends in Food Science & Technology*, *12*(11), 414–424. doi:10.1016/S0924-2244(02)00019-5

Elobeid, T., Savvaidis, I., & Ganji, V. (2019). Impact of food safety training on the knowledge, practice, and attitudes of food handlers working in fast-food restaurants. *British Food Journal*, *121*(4), 937–949. doi:10.1108/BFJ-01-2019-0066

Elshahat, S., Woodside, J. V., & Mckinley, M. C. (2019). Meat thermometer usage amongst European and North American consumers: A scoping review. *Food Control*, *106*, 106692. doi:10.1016/j.foodcont.2019.06.018

Emma, J., Sherrington, B. A., & Hones, P. A. (2001). The toxicity of styrene monomer. *Adverse Drug Reactions and Toxicological Reviews*, *20*, 9–35. PMID:11395942

Ergönül, B. (2013). Consumer awareness and perception to food safety: A consumer analysis. *Food Control*, *32*(2), 461–471. doi:10.1016/j.foodcont.2013.01.018

Eurofins. (2020). *Why you need a food defense plan.* https://www.eurofinsus.com/food-testing/resources/why-you-need-a-food-defense-plan/

European Food Safety Authority (EFSA). (n.d.). *About us.* https://www.efsa.europa.eu/en/aboutefsa

Fabisz-Kijowska, A., & Kijowski, J. (2006). Food safety requirements by new international standard. *Mieso Wedl*, *5*, 8–12.

Fang, T. J., Wei, Q., Liao, C., Hung, M., & Wang, T. (2003). Microbiological quality of 18°C ready-to-eat food products sold in Taiwan. *International Journal of Food Microbiology*, *80*(3), 241–250. doi:10.1016/S0168-1605(02)00172-1 PMID:12423926

FAO, Food and Agricultural Organization. (1997). *Hazard analysis and critical control point (HACCP) System and guidelines for its application.* https://www.fao.org/docrep/005/y1579e/y1579e03.htm

FAO, Food and Agricultural Organization. (2020). *The importance of food quality and safety for developing countries*. https://www.fao.org/docrep/meeting/x1845e.htm

Faraz, A., Lateef, M., Mustafa, M. I., Akhtar, P., & Yaqoob, M. (2013). Detection of adulteration, chemical composition and hygienic status of milk supplied to various canteens of educational institutes and public places in Faisalabad. *Journal Animal Plant Science*, *23*, 119–124.

Faridah, H. I., Chemah, T. C., Rosmaliza, M., & Norhayati, M. Y. (2016). Food safety knowledge and personal hygiene practices amongst mobile food handlers in Shah Alam, Selangor. *Procedia: Social and Behavioral Sciences*, *222*, 290–298. doi:10.1016/j.sbspro.2016.05.162

FDA. (1998). *Guide to minimize microbial food safety hazards for fresh fruits and vegetables.* Center for Food Safety and Applied Nutrition.

FDA. (2018). *Vulnerability assessments and key activity types.* https://www.fda.gov/food/food-defense/food-defense-programs

FDA. (2019). *Protecting the food supply from intentional adulteration, such as acts of terrorism.* https://www.fda.gov/food/conversations-expertsfood-topics/protecting-food-supply-intentional-adulteration-such-actsterrorism

Fiedler, G., Schneider, C., Igbinosa, E. O., Kabisch, J., Brinks, E., Becker, B., Stoll, D. A., Cho, G.-S., Huch, M., & Franz, C. (2019). Antibiotics resistance and toxin profiles of Bacillus cereus-group isolates from fresh vegetables from German retail markets. *BMC Microbiology*, *19*(1), 1–13. doi:10.118612866-019-1632-2 PMID:31706266

Fife-Schaw, C., & Rowe, G. (1996). Public perceptions of everyday food hazards: A psychometric study. *Risk Analysis*, *16*(4), 487–500. doi:10.1111/j.1539-6924.1996.tb01095.x PMID:8819341

Food Act 1983 and regulations 1985, Part VII Incidental Constituent: 39(2).

Food and Agriculture Organization & World Health Organization. (2010). *Protecting health, facilitating trade.* https://www.fao.org/fao-who-codexalimentarius/en/

Food and Drug Administration (FDA). (2010). *Program information manual retail food protection: recommendations for the temperature control of cut leafy greens during storage and display in retail food establishments.* https://www.fda.gov/food/retail-food-industryregulatory-assistance-training/ program-information-manual-retail-food-protection-recommendations-temperature-control-cut-leafy #prop

Food and Drug Administration (FDA). (2012). *Food allergies.* https://www.fda.gov/food/food-labeling-nutrition/food-allergies

Food and Drug Administration (FDA). (2013). *Food Code.* Public Health Service of Food and Drug Administration.

Food and Drug Administration (FDA). (2018). *Food allergen labeling and consumer protection act of 2004 (FALCPA).* https://www.fda.gov/food/food-allergensgluten-free-guidance-documents-regulatory-information/food-allergen-labeling-and-consumer-protection-act-2004-falcpa

Food and Drug Administration (FDA). (2018). *Serving up safe buffets.* https://www.fda.gov/food/buy-store-serve-safe-food/serving-safe-buffets

Food and Drug Administration. (2012). Bad Bug Book, Foodborne Pathogenic Microorganisms and Natural Toxins (2nd ed.). Author.

Food Safety and Quality Division. (2018). *Annual Report Food Safety And Quality 2018.* Author.

Food safety and standards authority of India (FSSAI). (2006). *Food Safety and Standards Act 2006.* Ministry of Law and Justice, New Delhi. https://fssai.gov.in/upload/uploadfiles/files/FOOD-ACT.pdf

Food Safety and Standards Authority of India. (2018). *About Codex.* https://archive.fssai.gov.in/home/fss-legislation/about-codex.html

Food Safety and Standards Authority of India. (2020). *Food law: Codex Alimentarius Commission.* https://fssai.gov.in/cms/codex.php

Food Safety Authority of Ireland (FSAI). (2014). Country markets: Guide to good hygiene practice. Food Safety Authority of Ireland.

Food Safety Authority of Ireland. (2017). *Hand washing and food safety.* https://www.fsai.ie/faq/hand_washing.html

Food Safety of Australia and New Zealand (FSANZ). (2019). *Preparing and cook food.* https://www.foodstandards.gov.au/consumer/safety/faqsafety/pages/foodsafetyfactsheets/charitiesandcommunityorganisationsfactsheets/preparingandcookingf1479.aspx

Food Standard Agency. (2016). *Understanding NI consumer needs around food labelling.* https://www.food.gov.uk/research/research-projects/understanding-northern-ireland-consumer-needs-around-food-labelling

Food Standards Act. (1999). https://www.legislation.gov.uk/ukpga/1999/28#reference-c1312451

Food Standards Agency. (2018). *Personal hygiene: Guidance on what you and your staff must do when handling food.* https://www.food.gov.uk/business-guidance/personal-hygiene

Food Standards Agency. (n.d.). *Homepage.* https://www.food.gov.uk/

Food Standards Australia New Zealand (FSAI). (2019). *Receiving food safely*. https://www.foodstandards.gov.au/consumer/safety/faqsafety/pages/foodsafetyfactsheets/receivingfoodsafely.aspx

Food Standards Australia New Zealand (FSANZ). (2001). *Food safety: An audit system*. https://www.foodstandards.gov.au/publications/Pages/anauditsystem.aspx

Food Standards Australia New Zealand (FSANZ). (2018). *Compendium of Microbiological Criteria for Food September 2018*. Author.

Food Standards Scotland. (n.d.). https://www.foodstandards.gov.scot

Frewer, L. J., & Miles, S. (2001). Risk perception, communication and trust. How might consumer confidence in the food supply be maintained? In *Food, people and society* (pp. 401–413). Springer. doi:10.1007/978-3-662-04601-2_24

Gadaga, T. H., Samende, B. K., Musuna, C., & Chibanda, D. (2008). The microbiological quality of informally vended foods in Harare, Zimbabwe. *Food Control*, *19*(8), 829–832. doi:10.1016/j.foodcont.2007.07.016

Galvin-King, P., Haughey, S. A., & Elliott, C. T. (2018). Herb and spice fraud; the drivers, challenges and detection. *Food Control*, *88*, 85–97. doi:10.1016/j.foodcont.2017.12.031

Garayoa, R., Abundancia, C., Díez-Leturia, M., & Vitas, A. I. (2017). Essential tools for food safety surveillance in catering services: On-site inspections and control of high risk cross-contamination surfaces. *Food Control*, *75*, 48–54. doi:10.1016/j.foodcont.2016.12.032

Gdoura-Ben Amor, M., Jan, S., Baron, F., Grosset, N., Culot, A., Gdoura, R., Gautier, M., & Techer, C. (2019). Toxigenic potential and antimicrobial susceptibility of Bacillus cereus group bacteria isolated from Tunisian foodstuffs. *BMC Microbiology*, *19*(1), 1–12. doi:10.118612866-019-1571-y PMID:31445510

Gendel, S. M. (2012). Comparison of international food allergen labeling regulations. *Regulatory Toxicology and Pharmacology*, *63*(2), 279–285. doi:10.1016/j.yrtph.2012.04.007 PMID:22565206

Ghasemi-Varnamkhasti, M., Mohtasebi, S. S., & Siadat, M. (2010). Biomimetic-based odor and taste sensing systems to food quality and safety characterization: An overview on basic principles and recent achievements. *Journal of Food Engineering*, *100*(3), 377–387. doi:10.1016/j.jfoodeng.2010.04.032

Ghosh, K., Degeneffe, D., Kinsey, J., & Stinson, T. (2007). How would Americans allocate anti-terrorism spending? Findings from a national survey of attitudes about terrorism. *Homeland Security Affairs*, *3*. https://www.hsaj.org/articles/148

Global Food Safety Resource (GFSR). (2016). *Food Safety Audits*. https://globalfoodsafetyresource.com/food-safety-audits/#

Gomez, M., Oliete, B., Garcia-Alvarez, J., Ronda, F., & Salazar, J. (2008). Characterization of cake batters by ultrasound measurements. *Journal of Food Engineering*, *89*(4), 408–413. doi:10.1016/j.jfoodeng.2008.05.024

Gormley, F. J., Rawal, N., & Little, C. L. (2012). Choose your menu wisely: Cuisine-associated food-poisoning risks in restaurants in England and Wales. *Epidemiology and Infection*, *140*(6), 997–1007. doi:10.1017/S0950268811001567 PMID:21854669

Gorris, L. G. (2005). Food safety objective: An integral part of food chain management. *Food Control*, *16*(9), 801–809. doi:10.1016/j.foodcont.2004.10.020

GOV.UK. (n.d.). *Food Standards Agency*. https://www.gov.uk/government/organisations/food-standards-agency

Gowen, A. A., Feng, Y., Gaston, E., & Valdramidis, V. (2015). Recent applications of hyperspectral imaging in microbiology. *Talanta*, *137*, 43–54. doi:10.1016/j.talanta.2015.01.012 PMID:25770605

Granda, C., Moreira, R. G., & Tichy, S. E. (2004). Reduction of acrylamide formation in potato chips by low-temperature vacuum frying. *Journal of Food Science*, *69*(8), E405–E411. doi:10.1111/j.1365-2621.2004.tb09903.x

Green, L. R., Selman, C. A., Radke, V., Ripley, D., Mack, J. C., Reimann, D. W., Stigger, T., Motsinger, M., & Bushnell, L. (2006). Food worker hand washing practices: An observation study. *Journal of Food Protection*, *69*(10), 2417–2423. doi:10.4315/0362-028X-69.10.2417 PMID:17066921

Grote, B., Zense, T., & Hitzmann, B. (2014). 2D-fluorescence and multivariate data analysis for monitoring of sourdough fermentation process. *Food Control*, *38*, 8–18. doi:10.1016/j.foodcont.2013.09.039

Guinebretiere, M. H., Girardin, H., Dargaignaratz, C., Carlin, F., & Nguyen-The, C. (2003). Contamination flows of Bacillus cereus and spore-forming aerobic bacteria in a cooked, pasteurized and chilled zucchini purée processing line. *International Journal of Food Microbiology*, *82*(3), 223–232. doi:10.1016/S0168-1605(02)00307-0 PMID:12593925

Guo, Q., Sun, D. W., Cheng, J. H., & Han, Z. (2017). Microwave processing techniques and their recent applications in the food industry. *Trends in Food Science & Technology*, *67*, 236–247. doi:10.1016/j.tifs.2017.07.007

Haqim, M. M. (2015). *Knowledge, attitude and practice of styrene relating to usage of polystyrene food packaging among Universiti Putra Malaysia students* [Unpublished doctoral dissertation]. Universiti Putra Malaysia, Serdang, Malaysia.

Haraminac, E. (2017). Cooling hot food, do it right to prevent bacterial growth. *Michigan State University Extension*. http://msue.anr.msu.edu/news/cooling_hot_food_do_it_right_to_prevent_bacterial_growth

Harris, L. (2019). *Auditor shortage slowly being addressed*. Global Food Safety Resource. https://globalfoodsafetyresource.com/auditor-shortage-slowly-addressed/

Harris, K. J., Ali, F., & Ryu, K. (2018). Foodborne illness outbreaks in restaurants and patrons' propensity to return. *International Journal of Contemporary Hospitality Management*, *30*(3), 1273–1292. doi:10.1108/IJCHM-12-2016-0672

Harris, L. J. (2002). *Guidelines for food safety during short-term power outages: Consumer fact sheet*. University of California, Department of agriculture and natural resources. doi:10.3733/ucanr.7264

Harrison, I. (2003). Non-destructive testing for fruit quality assurance. *Innovations in Food Technology*, *19*, 86–87.

Ha, T. M., Shakur, S., & Pham Do, K. H. (2019). Consumer concern about food safety in Hanoi, Vietnam. *Food Control*, *98*, 238–244. doi:10.1016/j.foodcont.2018.11.031

Havinga, T. (2006). Private regulation of food safety by supermarkets. *Law & Policy*, *28*(4), 515–533. doi:10.1111/j.1467-9930.2006.00237.x

Hedberg, C. W. (2013). Explaining the risk of foodborne illness associated with restaurants: The environmental health specialists' network (ehs-net). *Journal of Food Protection*, *76*(12), 2124–2125. doi:10.4315/0362-028X.JFP-13-270 PMID:24290690

Henson, S. (2003). *The economics of food safety in developing countries*. ESA Working Paper No. 03-19. Agricultural Development Economics Division, FAO. https://ftp.fao.org/docrep/fao/007/ae052e/ae052e00.pdf

Herron, M. (2018). Is predictive analytics the end of the annual audit? *AccountancyAge*. https://www.accountancyage.com/2018/05/21/is-predictive-analytics-the-end-of-the-annual-audit/

Hirri, A., Bassbasi, M., Platikanov, S., Tauler, R., & Oussama, A. (2016). FTIR spectroscopy and PLS-DA classification and prediction of four commercial grade virgin olive oils from Morocco. *Food Analytical Methods*, *9*(4), 974–981. doi:10.100712161-015-0255-y

Hogan, E., Kelly, A. L., & Sun, D. W. (2005). *High pressure processing of foods: An overview. In Emerging Technologies for Food Processing (pp. 3–31)*. Academic Press.

Hoogland, J. P., Jellema, A., & Jorgen, M. T. G. (1998). Quality assurance systems. In W. M. F. Jongen & M. T. G. Meulenberg (Eds.), *Innovative of Food Productions Systems: Product Quality And Consumer Acceptance*. Wageningen Pers.

Horigome, J., Kozuma, M., & Shirasaki, T. (2016). Fluorescence pattern analysis to assist food safety. *Hitachi Review*, *65*(7), 249.

Hubbard, M. R. (2012). *Statistical quality control for the food industry*. Springer Science and Business Media.

Hulland, K. R. S., Leontsini, E., Dreibelbis, R., Unicomb, L., Afroz, A., Dutta, N. C., Nizame, F. A., Luby, S. P., Ram, P. K., & Winch, P. J. (2013). Designing a handwashing station for infrastructure-restricted communities in Bangladesh using the integrated behavioural model for water, sanitation and hygiene interventions (IBM-WASH). *BMC Public Health*, *13*(1), 877. Advance online publication. doi:10.1186/1471-2458-13-877 PMID:24060247

Hwang, J. Y., & Park, J. H. (2015). Characteristics of enterotoxin distribution, hemolysis, lecithinase, and starch hydrolysis of *Bacillus cereus* isolated from infant formulas and ready-to-eat foods. *Journal of Dairy Science*, *98*(3), 1652–1660. doi:10.3168/jds.2014-9042 PMID:25597976

Hyman, F. N., Klontz, K. C., & Tollefson, L. (1993). Food and Drug Administration surveillance of the role of foreign objects in foodborne injuries. *Public Health Reports*, *108*(1), 54. PMID:8434098

IDFA, International Dairy Foundation. (2020). *Pasteurization.* https://www.idfa.org/pasteurization

Igual, M. G. M. E., García-Martínez, E., Camacho, M. M., & Martínez-Navarrete, N. (2010). Effect of thermal treatment and storage on the stability of organic acids and the functional value of grapefruit juice. *Food Chemistry*, *118*(2), 291–299. doi:10.1016/j.foodchem.2009.04.118

Insfran-Rivarola, A., Tlapa, D., Limon-Romero, J., Baez-Lopez, Y., Miranda-Ackerman, M., Arredondo-Soto, K., & Ontiveros, S. (2020). A systematic review and meta-analysis of the effects of food safety and hygiene training on food handlers. *Foods*, *9*(9), 1169. Advance online publication. doi:10.3390/foods9091169 PMID:32854221

International Organization for Standardization (ISO). (2018a). *International standard ISO 19011 - Guidelines for auditing management systems.* https://www.borhanjooyan.com/DL/ISO-19011-2018.pdf

International Organization for Standardization (ISO). (2018b). *Management systems for food safety.* https:// committee.iso.org/home/tc34sc17

Ishak, S., & Zabil, N. F. M. (2012). Impact of consumer awareness and knowledge to consumer effective behavior. *Journal of Asian Social Science*, *8*(13). Advance online publication. doi:10.5539/ass.v8n13p108

ISO 19011:2002. *Guidelines for Quality and/or Environmental Management Systems Auditing*. https://www.academia.edu/36204649/Guidelines_for_quality_and_or_environmental_management_systems_auditing

ISO 9000:2000. *Quality Management Systems – Fundamentals and Vocabulary*. https://www.iso.org/files/live/sites/isoorg/files/archive/pdf/en/watermarksample.pdf

ISO 9004:2000. *Quality Management Systems – Guidelines for Performance Improvements*. https://www.iso.org/standard/28692.html

ISO. (2021). *About us.* https://www.iso.org/about-us.html

Isoni Auad, L., Cortez Ginani, V., dos Santos Leandro, E., Stedefeldt, E., Costa Santos Nunes, A., Yoshio Nakano, E., & Puppin Zandonadi, R. (2019). Brazilian food truck consumers' profile, choices, preferences, and food safety importance perception. *Nutrients*, *11*(5), 1175. doi:10.3390/nu11051175 PMID:31130664

Jabeur, H., Zribi, A., Makni, J., Rebai, A., Abdelhedi, R., & Bouaziz, M. (2014). Detection of Chemlali extra-virgin olive oil adulteration mixed with soybean oil, corn oil, and sunflower oil by using GC and HPLC. *Journal of Agricultural and Food Chemistry*, *62*(21), 4893–4904. doi:10.1021/jf500571n PMID:24811341

Jayaraman, K., Haron, H., Sung, G. B., & Lin, S. K. (2011). Consumer reflections on the usage of plastic bags to parcel hot edible items: An empirical study in Malaysia. *Journal of Cleaner Production*, *19*(13), 1527–1535. doi:10.1016/j.jclepro.2011.03.019

Jenkins, J. P. (2020). Terrorism. *Encyclopedia Britannica.* https://www.britannica.com/topic/terrorism

Jeukendrup, A., & Gleeson, M. (2018). Sport nutrition. Human Kinetics, 18-30.

Jha, S. N. (2016). Food standards and permissible limits. *Rapid detection of food adulterants and contaminants*, 63–106.

Jones, T. L., Baxter, M. A. J., & Khanduja, V. (2013). A quick guide to survey research. *Annals of the Royal College of Surgeons of England*, *95*(1), 5–7. doi:10.1308/003588413X13511609956372 PMID:23317709

Kademi, H. I., Ulusoy, B. H., & Hecer, C. (2019). Applications of miniaturized and portable near infrared spectroscopy (NIRS) for inspection and control of meat and meat products. *Food Reviews International*, *35*(3), 201–220. doi:10.1080/87559129.2018.1514624

Kamal, M., & Karoui, R. (2017). Monitoring of mild heat treatment of camel milk by front-face fluorescence spectroscopy. *Lebensmittel-Wissenschaft + Technologie*, *79*, 586–593. doi:10.1016/j.lwt.2016.11.013

Kamidi, T. (2015). Styrofoam: The silent killer. *Borneo Post Online*. http://www.theborneopost.com /2015/02/18/styrofoam-the-silent-killer

Karayan, H. (2016). *Food defense: Protecting food against intentional adulteration.* https://www.sgs.com/en/news/2016/12/fsma-food-defenseprotecting-food-against-intentional-adulteration

Karmacharya, J. B. (2014). Good manufacturing practices (GMP) for medicinal products. *Promising Pharmaceuticals*, 101-148.

Katiyo, W., De Kock, H. L., Coorey, R., & Buys, E. M. (2019). Assessment of safety risks associated with handling chicken as based on practices and knowledge of a group of South African consumers. *Food Control*, *101*, 104–111. doi:10.1016/j.foodcont.2019.02.027

Kaur, N., & Singh, A. K. (2016). Ohmic heating: Concept and applications: A review. *Critical Reviews in Food Science and Nutrition*, *56*(14), 2338–2351. doi:10.1080/10408398.2013.835303 PMID:25830778

Keogh, J. G. (n.d.). Dubai leadership: The digitization of food safety management. *Global Food Safety Resource.* https://globalfoodsafetyresource.com/dubai-leadership-digitization-food-safety-management/

Khairuzzaman, M., Chowdhury, F. M., Zaman, S., Al Mamun, A., & Bari, M. L. (2014). Food safety challenges towards safe, healthy, and nutritious street foods in Bangladesh. *International Journal of Food Sciences*, *2014*, 1–9. Advance online publication. doi:10.1155/2014/483519 PMID:26904635

Khaksar, M. R., & Khansari, M. G. (2009). Determinants of migration monomer styrene from GPPS (General Purpose Polystyrene) and HIPS (High Impact Polystyrene) cups to hot drinks. *Toxicology Mechanisms and Methods*, *19*(3), 257–261. doi:10.1080/15376510802510299 PMID:19750020

Khan, K. M., Krishna, H., Majumder, S. K., & Gupta, P. K. (2015). Detection of urea adulteration in milk using near-infrared Raman spectroscopy. *Food Analytical Methods*, *8*(1), 93–102. doi:10.100712161-014-9873-z

Kim, B., Kim, D., Cho, D., & Cho, S. (2003). Bactericidal effect of TiO2 photocatalyst on selected food-borne pathogenic bacteria. *Chemosphere*, *52*(1), 277–281. doi:10.1016/S0045-6535(03)00051-1 PMID:12729712

Kim, K.-S., Song, S.-M., Kwon, S.-H., Jang, S.-E., Lee, B.-M., Kim, M.-H., Han, Y.-S., Hur, M.-J., & Kwon, M.-J. (2021). A Survey on the actual condition of products not labelled with allergens. *Journal of Food Hygiene and Safety*, *36*(3), 257–263. doi:10.13103/JFHS.2021.36.3.257

King, F. S., Burgess, A., Quinn, V. J., & Osei, A. K. (Eds.). (2015). *Nutrition for developing countries*. Oxford University Press. doi:10.1093/med/9780199685226.001.0001

Kinsey, J. D., Stinson, T. F., Degeneffe, D. J., Ghosh, K., & Busta, F. F. (2009). *Consumers response to a new food safety issue: Food terrorism.* https://iufost.edpsciences.org/articles/iufost/pdf/2006/01/iufost06000666.pdf

Knaflewska, J., & Pospiech, E. (2007). Quality assurance systems in food industry and health security of food. *Acta Scientiarum Polonorum. Technologia Alimentaria*, *6*(2), 75–85.

Knight, J. L. K., Joana Da Cunha Forte, G. M., & Gautam, O. P. (2020). *Technical Guide for handwashing facilities in public places and buildings*. WaterAid. https://washmatters.wateraid.org /sites/g/files/jkxoof256/files/technical-guide-for-handwashing-facilities-in-public-places-and-buildings.pdf

Kosa, K. M., Cates, S. C., Godwin, S., & Chabers, I. V. E. (2017). Barriers to Using a Food Thermometer When Cooking Poultry at Home: Results from a National Survey. *Food Protection Trends*, *37*(2), 116–125.

Kotiranta, A., Lounatmaa, K., & Haapasalo, M. (2000). Epidemiology and pathogenesis of *Bacillus cereus* infections. *Microbes and Infection*, *2*(2), 189–198. doi:10.1016/S1286-4579(00)00269-0 PMID:10742691

Kotsanopoulos, K. V., & Arvanitoyannis, I. S. (2017). The role of auditing, food safety, and food quality standards in the food industry: A review. *Comprehensive Reviews in Food Science and Food Safety*, *16*(5), 760–775. doi:10.1111/1541-4337.12293 PMID:33371608

Krejcie, R. V., & Morgan, D. W. (1970). Determining sample size for research activities. *Educational and Psychological Measurement*, *30*(3), 607–610. doi:10.1177/001316447003000308

Kwak, Y. H., & Anbari, F. T. (2006). Benefits, obstacles, and future of six sigma approach. *Technovation*, *26*(5-6), 708–715. doi:10.1016/j.technovation.2004.10.003

Lee, H. K., Abdul Halim, H., Thong, K. L., & Chai, L. C. (2017). Assessment of food safety knowledge, attitude, self-reported practices, and microbiological hand hygiene of food handlers. *International Journal of Environmental Research and Public Health*, *14*(1), 55. doi:10.3390/ijerph14010055 PMID:28098788

Lee, J., Jeong, S., Ko, G., Park, H., & Ko, Y. (2016). Development of a Food Safety and Nutrition Education Program for Adolescents by Applying Social Cognitive Theory. *Osong Public Health and Research Perspectives*, *7*(4), 248–260. doi:10.1016/j.phrp.2016.05.005 PMID:27635375

Lieberman, J. A., Gupta, R. S., Knibb, R. C., Haselkorn, T., Tilles, S., Mack, D. P., & Pouessel, G. (2021). The global burden of illness of peanut allergy: A comprehensive literature review. *Allergy: European Journal of Allergy and Clinical Immunology*, *76*(5), 1367–1384. doi:10.1111/all.14666 PMID:33216994

Lim, T.-P., Chye, F. Y., Mohd Rosni, S., Norazah, M. S., & Lee, J.-S. (2016). A structural modeling on food safety knowledge, attitude, and behaviour among Bum Bum Island community of Semporna, Sabah. *Food Control*, *60*, 241–246. doi:10.1016/j.foodcont.2015.07.042

Lin, C. T. J., Jensen, K. L., & Yen, S. T. (2005). Awareness of foodborne pathogens among US consumers. *Food Quality and Preference*, *16*(5), 401–412. doi:10.1016/j.foodqual.2004.07.001

Liu, A., & Niyongira, R. (2017). Chinese consumers food purchasing behaviors and awareness of food safety. *Food Control*, *79*, 185–191. doi:10.1016/j.foodcont.2017.03.038

Liu, D., Zeng, X. A., & Sun, D. W. (2015). Recent developments and applications of hyperspectral imaging for quality evaluation of agricultural products: A review. *Critical Reviews in Food Science and Nutrition*, *55*(12), 1744–1757. doi:10.1080/10408398.2013.777020 PMID:24915395

Liu, X., & Metzger, L. E. (2007). Application of fluorescence spectroscopy for monitoring changes in nonfat dry milk during storage. *Journal of Dairy Science*, *90*(1), 24–37. doi:10.3168/jds.S0022-0302(07)72605-X PMID:17183072

Liu, Y., Chao, K., Kim, M. S., Tuschel, D., Olkhovyk, O., & Priore, R. J. (2009). Potential of Raman spectroscopy and imaging methods for rapid and routine screening of the presence of melamine in animal feed and foods. *Applied Spectroscopy*, *63*(4), 477–480. doi:10.1366/000370209787944398 PMID:19366516

López-Gálvez, F., Gómez, P. A., Artés, F., Artés-Hernández, F., & Aguayo, E. (2021). Interactions between microbial food safety and environmental sustainability in the fresh produce supply chain. *Foods*, *10*(7), 1655. Advance online publication. doi:10.3390/foods10071655 PMID:34359525

Loutfi, A., Coradeschi, S., Mani, G. K., Shankar, P., & Rayappan, J. B. B. (2015). Electronic noses for food quality: A review. *Journal of Food Engineering*, *144*, 103–111. doi:10.1016/j.jfoodeng.2014.07.019

Low, W. Y., Jani, R., Halim, H. A., Alias, A. A., & Moy, F. M. (2016). Determinants of food hygiene knowledge among youths: A cross-sectional online study. *Food Control*, *59*, 88–93. doi:10.1016/j.foodcont.2015.04.032

Luo, J., Liu, T., & Liu, Y. (2011, October). FT-NIR and confocal microscope raman spectroscopic studies of sesame oil adulteration. In *International Conference on Computer and Computing Technologies in Agriculture* (pp. 24-31). Springer.

Madaki, M. Y., & Bavorova, M. (2019). Food safety knowledge of food vendors of higher educational institutions in Bauchi state, Nigeria. *Food Control*, *106*, 106703. doi:10.1016/j.foodcont.2019.06.029

Madhwal, S., & Sharma, S. (2017). Food Safety Issues and Risk Associated with Refrigerated Foods. *International Journal of Current Microbiology and Applied Sciences*, *6*(12), 4196–4203. doi:10.20546/ijcmas.2017.612.482

Mahendran, R., Jayashree, G. C., & Alagusundaram, K. (2012). Application of computer vision technique on sorting and grading of fruits and vegetables. *Journal of Food Processing & Technology*, *10*, 2157–7110.

Mala, D. M., Yoshimura, M., Kawasaki, S., Tsuta, M., Kokawa, M., Trivittayasil, V., ... Kitamura, Y. (2016). Fiber optics fluorescence fingerprint measurement for aerobic plate count prediction on sliced beef surface. *Lebensmittel-Wissenschaft + Technologie*, *68*, 14–20.

Maldives Food and Drug Authority. (2017). *National Food Safety Policy (2017-2026), Pub. L. No. Policy/23-MoH/2017/02, Ministry of Health, Maldives, 40*. http://www.health.gov.mv/Uploads/Downloads//Publications/Publication(29).pdf

Mallett, R. (2017). *HACCP International Food Safety Bulletin*. HACCP Europe., doi:10.18502/jfsh.v5i4.5700

Mamani-Linares, L. W., Gallo, C., & Alomar, D. (2012). Identification of cattle, llama and horse meat by near infrared reflectance or transflectance spectroscopy. *Meat Science*, *90*(2), 378–385. doi:10.1016/j.meatsci.2011.08.002 PMID:21889854

Mandal, M. D., & Mandal, S. (2011). Honey: Its medicinal property and antibacterial activity. *Asian Pacific Journal of Tropical Biomedicine*, *1*(2), 154–160. doi:10.1016/S2221-1691(11)60016-6 PMID:23569748

Manetta, A. C., Di Giuseppe, L., Giammarco, M., Fusaro, I., Simonella, A., Gramenzi, A., & Formigoni, A. (2005). High-performance liquid chromatography with post-column derivatisation and fluorescence detection for sensitive determination of aflatoxin M1 in milk and cheese. *Journal of Chromatography. A*, *1083*(1-2), 219–222. doi:10.1016/j.chroma.2005.06.039 PMID:16078711

Mangal, M., Bansal, S., & Sharma, M. (2014). Macro and micromorphological characterization of different *Aspergillus* isolates. *Legume Research*, *37*(4), 372–378. doi:10.5958/0976-0571.2014.00646.8

Manning, L. (2019). Food defence: Refining the taxonomy of food defence threats. *Trends in Food Science & Technology*, *85*, 107–115. doi:10.1016/j.tifs.2019.01.008

Marriott, N. G., Gravani, R. B., & Schilling, M. W. (2006). *Principles of food sanitation* (Vol. 22). Springer.

Marsh, M., & Bugusu, B. (2007). Food packaging: Roles, materials, and environmental issues. *Journal of Food Science*, *72*(3), 3. doi:10.1111/j.1750-3841.2007.00301.x PMID:17995809

Mathur, P. (2005). *Food microbiology and safety*. Indira Gandhi National Open University.

MATRADE. (2016). *FDA issues new regulation to protect food safety.* http://www.matrade.gov.my /en/about-matrade/media/news-clippings/146-press-releases/press-releases-2016/3526-fda-issues-new-regulation-toprotect-food-safety

Matthews, D. (2006). *A history of auditing: The changing audit process from the 19th century till date.* Routledge, Taylor & Francis Group.

McCurdy, S., Peutz, J., & Wittman, G. (2009). *Storing food for safety and quality*. Pacific Northwest Extension Publication. https://catalog.extension.oregonstate.edu/sites/catalog/files/project/pdf/ pnw612.pdf

McLoughlin, C., Miura, T., Junpei, N., Mendez, A., & Homel, W. (2017). *True Kaizen: Management's role in improving work climate and culture*. Productivity Press. doi:10.1201/9781315180373

McMillan, P. F., & Hofmeister, A. M. (2018). Infrared and Raman spectroscopy. In F. C. Hawthorne (Ed.), *Spectroscopic methods in mineralogy and geology* (pp. 99–160). De Gruyter.

Medina, S., Pereira, J. A., Silva, P., Perestrelo, R., & Câmara, J. S. (2019). Food fingerprints – A valuable tool to monitor food authenticity and safety. *Food Chemistry*, *278*, 144–162. doi:10.1016/j.foodchem.2018.11.046 PMID:30583355

Membré, J. M., Amézquita, A., Bassett, J., Giavedoni, P., Blackburn, C. D. W., & Gorris, L. G. M. (2006). A probabilistic modeling approach in thermal inactivation: Estimation of postprocess Bacillus cereus spore prevalence and concentration. *Journal of Food Protection*, *69*(1), 118–129. doi:10.4315/0362-028X-69.1.118 PMID:16416909

Mendes, T. O., da Rocha, R. A., Porto, B. L., de Oliveira, M. A., dos Anjos, V. D. C., & Bell, M. J. (2015). Quantification of extra-virgin olive oil adulteration with soybean oil: A comparative study of NIR, MIR, and Raman spectroscopy associated with chemometric approaches. *Food Analytical Methods*, *8*(9), 2339–2346. doi:10.100712161-015-0121-y

Mercado Solutions. (2018). *A report in food safety survey.* https://www.cfs.gov.hk/english/multimedia/multimedia_pub/files/foodsurvey.pdf

Merna, T., & Al-Thani, F. F. (2008). *Corporate risk management.* John Wiley & Sons.

Meulenbelt, S. (2018). Assessing chemical, biological, radiological and nuclear threats to the food supply chain. *Global Security: Health. Science and Policy*, *3*(1), 14–27.

Mignani, A. G., Ciaccheri, L., Ottevaere, H., Thienpont, H., Conte, L., Marega, M., Cichelli, A., Attilio, C., & Cimato, A. (2011). Visible and near-infrared absorption spectroscopy by an integrating sphere and optical fibers for quantifying and discriminating the adulteration of extra virgin olive oil from Tuscany. *Analytical and Bioanalytical Chemistry*, *399*(3), 1315–1324. doi:10.100700216-010-4408-y PMID:21107823

Millstone, E., & Van Zwanenberg, P. (2003). The evolution of food safety policy–making institutions in the UK, EU and Codex Alimentarius. *Social Policy and Administration*, *36*(6), 593–609. doi:10.1111/1467-9515.t01-1-00306

Ministry of Health Malaysia (MOHM). (2021). *Guide to Safe Food Storage – Portal My Health.* http://www.myhealth.gov.my/en/guide-to-safe-food-storage/

Ministry of Health Malaysia. (2014). *5 Guides To Safer Food.* Retrieved from www.myhealth.gov.my/en/5-guides-to-safer-food/%0D

Ministry of Health Malaysia. (2016). *Health Facts 2016 (Reference data for 2015).* Retrieved from www.moh.gov.my/moh/resources/Penerbitan/Penerbitan Utama/HEALTH FACTS/KKMHEALTH FACTS 2016.pdf%0D

Ministry of Health Malaysia. (2017). *Health Facts 2017 (Reference data for 2016).* Retrieved from https://myhdw.moh.gov.my/public/documents/20186/150084/HEALTH+FACTS+2017/98041185-ce34-4877-9ea1-4d5341e43187

Ministry of Health Malaysia. (2020). *Health facts 2020. Reference data for year 2019.* Ministry of Health Malaysia. Planning Division. Health Informatics Centre.

Ministry of Health. (2016). *Maldives health profile.* http://www.health.gov.mv/publications/50_Maldives_Health_Profile_2016_D13rdMay.pdf

Mjoka, J., & Selepe, M. (2018). Assessment of food hygiene knowledge and practices among food handlers in selected hotels around uMhlathuze Area. *African Journal of Hospitality, Tourism and Leisure*, *7*(4).

Moerman, F., & Wouters, P. C. (2016). Emerging Trends and Methods in Food Factory Design. In *Innovation and future trends in food manufacturing and supply chain technologies* (pp. 41–79). Woodhead Publishing. doi:10.1016/B978-1-78242-447-5.00003-4

Mohamad Fithri, A. O., Nurdiyana Syahirah, A. M., Siti Nurnajwa Nadhirah, M. S., Norsuhaida, Z., Nur Farhah Najwa, A., Siti Farah Alwani, M. N., & Zaini, M. Z. (2018). Determination of the Optimum Time for Preparation of Half-Boiled Eggs Free from Salmonella Enterica Serovar Enteritidis. *Journal of Clinical And Health Sciences*, *3*(1), 16–19. doi:10.24191/jchs.v3i1.6152

Mohamad Hapiz, A. R. (2019). A study on causal relationship between food handler's awareness and safe food handling practices in Malaysia. *Politeknik & Kolej Komuniti Journal of Social Sciences and Humanities*, *4*(1), 145–162.

Mohd Firdaus Siau, A., Son, R., Mohhiddin, O., Toh, P. S., & Chai, L. C. (2015). Food court hygiene assessment and food safety knowledge, attitudes and practices of food handlers in Putrajaya. *International Food Research Journal*, *22*(5), 1843–1854.

Mohd Halim, J., Mazni, S., Mohd Shazali, M. S., & Norazmir, M. N. (2016). Hygiene practices and food safety knowledge for biological, chemical and physical hazards. *Social Sciences*, *11*(19), 4633–4637.

Mohd Yusof, A. M., Rahman, N. A. A., & Haque, M. (2018). Knowledge, attitude, and practice toward food poisoning among food handlers and dietetic students in a public university in Malaysia. *Journal of Pharmacy & Bioallied Sciences*, *10*(4), 232–239. doi:10.4103/JPBS.JPBS_141_18 PMID:30568381

Mohd Zain, M., & Naing, N. N. (2002). Sociodemographic characteristics of food handlers and their knowledge, attitude and practice towards food sanitation: A preliminary report. *The Southeast Asian Journal of Tropical Medicine and Public Health*, *33*(2), 410–417. PMID:12236444

Moher, D., Liberati, A., Tetzlaff, J., & Altman, D. G.The Prisma Group. (2009). Preferred reporting items for systematic reviews and meta-analyses: The PRISMA statement. *PLoS Medicine*, *6*(7), e1000097. doi:10.1371/journal.pmed.1000097 PMID:19621072

Mozafari, M. R., Flanagan, J., Matia-Merino, L., Awati, A., Omri, A., Suntres, Z. E., & Singh, H. (2006). Recent trends in the lipid-based nanoencapsulation of antioxidants and their role in foods. *Journal of the Science of Food and Agriculture*, *86*(13), 2038–2045. doi:10.1002/jsfa.2576

Mun, S. G. (2020). The effects of ambient temperature changes on foodborne illness outbreaks associated with the restaurant industry. *International Journal of Hospitality Management*, *85*, 102432. Advance online publication. doi:10.1016/j.ijhm.2019.102432

Mustaffa, N. A., Rahman, R. A., Hassim, M. H., & Ngadi, N. (2017). Evaluation of Knowledge, Attitude and Practices of Food Handlers in Campus Cafeterias. *Chemical Engineering Transactions*, *56*, 1297–1302.

Nasreen, S., & Ahmed, T. (2014). Food adulteration and consumer awareness in Dhaka City, 1995-2011. *Journal of Health, Population and Nutrition*, *32*(3), 452. PMID:25395908

Nasrolahei, M., Mirshafiee, S., Kholdi, S., Salehian, M., & Nasrolahei, M. (2017). Bacterial assessment of food handlers in Sari City, Mazandaran Province, North of Iran. *Journal of Infection and Public Health*, *10*(2), 171–176. doi:10.1016/j.jiph.2016.03.006 PMID:27435639

National Bureau of Statistics. (2017). *Statistical yearbook of Maldives 2017.* Ministry of Finance and Treasury, Malé, Republic of Maldives. http://statisticsmaldives.gov.mv/to-share/yearbook/StatisticalYearbookofMal-dives2017.pdf

Ncube, F., Kanda, A., Chijokwe, M., Mabaya, G., & Nyamugure, T. (2020). Food safety knowledge, attitudes and practices of restaurant food handlers in a lower-middle-income country. *Food Science & Nutrition*, *8*(3), 1677–1687. doi:10.1002/fsn3.1454 PMID:32180975

Nee, S. O., & Abdullah Sani, N. (2011). Assessment of knowledge, attitudes and practices (KAP) among food handlers at residential colleges and canteen regarding food safety. *Sains Malaysiana*, *40*(4), 403–410.

Nesbitt, A., Thomas, M. K., Marshall, B., Snedeker, K., Meleta, K., Watson, B., & Bienefeld, M. (2014). Baseline for consumer food safety knowledge and behaviour in Canada. *Food Control*, *38*, 157–173. doi:10.1016/j.foodcont.2013.10.010

Neuman, W. L. (2014). *Social research methods: Qualitative and quantitative approaches* (7th ed.). Pearson.

New Straits Times. (2019, February 15). 49 Keramat students struck down with food poisoning; canteen ordered shut. *New Straits Times.*

Newslow, D. (2013). *Food safety management programs: Applications, best practices, and compliance*. CRC Press. doi:10.1201/b16231

Ngo, H. M., Liu, R., Moritaka, M., & Fukuda, S. (2020). Urban consumer trust in safe vegetables in Vietnam: The role of brand trust and the impact of consumer worry about vegetable safety. *Food Control*, *108*, 106856. doi:10.1016/j.foodcont.2019.106856

Nguz, K. (2007). Assessing food safety system in sub-Saharan countries: An overview of key issues. *Food Control*, *18*(2), 131–134. doi:10.1016/j.foodcont.2005.09.003

Nicolai, B. M., Beullens, K., Bobelyn, E., Peirs, A., Saeys, W., Theron, K. I., & Lammertyn, J. (2007). Nondestructive measurement of fruit and vegetable quality by means of NIR spectroscopy: A review. *Postharvest Biology and Technology*, *46*(2), 99–118. doi:10.1016/j.postharvbio.2007.06.024

Nicoli, M. C., & Calligaris, S. (2018). Secondary shelf life: An underestimated issue. *Food Engineering Reviews*, *10*(2), 57–65. doi:10.100712393-018-9173-2

Norazmir, M. N., Noor Hasyimah, M. A., Siti Shafurah, A., Siti Sabariah, B., Ajau, D., & Norazlanshah, H. (2012). Knowledge and Practices on Food Safety among Secondary School Students in Johor Bahru, Johor, Malaysia. *Pakistan Journal of Nutrition*, *11*(2), 110–115. doi:10.3923/pjn.2012.110.115

Nur Izyan, F. R., Zuraini, M. I., Maria, M. S., Lovelyna, B. J., Maimunah, M., & Saidatul Afzan, A. A. (2019). A preliminary study on food safety knowledge, attitude and practices among home-based food providers in Klang Valley, Malaysia. *Malaysian Applied Biology*, *48*(2), 157–160.

Nychas, G. J. E., Skandamis, P. N., Tassou, C. C., & Koutsoumanis, K. P. (2008). Meat spoilage during distribution. *Meat Science*, *78*(1-2), 77–89. doi:10.1016/j.meatsci.2007.06.020 PMID:22062098

Odeyemi, O. A., Sani, N. A., Obadina, A. O., Saba, C. K. S., Bamidele, F. A., Abughoush, M., Ashgar, A., Dongmo, F. F. D., Macer, D., & Aberoumand, A. (2019). Food safety knowledge, attitudes and practices among consumers in developing countries: An international survey. *Food Research International*, *116*, 1386–1390. doi:10.1016/j.foodres.2018.10.030 PMID:30716930

Odumeru, J. A. (2012). Microbial safety of food and food products. In B. K. Simpson (Ed.), *Food Biochemistry and Food Processing* (pp. 785–797). Wiley Blackwell. doi:10.1002/9781118308035.ch41

Okareh, O. T., & Erhahon, O. O. (2015). Microbiological Assessment of Food and Hand-Swabs Samples of School Food Vendors in Benin City, Nigeria. *Food and Public Heath*, *5*(1), 23–28.

Okazaki, S., Hiramatsu, M., Gonmori, K., Suzuki, O., & Tu, A. T. (2009). Rapid nondestructive screening for melamine in dried milk by Raman spectroscopy. *Forensic Toxicology*, *27*(2), 94–97. doi:10.100711419-009-0072-3

Okour, A. M., Alzein, E., Saadeh, R., & Alfaqih, M. (2020). Food safety knowledge among Jordanians: A national study. *Food Control*, *114*, 107216. doi:10.1016/j.foodcont.2020.107216

Okoye, J., & Oni, K. (2017). Promotion of indigenous food preservation and processing knowledge and the challenge of food security in Africa. *Journal of Food Security*, *5*(3), 75–87.

Olsen, A. R. (1998). Regulatory action criteria for filth and other extraneous materials: I. Review of hard or sharp foreign objects as physical hazards in food. *Regulatory Toxicology and Pharmacology*, *28*(3), 181–189. doi:10.1006/rtph.1998.1249 PMID:10049789

Olu-Taiwo, M., De-Graft, B. M., & Forson, A. O. (2021). Microbial quality of sliced pawpaw (*Carica papaya*) and watermelon (*Citrullus lanatus*) sold on some streets of Accra Metropolis, Ghana. *International Journal of Microbiology*, *2021*, 1–8. Advance online publication. doi:10.1155/2021/6695957 PMID:33574850

Oroian, M., Ropciuc, S., & Paduret, S. (2018). Honey adulteration detection using Raman spectroscopy. *Food Analytical Methods*, *11*(4), 959–968. doi:10.100712161-017-1072-2

Osaili, T. M., Jamous, D. O. A., Obeidat, B. A., Bawadi, H. A., Tayyem, R. F., & Subih, H. S. (2013). Food safety knowledge among food workers in restaurants in Jordan. *Food Control*, *31*(1), 145–150. doi:10.1016/j.foodcont.2012.09.037

Oto, N., Oshita, S., Makino, Y., Kawagoe, Y., Sugiyama, J., & Yoshimura, M. (2013). Non-destructive evaluation of ATP content and plate count on pork meat surface by fluorescence spectroscopy. *Meat Science*, *93*(3), 579–585. doi:10.1016/j.meatsci.2012.11.010 PMID:23273467

Oxford Reference. (2021). *Terrorism*. https://www.oxfordreference.com/view/10.1093/oi / authority.20110803103209420

Ozogul, F., & Hamed, I. (2018). The importance of lactic acid bacteria for the prevention of bacterial growth and their biogenic amines formation: A review. *Critical Reviews in Food Science and Nutrition*, *58*(10), 1660–1670. doi:10.1080/10408398.2016.1277972 PMID:28128651

Paek, H., & Hove, T. (2017). *Risk perceptions and risk characteristics*. Oxford Research Encyclopaedia of Communication. doi:10.1093/acrefore/9780190228613.013.283

Paiva, C. L. (2013). Quality management: Important aspects for the food industry. *Food Industries*, 191–218.

Papadopoulou, O., Panagou, E. Z., Tassou, C. C., & Nychas, G. J. (2011). Contribution of Fourier transform infrared (FTIR) spectroscopy data on the quantitative determination of minced pork meat spoilage. *Food Research International*, *44*(10), 3264–3271. doi:10.1016/j.foodres.2011.09.012

Pardeshi, S. (2019). Food adulteration: Injurious adulterants and contaminants in foods and their health effects and its safety measures in India. *International Journal of Scientific Development and Research*, *4*(6), 231–232.

Parry-Hanson Kunadu, A., Ofosu, D. B., Aboagye, E., & Tano-Debrah, K. (2016). Food safety knowledge, attitudes and self-reported practices of food handlers in institutional foodservice in Accra, Ghana. *Food Control*, *69*, 324–330. doi:10.1016/j.foodcont.2016.05.011

Patel, K. T., & Chotai, N. P. (2013). GMP requirements for "buildings and facilities" for api-comparison of schedule m, india and ich guideline and approach for compliance to different regulatory expectations. *Pharma Science Monitor*, *4*(1).

Patterson, M. (1999). High-pressure treatment of foods. In The encyclopaedia of food microbiology (pp. 1059–1065). New York: Academic Press.

Penido, A., Mendes, P., Campos, I., & Mendes, L. (2013). Enterotoxigenic Bacillus cereus from cooked chicken meat: A potential public health hazard. *Malaysian Journal of Microbiology*, *9*(2), 166–175.

Pérez, J. S. (2013). *Guidance on Food Allergen Management for Food Manufacturers*. Food Drink Europe.

Peri, C. (2006). The universe of food quality. *Food Quality and Preference*, *17*(1-2), 3–8. doi:10.1016/j.foodqual.2005.03.002

Peter, R., Mateja, A., & Mojca, J. (2013). Food chain safety management systems: The impact of good practices. In *Advances in Food Process Engineering Research and Applications* (pp. 607–625). Springer. doi:10.1007/978-1-4614-7906-2_30

Poonia, A., Jha, A., Sharma, R., Singh, H. B., Rai, A. K., & Sharma, N. (2016). Detection of adulteration in milk: A review. *International Journal of Dairy Technology*, *70*(1), 23–42. doi:10.1111/1471-0307.12274

Pretty, J. (2012). *The pesticide detox: towards a more sustainable agriculture*. Earthscan. doi:10.4324/9781849773188

Prieto, N., Andrés, S., Giráldez, F. J., Mantecón, A. R., & Lavín, P. (2008). Discrimination of adult steers (oxen) and young cattle ground meat samples by near infrared reflectance spectroscopy (NIRS). *Meat Science*, *79*(1), 198–201. doi:10.1016/j.meatsci.2007.08.001 PMID:22062613

Prüss-Ustün, A., Vickers, C., Haefliger, P., & Bertollini, R. (2011). Knowns and unknowns on burden of disease due to chemicals: A systematic review. *Environmental Health*, *10*(1), 1–15. doi:10.1186/1476-069X-10-9 PMID:21255392

Purna, S. G., Prow, L. A., & Metzger, L. E. (2005). Utilization of front-face fluorescence spectroscopy for analysis of process cheese functionality. *Journal of Dairy Science*, *88*(2), 470–477. doi:10.3168/jds.S0022-0302(05)72708-9 PMID:15653511

Qin, J., Chao, K., & Kim, M. S. (2010, June). *Development of a Raman chemical imaging system for food safety inspection*. American Society of Agricultural and Biological Engineers.

Queirós, A., Faria, D., & Almeida, F. (2017). Strengths and limitations of qualitative and quantitative research methods. *European Journal of Education Studies*, *3*(9), 369–387.

Qu, J. H., Liu, D., Cheng, J. H., Sun, D. W., Ma, J., Pu, H., & Zeng, X. A. (2015). Applications of near-infrared spectroscopy in food safety evaluation and control: A review of recent research advances. *Critical Reviews in Food Science and Nutrition*, *55*(13), 1939–1954. doi:10.1080/10408398.2013.871693 PMID:24689758

Rai, P. K., Lee, S. S., Zhang, M., Tsang, Y. F., & Kim, K. H. (2019). Heavy metals in food crops: Health risks, fate, mechanisms, and management. *Environment International*, *125*, 365–385. doi:10.1016/j.envint.2019.01.067 PMID:30743144

Ram, A. (2011). *Indian food laws in global context a critique*. http://hdl.handle.net/10603/87550

Ramasamy, I., Law, M., Collins, S., & Brooke, F. (2003). Organ distribution of prion proteins in variant Creutzfeldt-Jakob disease. *The Lancet. Infectious Diseases*, *3*(4), 214–222. doi:10.1016/S1473-3099(03)00578-4 PMID:12679264

Rebouças, L. T., Santiago, L. B., Martins, L. S., Menezes, A. C. R., & Araújo, M. (2017). Food safety knowledge and practices of food handlers, head chefs and managers in hotels' restaurants of Salvador, Brazil. *Food Control*, *73*, 372–381. doi:10.1016/j.foodcont.2016.08.026

Reddy, D., Venkatesh, K., & Reddy, C. (2017). Adulteration of milk and its detection: A review. *International Journal of Chemical Studies*, *5*(4), 613–617.

Redmond, E. C., & Griffith, C. J. (2003). Consumer food handling in the home: A review of food safety studies. *Journal of Food Protection*, *66*(1), 130–161. doi:10.4315/0362-028X-66.1.130 PMID:12540194

Rennie, D. (1995). Health education models and food hygiene education. *Journal of the Royal Society of Health*, *1*(2), 75–79. doi:10.1177/146642409511500203 PMID:7738994

Ricci, A., Chemaly, M., Davies, R., Fernández Escámez, P. S., Girones, R., Herman, L., Lindqvist, R., Nørrung, B., Robertson, L., Ru, G., Simmons, M., Skandamis, P., Snary, E., Speybroeck, N., Ter Kuile, B., Threlfall, J., Wahlström, H., Allende, A., Barregård, L., & Bolton, D. (2017). Hazard analysis approaches for certain small retail establishments in view of the application of their food safety management systems. *EFSA Journal*, *15*(3). Advance online publication. doi:10.2903/j.efsa.2017.4697 PMID:32625423

Roberts, K. R., Olds, D. A., Shanklin, C., Sauer, K., & Sneed, J. (2013). Cooling of foods in retail foodservice operations. *Food Protection Trends*, *33*(1), 27–31.

Rodriguez-Martinez, V., Velázquez, G., & Altaif, R. (2020). Deterministic and probabilistic predictive microbiology-based indicator of the listeriosis and microbial spoilage risk of pasteurized milk stored in residential refrigerators. *Lebensmittel-Wissenschaft + Technologie*, *117*, 108650. doi:10.1016/j.lwt.2019.108650

Röhr, A., Lüddecke, K., Drusch, S., Müller, M. J., & Alvensleben, R. V. (2005). Food quality and safety: Consumer perception and public health concern. *Food Control*, *16*(8), 649–655. doi:10.1016/j.foodcont.2004.06.001

Rosnani, A. H., Son, R., Mohhidin, O., Toh, P. S., & Chai, L. C. (2014). Assessment of knowledge, attitude and practices concerning food safety among restaurant workers in Putrajaya, Malaysia. *Food Science and Quality Management*, *32*, 20–28.

Ruby, G. E., Ungku Zainal Abidin, U. F., Lihan, S., Jambari, N. N., & Radu, S. (2019). A cross sectional study on food safety knowledge among adult consumers. *Food Control*, *99*, 98–105. doi:10.1016/j.foodcont.2018.12.045

Sadiku, M. N. O., Musa, S. M., & Ashaolu, T. J. (2019). Food terrorism. *International Journal of Trend in Scientific Research and Development*, *3*(4), 134–135. doi:10.31142/ijtsrd23642

Safe Food Alliance. (n.d.). *What is GFSI? Getting started with the Global Food Safety Initiative (GFSI).* https://safefoodalliance.com/food-safety-resources/what-is-gfsi/

Safe Quality Food Institute. (2008). https://www.sqfi.com/

Sahar, A., Rahman, U., Kondjoyan, A., Portanguen, S., & Dufour, E. (2016). Monitoring of thermal changes in meat by synchronous fluorescence spectroscopy. *Journal of Food Engineering*, *168*, 160–165. doi:10.1016/j.jfoodeng.2015.07.038

Saira Banu, M. R., Humairak, S., Zakiah, M. D., & Siti Adila, Z. A. (2019). Retrospective study on persistent salmonella serotypes in meat samples tested in the veterinary public health section, regional of veterinary laboratory Bukit Tengah, Penang. *Malaysian Journal of Veterinary Research*, *10*(1), 34–42.

Saldaña, E., Siche, R., Luján, M., & Quevedo, R. (2013). Computer vision applied to the inspection and quality control of fruits and vegetables. *Brazilian Journal of Food Technology*, *16*(4), 254–272. doi:10.1590/S1981-67232013005000031

Salguero-Chaparro, L., Gaitán-Jurado, A. J., Ortiz-Somovilla, V., & Peña-Rodríguez, F. (2013). Feasibility of using NIR spectroscopy to detect herbicide residues in intact olives. *Food Control, 30*(2), 504–509. doi:10.1016/j.foodcont.2012.07.045

Samapundo, S., Cam Thanh, T. N., Xhaferi, R., & Devlieghere, F. (2016). Food safety knowledge, attitudes and practices of street food vendors and consumers in Ho Chi Minh city, Vietnam. *Food Control, 70*, 79–89. doi:10.1016/j.foodcont.2016.05.037

Sánchez, M. T., Pérez-Marín, D., Flores-Rojas, K., Guerrero, J. E., & Garrido-Varo, A. (2009). Use of near-infrared reflectance spectroscopy for shelf-life discrimination of green asparagus stored in a cool room under controlled atmosphere. *Talanta, 78*(2), 530–536. doi:10.1016/j.talanta.2008.12.004 PMID:19203619

Sánchez, M., Neira, C., Laca, A., Laca, A., & Díaz, M. (2019). Survival and development of Staphylococcus in egg products. *Lebensmittel-Wissenschaft + Technologie, 101*, 685–693. doi:10.1016/j.lwt.2018.11.092

Sandra, A., Afsah-Hejri, L., Tunung, R., Tuan Zainazor, T. T. C., Tang, J. Y. H., Ghazali, F. M., & Son, R. (2012). *Bacillus cereus* and *Bacillus thuringiensis* in ready-to-eat cooked rice in Malaysia. *International Food Research Journal, 19*(3), 829–836.

Saw, S. H., Mak, J. L., Tan, M. H., Teo, S. T., Tan, T. Y., Cheow, M. Y. K., Ong, C. A., Chen, S. N., Yeo, S. K., Kuan, C. S., New, C. Y., Radu, S., Phuah, E. T., Thung, T. Y., & Kuan, C. H. (2020). Detection and quantification of Salmonella in fresh vegetables in Perak, Malaysia. *Food Research, 4*(2), 441–448. doi:10.26656/fr.2017.4(2).316

Schaffner, D. W., Brown, L. G., Ripley, D., Reimann, D., Koktavy, N., Blade, H., & Nicholas, D. (2015). Quantitative data analysis to determine best food cooling practices in U.S. restaurants. *Journal of Food Protection, 78*(4), 778–783. doi:10.4315/0362-028X.JFP-14-252 PMID:25836405

Schilling, E. G., & Neubauer, D. V. (2017). *Acceptance sampling in quality control*. CRC Press. doi:10.1201/9781315120744

Schneider, K. R., Schneider, R. G., Silverberg, R., Kurdmongkoltham, P., & Bertoldi, B. (2017). *Preventing Foodborne Illness: Bacillus cereus and Bacillus anthracis*. Institute of Food and Agricultural Sciences (IFAS), University of Florida. FSHN04-05:1–5.

SCV, The Foundation for the Certification of Food Safety Systems. (2008). *HACCP requirements*. http://www.foodsafetymanagement. info

Sebastian, J. (n.d.). Effectively digitizing food safety audits. *Global Food Safety Resource*. https://globalfoodsafetyresource.com/effectively-digitizing-food-safety-audits/

Severi, E., Booth, L., Johnson, S., Cleary, P., Rimington, M., Saunders, D., Cockcroft, P., & Ihekweazu, C. (2012). Large outbreak of *Salmonella Enteritidis* PT8 in Portsmouth, UK, associated with a restaurant. *Epidemiology and Infection, 140*(10), 1748–1756. doi:10.1017/S0950268811002615 PMID:22166322

Shafie, A. A., & Azman, A. W. (2015). Assessment of knowledge, attitude and practice of food allergies among food handlers in the state of Penang, Malaysia. *Public Health, 129*(9), 1278–1284. doi:10.1016/j.puhe.2015.03.016 PMID:25931434

Shaikh, S., & O'Donnell, C. (2017). Applications of fluorescence spectroscopy in dairy processing: A review. *Current Opinion in Food Science, 17*, 16–24. doi:10.1016/j.cofs.2017.08.004

Shende, C., Gift, A., Inscore, F., Maksymiuk, P., & Farquharson, S. (2004). Inspection of pesticide residues on food by surface-enhanced Raman spectroscopy. In *Monitoring food safety, agriculture, and plant health* (Vol. 5271, pp. 28–34). International Society for Optics and Photonics. doi:10.1117/12.511941

Sheridan, M. (2003). *Packing up profits. Restaurants and institutions.* Retrieved from www.rimag.com/archives/2003/06b/ops1.asp

Shilla, J. (2011). Survey beta lactamase production and resistance pattern into beta lactame antibiotics in *Bacillus cereus* strain isolated from staff hands and hospital environment in Iran. *African Journal of Microbiological Research, 5*(19), 2980–2985. doi:10.5897/AJMR11.515

Siddiqui, M. W., Longkumer, M., Ahmad, M. S., Barman, K., Thakur, P. K., & Kabir, J. (2014). Postharvest biology and technology of sapota: A concise review. *Acta Physiologiae Plantarum, 36*(12), 3115–3122. doi:10.100711738-014-1696-4

Sileyew, K. J. (2019). *Research design and methodology.* https://www.intechopen.com/online-first/research-design-and-methodology

Sinelli, N., Cerretani, L., Di Egidio, V., Bendini, A., & Casiraghi, E. (2010). Application of near (NIR) infrared and mid (MIR) infrared spectroscopy as a rapid tool to classify extra virgin olive oil on the basis of fruity attribute intensity. *Food Research International, 43*(1), 369–375. doi:10.1016/j.foodres.2009.10.008

Singapore Food Agency (SFA). (2020). *Guidelines on food safety & hygiene practices for residents preparing food under the HDB/URA's home-based small-scale business scheme.* Author.

Singuluri, H., & Sukumaran, M. K. (2014). Milk adulteration in Hyderabad, India–A comparative study on the levels of different adulterants present in milk. *Journal of Chromatography & Separation Techniques, 5*(1), 212. doi:10.4172/2157-7064.1000212

Sirisomboon, C. D., Putthang, R., & Sirisomboon, P. (2013). Application of near infrared spectroscopy to detect aflatoxigenic fungal contamination in rice. *Food Control, 33*(1), 207–214. doi:10.1016/j.foodcont.2013.02.034

Siti Nurul Ain, S., Sahilah, A. M., & Razalee, S. (2018). Knowledge, attitude and practice of food utensils hygiene amongst food handlers in Kuala Pilah, Negeri Sembilan, Malaysia. *Sains Malaysiana, 47*(7), 1527–1533. doi:10.17576/jsm-2018-4707-21

Sivertsen, A. H., Kimiya, T., & Heia, K. (2011). Automatic freshness assessment of cod (Gadus morhua) fillets by Vis/Nir spectroscopy. *Journal of Food Engineering, 103*(3), 317–323. doi:10.1016/j.jfoodeng.2010.10.030

Solar, L. (n.d.). BRC Global Standards V.8: New changes for improving compliance. *Global Food Safety Resource.* https://globalfoodsafetyresource.com/brc-global-standards-v-8-new-changes-improving-compliance/

Sonnemann, G., Tsang, M., & Schuhmacher, M. (2018). *Integrated lifecycle and risk assessment for industrial processes and products*. CRC Press. doi:10.1201/9780429436949

Sornchuer, P., & Tiengtip, R. (2021). Prevalence, virulence genes, and antimicrobial resistance of *Bacillus cereus* isolated from foodstuffs in Pathum Thani Province, Thailand. *Pharmaceutical Sciences Asia, 48*(2), 194–203. doi:10.29090/psa.2021.02.19.119

Souza, C., Azevedo, P. R. M., & Seabra, L. M. A. J. (2018). Food safety in Brazilian popular public restaurants: Food handlers' knowledge and practices. *Journal of Food Safety, 38*(5), e12512. Advance online publication. doi:10.1111/jfs.12512 PMID:30449912

Spink, J. (2014). *GFSI direction on food fraud and vulnerability assessment (VACCP).* http://foodfraud.msu.edu/2014/05/08/gfsidirection-on-food-fraud-and-vulnerability-as

Spink, J., & Moyer, D. C. (2011). Defining the public health threat of food fraud. *Journal of Food Science, 76*(9), R157–R163. doi:10.1111/j.1750-3841.2011.02417.x PMID:22416717

Spink, J., Moyer, D. C., & Whelan, P. (2016). The role of the public private partnership in food fraud prevention—Includes implementing the strategy. *Current Opinion in Food Science, 10*, 68–75. doi:10.1016/j.cofs.2016.10.002

Srivastava, S. (2015). Food adulteration affects the nutrition and health of human beings. *Journal of Biological Sciences and Medicine, 1*(1), 68.

Stankovic, I. (2016). Codex Alimentarius. Encyclopedia of Food and Health, 191–196.

Stier, R. F. (2020). The evolution of food safety management systems—from early 1900s to audits today. *Food Engineering.* https://www.foodengineeringmag.com/articles/98678-the-evolution-of-food-safety-management-systemsfrom-early-1900s-to-audits-today

Stinson, T. F., Albisu, L. M., Canavari, M., Larson, R., & Gracia, A. (2011). Differences in household attitudes on food defence and food safety: An international comparison. *International Journal of Food Safety, Nutrition and Public Health, 4*(1), 29–44. doi:10.1504/IJFSNPH.2011.042573

Sun, Y., Wu, S., & Gong, G. (2019). Trends of research on polycyclic aromatic hydrocarbons in food: A 20-year perspective from 1997 to 2017. *Trends in Food Science & Technology, 83*, 86–98. doi:10.1016/j.tifs.2018.11.015

Surak, J. G., & Lorca, T. A. (2007). Process auditing for food safety. *Food Safety Magazine.* http://www.foodsafetymagazine.com/magazinearchive1/augustseptember2007/process-auditing-for-food-safety/

Su, W. H., Bakalis, S., & Sun, D. W. (2018). Fourier transform mid-infrared-attenuated total reflectance (FTMIR-ATR) microspectroscopy for determining textural property of microwave baked tuber. *Journal of Food Engineering, 218*, 1–13. doi:10.1016/j.jfoodeng.2017.08.016

Su, W. H., & Sun, D. W. (2019). Mid-infrared (MIR) spectroscopy for quality analysis of liquid foods. *Food Engineering Reviews*, *11*(3), 142–158. doi:10.100712393-019-09191-2

Syahira, S., Huda, B. Z., & Mohd Rafee, B. B. (2019). Factors associated with level of food safety knowledge among form four students In Hulu Langat District, Selangor. *International Journal of Public Health and Clinical Sciences*, *6*(2), 252–265. doi:10.32827/ijphcs.6.2.252

Taha, S., Osaili, T. M., Saddal, N. K., Al-Nabulsi, A. A., Ayyash, M. M., & Obaid, R. S. (2020). Food safety knowledge among food handlers in food service establishments in United Arab Emirates. *Food Control*, *110*, 106968. doi:10.1016/j.foodcont.2019.106968

Tan, S. L., Cheng, P. L., Soon, H. K., Ghazali, H., & Mahyudin, N. A. (2013). A qualitative study on personal hygiene knowledge and practices among food handlers at selected primary schools in Klang valley area, Selangor, Malaysia. *International Food Research Journal*, *20*(1), 71–76.

Tan, S. L., Cheng, P. L., Soon, H. K., Ghazali, H., & Mahyudin, N. A. (2013). A qualitative study on personal hygiene knowledge and practices among food handlers at selected primary schools in Klang Valley area, Selangor, Malaysia. *International Food Research Journal*, *20*(1), 71–76.

Tan, S. L., Fatimah, A. B., Muhammad Shahrim, A. K., Lee, H. Y., & Nor Ainy, M. (2013). Hand hygiene knowledge, attitudes and practices among food handlers at primary schools in Hulu Langat district, Selangor (Malaysia). *Food Control*, *34*(2), 428–435. doi:10.1016/j.foodcont.2013.04.045

TCEC. (2011). *Pilot testing data collection instruments*. https://tcec.sf.ucdavis.edu/pilot-testing

Tegegne, H. A., & Phyo, H. W. W. (2017). Food safety knowledge, attitude and practices of meat handler in abattoir and retail meat shops of Jigjiga Town, Ethiopia. *Journal of Preventive Medicine and Hygiene*, *58*, 320–327. PMID:29707664

Teh, W. S. (2016). *Silent killer: Polystyrene containers still widely used despite Ramadan night market ban*. https://malaysiandigest.com/news/618001silent-killer-polystyrene-containers-still-widely-used-despite-ramadan-night-market-ban.html

The Malaysian National News Agency (2018, October 13). Three more food poisoning cases reported in Perak. *New Straits Times*.

The Malaysian National News Agency (2019, June 19). 110 college students down with food poisoning. *New Straits Times*.

The Malaysian National News Agency (2020, January 6). "Ayam masak lemak" dish downs 137 kindergarten teachers. *Malaysiakini*.

The National Council of Educational Research and Training (NCERT). (2021-22). *Food safety and food quality*. https://ncert.nic.in/textbook/pdf/lehe106.pdf

Tibola, C. S., da Silva, S. A., Dossa, A. A., & Patrício, D. I. (2018). Economically motivated food fraud and adulteration in Brazil: Incidents and alternatives to minimize occurrence. *Journal of Food Science*, *83*(8), 2028–2038. doi:10.1111/1750-3841.14279 PMID:30020548

Tirloni, E., Bernardi, C., Ghelardi, E., Celandroni, F., Cattaneo, P., & Stella, S. (2019). *Bacillus cereus* in fried rice meals: Natural occurrence, strain dependent growth and haemolysin (HBL) production. *Lebensmittel-Wissenschaft + Technologie, 114*, 108393. doi:10.1016/j.lwt.2019.108393

Tjahjono, B., Ball, P., Vitanov, V. I., Scorzafave, C., Nogueira, J., Calleja, J., Minguet, M., Narasimha, L., Rivas, A., Srivastava, A., Srivastava, S., & Yadav, A. (2010). Six Sigma: A literature review. *International Journal of Lean Six Sigma, 1*(3), 216–233. doi:10.1108/20401461011075017

Todt, O., Müoz, E., González, M., Ponce, G., & Estévez, B. (2009). Consumer attitudes and the governance of food safety. *Public Understanding of Science (Bristol, England), 18*(1), 103–114. doi:10.1177/0963662507078019 PMID:19579538

Tomaszewska, M., Trafialek, J., Suebpongsang, P., & Kolanowski, W. (2018). Food hygiene knowledge and practice of consumers in Poland and in Thailand: A survey. *Food Control, 85*, 76–84. doi:10.1016/j.foodcont.2017.09.022

Torkar, K. G., & Bedenić, B. (2018). Antimicrobial susceptibility and characterization of metallo-β-lactamases, extended-spectrum β-lactamases, and carbapenemases of *Bacillus cereus* isolates. *Microbial Pathogenesis, 118*(March), 140–145. doi:10.1016/j.micpath.2018.03.026 PMID:29551437

Torso, L. M., Voorhees, R. E., Forest, S. A., Gordon, A. Z., Silvestri, S. A., Kissler, B., Schlackman, J., Sandt, C. H., Toma, P., Bachert, J., Mertz, K. J., & Harrison, L. H. (2015). *Escherichia coli* O157:H7 outbreak associated with restaurant beef grinding. *Journal of Food Protection, 78*(7), 1272–1279. doi:10.4315/0362-028X.JFP-14-545 PMID:26197277

Trafialek, J. (2019). The role of suppliers of raw materials in ensuring food safety. *Global Journal of Nutrition & Food Science, 1*(4). Advance online publication. doi:10.33552/GJNFS.2019.01.000517

Trafialek, J., & Kołożyn-Krajewska, D. (2011). Implementation of safety assurance system in food production in Poland. *Polish Journal of Food and Nutrition Sciences, 61*(2), 115–124. doi:10.2478/v10222-011-0012-x

Tricker, R. (2019). *Quality management systems: A practical guide to standards implementation*. Routledge. doi:10.4324/9780429274473

Troedsson, H. (2009). *Forging shared international standards for food safety*. WHO speech. http://www.wpro.who.int/china/media_centre/speeches/speech_20090621.htm

Tucker, M., Whaley, S. R., & Sharp, S. S. (2006). Consumer perceptions of food-related risks. *International Journal of Food Science & Technology, 41*(2), 135–146. doi:10.1111/j.1365-2621.2005.01010.x

Tumbarski, Y. D. (2020). Foodborne zoonotic agents and their food bioterrorism potential: A review. *Bulgarian Journal of Veterinary Medicine, 23*(2), 147–159. doi:10.15547//bjvm.2232

Turlejska, H. (2003). Zasady GHP/GMP oraz systemu HACCP jako narzedzia zapewnienia bezpieczenstwa zdrowotnego _ywnosci [GHP/GMP principles and HACCP system as a tolls for food safety assurance]. *Poradnik Dla przedsiebiorcy*.

Tutu, B. O., Hushie, C., Asante, R., & Egyakwa-Amusah, J. A. (2020). Food safety knowledge and self-reported practices among school children in the Ga West Municipality in Ghana. *Food Control, 110*, 107012. doi:10.1016/j.foodcont.2019.107012

U.S. National Library of Medicine. (2014). *HSDB: Styrene. Toxicology data network.* https://toxnet.nlm.nih.gov/cgibin/sis/search2/

Uçar, A., Yilmaz, M. V., & Çakiroglu, F. P. (2016). Food safety: Problems and solutions. In Significance, Prevention and Control of Food Related Diseases. InTech. doi:10.5772/63176

Uddin, M., Okazaki, E., Turza, S., Yumiko, Y., Tanaka, M., & Fukuda, Y. (2005). Non-destructive visible/NIR spectroscopy for differentiation of fresh and frozen-thawed fish. *Journal of Food Science, 70*(8), c506–c510. doi:10.1111/j.1365-2621.2005.tb11509.x

Uddin, M., Turza, S., & Okazaki, E. (2007). Rapid determination of intact sardine fat by NIRS using surface interactance fibre probe. *International Journal of Food Engineering, 3*(6). Advance online publication. doi:10.2202/1556-3758.1248

UNICEF (United Nation's Children's Fund). (2010). *Progress for children: Achieving the MDGs with equity.* UNICEF.

United States Department of Agriculture. (2012). Food safety plan HACCP-based standard operating procedures. Mississippi: National Food Service Management Institute.

United States Department of Agriculture. (2017). *Refrigerator thermometers - cold facts about food safety.* https://www.fda.gov/food/buy-store-serve-safe-food/refrigerator-thermometers-cold-facts-about-food-safety

United States Department of Agriculture. (2019). *Cold food storage chart.* https://www.foodsafety.gov/food-safety-charts/cold-food-storage-charts

United States Department of Agriculture. (2019). *What methods of reheating food are safe?* https://ask.usda.gov/s/article/What-methods-of-reheating-food-are-safe

United States Department of Agriculture. (2021). *Food safety during power outage.* https://www.foodsafety.gov/food-safety-charts/food-safety-during-power-outage

Unnevehr, L. J., & Huirne, R. B. M. (2002). *New approaches to food safety economics: Overview and new research directions.* Summary of papers presented at a Frontis workshop on New Approaches to Food Safety Economics, Wageningen, The Netherlands.

Unusan, N. (2007). Consumer food safety knowledge and practices in the home in Turkey. *Food Control, 18*(1), 45–51. doi:10.1016/j.foodcont.2005.08.006

Ursachi, G., Horodnic, I. A., & Zait, A. (2015). How reliable are measurement scales? External factors with indirect influence on reliability estimators. *Procedia Economics and Finance, 20*, 679–686. doi:10.1016/S2212-5671(15)00123-9

US Department of Health & Human Services. (2019). *Antibiotic resistance threats in the United States.* Centers for Disease Control and Prevention. https://www.cdc.gov/drugresistance/biggest_threats.html

Van der Meulen, B. (2014). Food law. Encyclopedia of Agriculture and Food Systems, 186–195.

Van der Meulen, B. (2010). *Development of food legislation around the world. In Ensuring Global Food Safety Exploring Global Harmonization.* Academic Press. doi:10.1016/B978-0-12-374845-4.00002-3

Van der Spiegel, M., Luning, P. A., Ziggers, G. W., & Jongen, W. M. F. (2003). Towards a conceptual model to measure effectiveness of food quality systems. *Trends in Food Science & Technology, 14*(10), 424–431. doi:10.1016/S0924-2244(03)00058-X

Vandenabeele, P. (2013). *Practical Raman spectroscopy: An introduction.* Wiley. doi:10.1002/9781119961284

Vašková, H. (2011). A powerful tool for material identification: Raman spectroscopy. *International Journal of Mathematical Models and Methods in Applied Sciences, 5*, 1205–1212.

Venture-Lucas, M. R. (2004). *Consumer perceptions and attitudes towards food safety in Portugal.* The 84th EAAE Seminar 'Food Safety in a Dynamic World'. 10.22004/ag.econ.24986

Verhoeckx, K. C. M., Vissers, Y. M., Baumert, J. L., Faludi, R., Feys, M., Flanagan, S., Herouet-Guicheney, C., Holzhauser, T., Shimojo, R., van der Bolt, N., Wichers, H., & Kimber, I. (2015). Food processing and allergenicity. *Food and Chemical Toxicology, 80*, 223–240. doi:10.1016/j.fct.2015.03.005 PMID:25778347

Wallace, C. A., Sperber, W. H., & Mortimore, S. E. (2018). *Food safety for the 21st century: Managing HACCP and food safety throughout the global supply chain.* John Wiley and Sons. doi:10.1002/9781119053569

Wang, L., Lee, F. S., Wang, X., & He, Y. (2006). Feasibility study of quantifying and discriminating soybean oil adulteration in camellia oils by attenuated total reflectance MIR and fiber optic diffuse reflectance NIR. *Food Chemistry, 95*(3), 529–536. doi:10.1016/j.foodchem.2005.04.015

Wang, S., Guo, Q., Wang, L., Lin, L., Shi, H., Cao, H., & Cao, B. (2015). Detection of honey adulteration with starch syrup by high performance liquid chromatography. *Food Chemistry, 172*, 669–674. doi:10.1016/j.foodchem.2014.09.044 PMID:25442605

Wang, W., & Paliwal, J. (2007). Near-infrared spectroscopy and imaging in food quality and safety. *Sensing and Instrumentation for Food Quality and Safety, 1*(4), 193–207. doi:10.100711694-007-9022-0

Werkema, C. (2004). *Criando a cultura seis sigma.* Nova Lima: Werkema Editora.

Whited, T., Feng, Y., & Bruhn, C. M. (2019). Evaluation of the high school food safety curriculum using a positive deviance model. *Food Control, 96*, 324–328. doi:10.1016/j.foodcont.2018.09.004

WHO. (2004). Food safety in developing countries: Building capacity. *Weekly Epidemiological Record*, *79*(18), 173–180. PMID:15168565

WHO. (2007). *Countries urged to be more vigilant about food safety*. www.who.int/mediacentre/news/releases/2007/pr39/en/

WHO. (2019). *Food Safety*. https://www.who.int/news-room/fact-sheets/detail/food-safety

WHO/FAO. (2011). *The international food safety authorities network (INFOSAN) progress report 2004-2010*. http://whqlibdoc.who.int/publications/2011/9789241501286_eng.pdf

Wierzbicki, S. (2003). AQAP w Sokołów S.A [AQAP at Sokołów S.A.]. *Bezp. Hig. żywn., 6*(1), 3.

Wilna, H. O., & Abdulkadir, A. E. (2016). Food quality and food safety. In N. J. Temple & N. Stey (Eds.), *Community Nutrition for Developing Countries*. Athabasca University Press and UNISA Press.

Wilson, A. D. (2013). Diverse applications of electronic-nose technologies in agriculture and forestry. *Sensors (Basel)*, *13*(2), 2295–2348. doi:10.3390130202295 PMID:23396191

Wong, J. X., Ramli, S., & Chen, S. N. (2021a). Assessment of toxic effect of *Centella asiatica* extract and its application as natural preservative in fresh-cut mango, pear and cabbage. *Journal of Food Processing and Preservation*, *45*(10), e15824. doi:10.1111/jfpp.15824

Wong, J. X., Ramli, S., Desa, S., & Chen, S. N. (2021b). Use of *Centella asiatica* extract in reducing microbial contamination and browning effect in fresh cut fruits and vegetables during storage: A potential alternative of synthetic preservatives. *Lebensmittel-Wissenschaft + Technologie*, *151*, 112229. Advance online publication. doi:10.1016/j.lwt.2021.112229

Wood, M. M., Mileto, D. S., Kano, M., Kelley, M. M., Regan, R., & Bourque, L. B. (2012). Communicating actionable risk for terrorism and other hazards. *An International Journal: Risk Analysis*, *32*(4), 601–615. doi:10.1111/j.1539-6924.2011.01645.x PMID:21689127

World Bank. (2018). *Food-borne illnesses cost US$ 110 billion per year in low- and middle-income countries*. https://www.worldbank.org/en/news/press-release/2018/10/23/food-borne-illnesses-cost-us-110-billion-per-year-in-low-and-middle-income-countries

World Health Organization. (1998). *Members and associate members of WHO*. https://www.who.int/whr/1998/en/whr98_annex.pdf

World Health Organization. (2002). *Potential effects of food terrorism. Terrorist threats to food: Guidance for establishing and strengthening prevention and response systems*. https://apps.who.int/iris/bitstream/handle/10665/42619/9241545844.pdf?sequence=1&isAllowed=y

World Health Organization. (2002). Statistical information on food-borne disease in europe microbiological and chemical hazards. *Pan European Conference on Food Safety and Quality*, 60–85.

World Health Organization. (2002). *WHO global strategy for food safety : Safer food for better health.* https://www.who.int/fsf

World Health Organization. (2004). *Annex 5: Precautions against the sabotage of drinking-water, food, and other products. Public health response to biological and chemical weapons: WHO guidance.* https://www.who.int/csr/delibepidemics/annex5.pdf?ua

World Health Organization. (2013). *Regional food safety strategy 2013 - 2017.* www.searo.who.int

World Health Organization. (2015). *WHO estimates of the global burden of foodborne diseases.* https://apps.who.int/iris/bitstream/handle/10665/199350/9789241565165_eng.pdf?sequence=1

World Health Organization. (2018, May 9). *Mycotoxins.* https://www.who.int/news-room/fact-sheets/detail/mycotoxins

Xue, J., & Zhang, W. (2013). Understanding China's food safety problem: An analysis of 2387 incidents of acute foodborne illness. *Food Control, 30*(1), 311–317. doi:10.1016/j.foodcont.2012.07.024

Xue, L., Cai, J., Li, J., & Liu, M. (2012). Application of particle swarm optimization (PSO) algorithm to determine dichlorvos residue on the surface of navel orange with Vis-NIR spectroscopy. *Procedia Engineering, 29*, 4124–4128. doi:10.1016/j.proeng.2012.01.631

Xu, Y., & Fisher, G. J. (2005). Ultraviolet (UV) light irradiation induced signal transduction in skin photoaging. *Journal of Dermatological Science. Supplement, 1*(2), S1–S8. doi:10.1016/j.descs.2005.06.002

Yang, D., & Ying, Y. (2011). Applications of Raman spectroscopy in agricultural products and food analysis: A review. *Applied Spectroscopy Reviews, 46*(7), 539–560. doi:10.1080/05704928.2011.593216

Yang, H., & Irudayaraj, J. (2003). Rapid detection of foodborne microorganisms on food surface using Fourier transform Raman spectroscopy. *Journal of Molecular Structure, 646*(1-3), 35–43. doi:10.1016/S0022-2860(02)00575-6

Yang, H., Irudayaraj, J., & Paradkar, M. M. (2005). Discriminant analysis of edible oils and fats by FTIR, FT-NIR and FT-Raman spectroscopy. *Food Chemistry, 93*(1), 25–32. doi:10.1016/j.foodchem.2004.08.039

Yang, W., Li, D., & Mugambi, A. (2017). *Spoilage Microorganisms in Cereal Products. In Food Spoilage Microorganisms: Ecology and Control.* CRC Press.

Young, I., Waddell, L. A., Wilhelm, B. J., & Greig, J. (2020). A systematic review and meta-regression of single group, pre-post studies evaluating food safety education and training interventions for food handlers. *Food Research International, 128*, 108711. doi:10.1016/j.foodres.2019.108711 PMID:31955782

Yu, H., Huang, Y., & Huang, Q. (2009). Synthesis and characterization of novel antimicrobial emulsifiers from polylysine. *Journal of Agricultural and Food Chemistry*, *58*(2), 1290–1295. doi:10.1021/jf903300m PMID:20020765

Yu, H., Sirsat, S. A., & Madera, J. M. (2018). Enhancing hospitality students' motivation to learn food safety knowledge using tablet personal computer (TPC)-based simulation game class activity. *Journal of Hospitality, Leisure, Sport and Tourism Education*, *23*, 82–94. doi:10.1016/j.jhlste.2018.08.002

Yu, S., Yu, P., Wang, J., Li, C., Guo, H., Liu, C., Kong, L., Yu, L., Wu, S., Chen, M., Zeng, H., Rui, P., Zheng, Y., Wei, X., Zhang, J., Wu, Q., & Ding, Y. (2020). A study on prevalence and characterization of Bacillus cereus in Ready-to-Eat foods in China. *Frontiers in Microbiology*, *10*(January), 1–11. doi:10.3389/fmicb.2019.03043 PubMed

Zhang, P. X., Zhou, X., Cheng, A. Y., & Fang, Y. (2006). Raman spectra from pesticides on the surface of fruits. Journal of Physics: Conference Series, 28(1), 2. doi:10.1088/1742-6596/28/1/002

Zhang, X., Wu, D., Zhou, X., Yu, Y., Liu, J., Hu, N., Wang, H., Li, G., & Wu, Y. (2019). Recent progress in the construction of nanozyme-based biosensors and their applications to food safety assay. *Trends in Analytical Chemistry*, *121*, 115668. doi:10.1016/j.trac.2019.115668

Related References

To continue our tradition of advancing information science and technology research, we have compiled a list of recommended IGI Global readings. These references will provide additional information and guidance to further enrich your knowledge and assist you with your own research and future publications.

Abtahi, M. S., Behboudi, L., & Hasanabad, H. M. (2017). Factors Affecting Internet Advertising Adoption in Ad Agencies. *International Journal of Innovation in the Digital Economy*, *8*(4), 18–29. doi:10.4018/IJIDE.2017100102

Agrawal, S. (2017). The Impact of Emerging Technologies and Social Media on Different Business(es): Marketing and Management. In O. Rishi & A. Sharma (Eds.), *Maximizing Business Performance and Efficiency Through Intelligent Systems* (pp. 37–49). Hershey, PA: IGI Global. doi:10.4018/978-1-5225-2234-8.ch002

Alnoukari, M., Razouk, R., & Hanano, A. (2016). BSC-SI: A Framework for Integrating Strategic Intelligence in Corporate Strategic Management. *International Journal of Social and Organizational Dynamics in IT*, *5*(2), 1–14. doi:10.4018/IJSODIT.2016070101

Alnoukari, M., Razouk, R., & Hanano, A. (2016). BSC-SI, A Framework for Integrating Strategic Intelligence in Corporate Strategic Management. *International Journal of Strategic Information Technology and Applications*, *7*(1), 32–44. doi:10.4018/IJSITA.2016010103

Altındağ, E. (2016). Current Approaches in Change Management. In A. Goksoy (Ed.), *Organizational Change Management Strategies in Modern Business* (pp. 24–51). Hershey, PA: IGI Global. doi:10.4018/978-1-4666-9533-7.ch002

Alvarez-Dionisi, L. E., Turner, R., & Mittra, M. (2016). Global Project Management Trends. *International Journal of Information Technology Project Management, 7*(3), 54–73. doi:10.4018/IJITPM.2016070104

Anantharaman, R. N., Rajeswari, K. S., Angusamy, A., & Kuppusamy, J. (2017). Role of Self-Efficacy and Collective Efficacy as Moderators of Occupational Stress Among Software Development Professionals. *International Journal of Human Capital and Information Technology Professionals, 8*(2), 45–58. doi:10.4018/IJHCITP.2017040103

Aninze, F., El-Gohary, H., & Hussain, J. (2018). The Role of Microfinance to Empower Women: The Case of Developing Countries. *International Journal of Customer Relationship Marketing and Management, 9*(1), 54–78. doi:10.4018/IJCRMM.2018010104

Arsenijević, O. M., Orčić, D., & Kastratović, E. (2017). Development of an Optimization Tool for Intangibles in SMEs: A Case Study from Serbia with a Pilot Research in the Prestige by Milka Company. In M. Vemić (Ed.), *Optimal Management Strategies in Small and Medium Enterprises* (pp. 320–347). Hershey, PA: IGI Global. doi:10.4018/978-1-5225-1949-2.ch015

Aryanto, V. D., Wismantoro, Y., & Widyatmoko, K. (2018). Implementing Eco-Innovation by Utilizing the Internet to Enhance Firm's Marketing Performance: Study of Green Batik Small and Medium Enterprises in Indonesia. *International Journal of E-Business Research, 14*(1), 21–36. doi:10.4018/IJEBR.2018010102

Atiku, S. O., & Fields, Z. (2017). Multicultural Orientations for 21st Century Global Leadership. In N. Baporikar (Ed.), *Management Education for Global Leadership* (pp. 28–51). Hershey, PA: IGI Global. doi:10.4018/978-1-5225-1013-0.ch002

Atiku, S. O., & Fields, Z. (2018). Organisational Learning Dimensions and Talent Retention Strategies for the Service Industries. In N. Baporikar (Ed.), *Global Practices in Knowledge Management for Societal and Organizational Development* (pp. 358–381). Hershey, PA: IGI Global. doi:10.4018/978-1-5225-3009-1.ch017

Ávila, L., & Teixeira, L. (2018). The Main Concepts Behind the Dematerialization of Business Processes. In M. Khosrow-Pour, D.B.A. (Ed.), Encyclopedia of Information Science and Technology, Fourth Edition (pp. 888-898). Hershey, PA: IGI Global. doi:10.4018/978-1-5225-2255-3.ch076

Bartens, Y., Chunpir, H. I., Schulte, F., & Voß, S. (2017). Business/IT Alignment in Two-Sided Markets: A COBIT 5 Analysis for Media Streaming Business Models. In S. De Haes & W. Van Grembergen (Eds.), *Strategic IT Governance and Alignment in Business Settings* (pp. 82–111). Hershey, PA: IGI Global. doi:10.4018/978-1-5225-0861-8.ch004

Bashayreh, A. M. (2018). Organizational Culture and Organizational Performance. In W. Lee & F. Sabetzadeh (Eds.), *Contemporary Knowledge and Systems Science* (pp. 50–69). Hershey, PA: IGI Global. doi:10.4018/978-1-5225-5655-8.ch003

Bedford, D. A. (2018). Sustainable Knowledge Management Strategies: Aligning Business Capabilities and Knowledge Management Goals. In N. Baporikar (Ed.), *Global Practices in Knowledge Management for Societal and Organizational Development* (pp. 46–73). Hershey, PA: IGI Global. doi:10.4018/978-1-5225-3009-1.ch003

Benmoussa, F., Nakara, W. A., & Jaouen, A. (2016). The Use of Social Media by SMEs in the Tourism Industry. In I. Lee (Ed.), *Encyclopedia of E-Commerce Development, Implementation, and Management* (pp. 2159–2170). Hershey, PA: IGI Global. doi:10.4018/978-1-4666-9787-4.ch155

Berger, R. (2016). Indigenous Management and Bottom of Pyramid Countries: The Role of National Institutions. In U. Aung & P. Ordoñez de Pablos (Eds.), *Managerial Strategies and Practice in the Asian Business Sector* (pp. 107–123). Hershey, PA: IGI Global. doi:10.4018/978-1-4666-9758-4.ch007

Bharwani, S., & Musunuri, D. (2018). Reflection as a Process From Theory to Practice. In M. Khosrow-Pour, D.B.A. (Ed.), Encyclopedia of Information Science and Technology, Fourth Edition (pp. 1529-1539). Hershey, PA: IGI Global. doi:10.4018/978-1-5225-2255-3.ch132

Bhatt, G. D., Wang, Z., & Rodger, J. A. (2017). Information Systems Capabilities and Their Effects on Competitive Advantages: A Study of Chinese Companies. *Information Resources Management Journal*, *30*(3), 41–57. doi:10.4018/IRMJ.2017070103

Bhushan, M., & Yadav, A. (2017). Concept of Cloud Computing in ESB. In R. Bhadoria, N. Chaudhari, G. Tomar, & S. Singh (Eds.), *Exploring Enterprise Service Bus in the Service-Oriented Architecture Paradigm* (pp. 116–127). Hershey, PA: IGI Global. doi:10.4018/978-1-5225-2157-0.ch008

Bhushan, S. (2017). System Dynamics Base-Model of Humanitarian Supply Chain (HSCM) in Disaster Prone Eco-Communities of India: A Discussion on Simulation and Scenario Results. *International Journal of System Dynamics Applications*, *6*(3), 20–37. doi:10.4018/IJSDA.2017070102

Biswas, A., & De, A. K. (2017). On Development of a Fuzzy Stochastic Programming Model with Its Application to Business Management. In S. Trivedi, S. Dey, A. Kumar, & T. Panda (Eds.), *Handbook of Research on Advanced Data Mining Techniques and Applications for Business Intelligence* (pp. 353–378). Hershey, PA: IGI Global. doi:10.4018/978-1-5225-2031-3.ch021

Bücker, J., & Ernste, K. (2018). Use of Brand Heroes in Strategic Reputation Management: The Case of Bacardi, Adidas, and Daimler. In A. Erdemir (Ed.), *Reputation Management Techniques in Public Relations* (pp. 126–150). Hershey, PA: IGI Global. doi:10.4018/978-1-5225-3619-2.ch007

Bureš, V. (2018). Industry 4.0 From the Systems Engineering Perspective: Alternative Holistic Framework Development. In R. Brunet-Thornton & F. Martinez (Eds.), *Analyzing the Impacts of Industry 4.0 in Modern Business Environments* (pp. 199–223). Hershey, PA: IGI Global. doi:10.4018/978-1-5225-3468-6.ch011

Buzady, Z. (2017). Resolving the Magic Cube of Effective Case Teaching: Benchmarking Case Teaching Practices in Emerging Markets – Insights from the Central European University Business School, Hungary. In D. Latusek (Ed.), *Case Studies as a Teaching Tool in Management Education* (pp. 79–103). Hershey, PA: IGI Global. doi:10.4018/978-1-5225-0770-3.ch005

Campatelli, G., Richter, A., & Stocker, A. (2016). Participative Knowledge Management to Empower Manufacturing Workers. *International Journal of Knowledge Management, 12*(4), 37–50. doi:10.4018/IJKM.2016100103

Căpusneanu, S., & Topor, D. I. (2018). Business Ethics and Cost Management in SMEs: Theories of Business Ethics and Cost Management Ethos. In I. Oncioiu (Ed.), *Ethics and Decision-Making for Sustainable Business Practices* (pp. 109–127). Hershey, PA: IGI Global. doi:10.4018/978-1-5225-3773-1.ch007

Carneiro, A. (2016). Maturity in Health Organization Information Systems: Metrics and Privacy Perspectives. *International Journal of Privacy and Health Information Management, 4*(2), 1–18. doi:10.4018/IJPHIM.2016070101

Chan, R. L., Mo, P. L., & Moon, K. K. (2018). Strategic and Tactical Measures in Managing Enterprise Risks: A Study of the Textile and Apparel Industry. In K. Strang, M. Korstanje, & N. Vajjhala (Eds.), *Research, Practices, and Innovations in Global Risk and Contingency Management* (pp. 1–19). Hershey, PA: IGI Global. doi:10.4018/978-1-5225-4754-9.ch001

Chandan, H. C. (2016). Motivations and Challenges of Female Entrepreneurship in Developed and Developing Economies. In N. Baporikar (Ed.), *Handbook of Research on Entrepreneurship in the Contemporary Knowledge-Based Global Economy* (pp. 260–286). Hershey, PA: IGI Global. doi:10.4018/978-1-4666-8798-1.ch012

Charlier, S. D., Burke-Smalley, L. A., & Fisher, S. L. (2018). Undergraduate Programs in the U.S: A Contextual and Content-Based Analysis. In J. Mendy (Ed.), *Teaching Human Resources and Organizational Behavior at the College Level* (pp. 26–57). Hershey, PA: IGI Global. doi:10.4018/978-1-5225-2820-3.ch002

Chaudhuri, S. (2016). Application of Web-Based Geographical Information System (GIS) in E-Business. In U. Panwar, R. Kumar, & N. Ray (Eds.), *Handbook of Research on Promotional Strategies and Consumer Influence in the Service Sector* (pp. 389–405). Hershey, PA: IGI Global. doi:10.4018/978-1-5225-0143-5.ch023

Choudhuri, P. S. (2016). An Empirical Study on the Quality of Services Offered by the Private Life Insurers in Burdwan. In U. Panwar, R. Kumar, & N. Ray (Eds.), *Handbook of Research on Promotional Strategies and Consumer Influence in the Service Sector* (pp. 31–55). Hershey, PA: IGI Global. doi:10.4018/978-1-5225-0143-5.ch002

Dahlberg, T., Kivijärvi, H., & Saarinen, T. (2017). IT Investment Consistency and Other Factors Influencing the Success of IT Performance. In S. De Haes & W. Van Grembergen (Eds.), *Strategic IT Governance and Alignment in Business Settings* (pp. 176–208). Hershey, PA: IGI Global. doi:10.4018/978-1-5225-0861-8.ch007

Damnjanović, A. M. (2017). Knowledge Management Optimization through IT and E-Business Utilization: A Qualitative Study on Serbian SMEs. In M. Vemić (Ed.), *Optimal Management Strategies in Small and Medium Enterprises* (pp. 249–267). Hershey, PA: IGI Global. doi:10.4018/978-1-5225-1949-2.ch012

Daneshpour, H. (2017). Integrating Sustainable Development into Project Portfolio Management through Application of Open Innovation. In M. Vemić (Ed.), *Optimal Management Strategies in Small and Medium Enterprises* (pp. 370–387). Hershey, PA: IGI Global. doi:10.4018/978-1-5225-1949-2.ch017

Daniel, A. D., & Reis de Castro, V. (2018). Entrepreneurship Education: How to Measure the Impact on Nascent Entrepreneurs. In A. Carrizo Moreira, J. Guilherme Leitão Dantas, & F. Manuel Valente (Eds.), *Nascent Entrepreneurship and Successful New Venture Creation* (pp. 85–110). Hershey, PA: IGI Global. doi:10.4018/978-1-5225-2936-1.ch004

David, F., van der Sijde, P., & van den Besselaar, P. (2016). Enterpreneurial Incentives, Obstacles, and Management in University-Business Co-Operation: The Case of Indonesia. In J. Saiz-Álvarez (Ed.), *Handbook of Research on Social Entrepreneurship and Solidarity Economics* (pp. 499–518). Hershey, PA: IGI Global. doi:10.4018/978-1-5225-0097-1.ch024

David, R., Swami, B. N., & Tangirala, S. (2018). Ethics Impact on Knowledge Management in Organizational Development: A Case Study. In N. Baporikar (Ed.), *Global Practices in Knowledge Management for Societal and Organizational Development* (pp. 19–45). Hershey, PA: IGI Global. doi:10.4018/978-1-5225-3009-1.ch002

Delias, P., & Lakiotaki, K. (2018). Discovering Process Horizontal Boundaries to Facilitate Process Comprehension. *International Journal of Operations Research and Information Systems, 9*(2), 1–31. doi:10.4018/IJORIS.2018040101

Denholm, J., & Lee-Davies, L. (2018). Success Factors for Games in Business and Project Management. In *Enhancing Education and Training Initiatives Through Serious Games* (pp. 34–68). Hershey, PA: IGI Global. doi:10.4018/978-1-5225-3689-5.ch002

Deshpande, M. (2017). Best Practices in Management Institutions for Global Leadership: Policy Aspects. In N. Baporikar (Ed.), *Management Education for Global Leadership* (pp. 1–27). Hershey, PA: IGI Global. doi:10.4018/978-1-5225-1013-0.ch001

Deshpande, M. (2018). Policy Perspectives for SMEs Knowledge Management. In N. Baporikar (Ed.), *Knowledge Integration Strategies for Entrepreneurship and Sustainability* (pp. 23–46). Hershey, PA: IGI Global. doi:10.4018/978-1-5225-5115-7.ch002

Dezdar, S. (2017). ERP Implementation Projects in Asian Countries: A Comparative Study on Iran and China. *International Journal of Information Technology Project Management, 8*(3), 52–68. doi:10.4018/IJITPM.2017070104

Domingos, D., Martinho, R., & Varajão, J. (2016). Controlled Flexibility in Healthcare Processes: A BPMN-Extension Approach. In M. Cruz-Cunha, I. Miranda, R. Martinho, & R. Rijo (Eds.), *Encyclopedia of E-Health and Telemedicine* (pp. 521–535). Hershey, PA: IGI Global. doi:10.4018/978-1-4666-9978-6.ch040

Domingos, D., Respício, A., & Martinho, R. (2017). Reliability of IoT-Aware BPMN Healthcare Processes. In C. Reis & M. Maximiano (Eds.), *Internet of Things and Advanced Application in Healthcare* (pp. 214–248). Hershey, PA: IGI Global. doi:10.4018/978-1-5225-1820-4.ch008

Dosumu, O., Hussain, J., & El-Gohary, H. (2017). An Exploratory Study of the Impact of Government Policies on the Development of Small and Medium Enterprises in Developing Countries: The Case of Nigeria. *International Journal of Customer Relationship Marketing and Management*, *8*(4), 51–62. doi:10.4018/IJCRMM.2017100104

Durst, S., Bruns, G., & Edvardsson, I. R. (2017). Retaining Knowledge in Smaller Building and Construction Firms. *International Journal of Knowledge and Systems Science*, *8*(3), 1–12. doi:10.4018/IJKSS.2017070101

Edvardsson, I. R., & Durst, S. (2017). Outsourcing, Knowledge, and Learning: A Critical Review. *International Journal of Knowledge-Based Organizations*, *7*(2), 13–26. doi:10.4018/IJKBO.2017040102

Edwards, J. S. (2018). Integrating Knowledge Management and Business Processes. In M. Khosrow-Pour, D.B.A. (Ed.), Encyclopedia of Information Science and Technology, Fourth Edition (pp. 5046-5055). Hershey, PA: IGI Global. doi:10.4018/978-1-5225-2255-3.ch437

Ejiogu, A. O. (2018). Economics of Farm Management. In *Agricultural Finance and Opportunities for Investment and Expansion* (pp. 56–72). Hershey, PA: IGI Global. doi:10.4018/978-1-5225-3059-6.ch003

Ekanem, I., & Abiade, G. E. (2018). Factors Influencing the Use of E-Commerce by Small Enterprises in Nigeria. *International Journal of ICT Research in Africa and the Middle East*, *7*(1), 37–53. doi:10.4018/IJICTRAME.2018010103

Ekanem, I., & Alrossais, L. A. (2017). Succession Challenges Facing Family Businesses in Saudi Arabia. In P. Zgheib (Ed.), *Entrepreneurship and Business Innovation in the Middle East* (pp. 122–146). Hershey, PA: IGI Global. doi:10.4018/978-1-5225-2066-5.ch007

El Faquih, L., & Fredj, M. (2017). Ontology-Based Framework for Quality in Configurable Process Models. *Journal of Electronic Commerce in Organizations*, *15*(2), 48–60. doi:10.4018/JECO.2017040104

El-Gohary, H., & El-Gohary, Z. (2016). An Attempt to Explore Electronic Marketing Adoption and Implementation Aspects in Developing Countries: The Case of Egypt. *International Journal of Customer Relationship Marketing and Management*, *7*(4), 1–26. doi:10.4018/IJCRMM.2016100101

Entico, G. J. (2016). Knowledge Management and the Medical Health Librarians: A Perception Study. In J. Yap, M. Perez, M. Ayson, & G. Entico (Eds.), *Special Library Administration, Standardization and Technological Integration* (pp. 52–77). Hershey, PA: IGI Global. doi:10.4018/978-1-4666-9542-9.ch003

Faisal, M. N., & Talib, F. (2017). Building Ambidextrous Supply Chains in SMEs: How to Tackle the Barriers? *International Journal of Information Systems and Supply Chain Management, 10*(4), 80–100. doi:10.4018/IJISSCM.2017100105

Fernandes, T. M., Gomes, J., & Romão, M. (2017). Investments in E-Government: A Benefit Management Case Study. *International Journal of Electronic Government Research, 13*(3), 1–17. doi:10.4018/IJEGR.2017070101

Fouda, F. A. (2016). A Suggested Curriculum in Career Education to Develop Business Secondary Schools Students' Career Knowledge Management Domains and Professional Thinking. *International Journal of Technology Diffusion, 7*(2), 42–62. doi:10.4018/IJTD.2016040103

Gallardo-Vázquez, D., & Pajuelo-Moreno, M. L. (2016). How Spanish Universities are Promoting Entrepreneurship through Your Own Lines of Teaching and Research? In L. Carvalho (Ed.), *Handbook of Research on Entrepreneurial Success and its Impact on Regional Development* (pp. 431–454). Hershey, PA: IGI Global. doi:10.4018/978-1-4666-9567-2.ch019

Gao, S. S., Oreal, S., & Zhang, J. (2018). Contemporary Financial Risk Management Perceptions and Practices of Small-Sized Chinese Businesses. In I. Management Association (Ed.), Global Business Expansion: Concepts, Methodologies, Tools, and Applications (pp. 917-931). Hershey, PA: IGI Global. doi:10.4018/978-1-5225-5481-3.ch041

Garg, R., & Berning, S. C. (2017). Indigenous Chinese Management Philosophies: Key Concepts and Relevance for Modern Chinese Firms. In B. Christiansen & G. Koc (Eds.), *Transcontinental Strategies for Industrial Development and Economic Growth* (pp. 43–57). Hershey, PA: IGI Global. doi:10.4018/978-1-5225-2160-0.ch003

Gencer, Y. G. (2017). Supply Chain Management in Retailing Business. In U. Akkucuk (Ed.), *Ethics and Sustainability in Global Supply Chain Management* (pp. 197–210). Hershey, PA: IGI Global. doi:10.4018/978-1-5225-2036-8.ch011

Giacosa, E. (2016). Innovation in Luxury Fashion Businesses as a Means for the Regional Development. In L. Carvalho (Ed.), *Handbook of Research on Entrepreneurial Success and its Impact on Regional Development* (pp. 206–222). Hershey, PA: IGI Global. doi:10.4018/978-1-4666-9567-2.ch010

Giacosa, E. (2018). The Increasing of the Regional Development Thanks to the Luxury Business Innovation. In L. Carvalho (Ed.), *Handbook of Research on Entrepreneurial Ecosystems and Social Dynamics in a Globalized World* (pp. 260–273). Hershey, PA: IGI Global. doi:10.4018/978-1-5225-3525-6.ch011

Gianni, M., & Gotzamani, K. (2016). Integrated Management Systems and Information Management Systems: Common Threads. In P. Papajorgji, F. Pinet, A. Guimarães, & J. Papathanasiou (Eds.), *Automated Enterprise Systems for Maximizing Business Performance* (pp. 195–214). Hershey, PA: IGI Global. doi:10.4018/978-1-4666-8841-4.ch011

Gianni, M., Gotzamani, K., & Linden, I. (2016). How a BI-wise Responsible Integrated Management System May Support Food Traceability. *International Journal of Decision Support System Technology*, *8*(2), 1–17. doi:10.4018/IJDSST.2016040101

Glykas, M., & George, J. (2017). Quality and Process Management Systems in the UAE Maritime Industry. *International Journal of Productivity Management and Assessment Technologies*, *5*(1), 20–39. doi:10.4018/IJPMAT.2017010102

Glykas, M., Valiris, G., Kokkinaki, A., & Koutsoukou, Z. (2018). Banking Business Process Management Implementation. *International Journal of Productivity Management and Assessment Technologies*, *6*(1), 50–69. doi:10.4018/IJPMAT.2018010104

Gomes, J., & Romão, M. (2017). The Balanced Scorecard: Keeping Updated and Aligned with Today's Business Trends. *International Journal of Productivity Management and Assessment Technologies*, *5*(2), 1–15. doi:10.4018/IJPMAT.2017070101

Gomes, J., & Romão, M. (2017). Aligning Information Systems and Technology with Benefit Management and Balanced Scorecard. In S. De Haes & W. Van Grembergen (Eds.), *Strategic IT Governance and Alignment in Business Settings* (pp. 112–131). Hershey, PA: IGI Global. doi:10.4018/978-1-5225-0861-8.ch005

Grefen, P., & Turetken, O. (2017). Advanced Business Process Management in Networked E-Business Scenarios. *International Journal of E-Business Research*, *13*(4), 70–104. doi:10.4018/IJEBR.2017100105

Haider, A., & Saetang, S. (2017). Strategic IT Alignment in Service Sector. In S. Rozenes & Y. Cohen (Eds.), *Handbook of Research on Strategic Alliances and Value Co-Creation in the Service Industry* (pp. 231–258). Hershey, PA: IGI Global. doi:10.4018/978-1-5225-2084-9.ch012

Haider, A., & Tang, S. S. (2016). Maximising Value Through IT and Business Alignment: A Case of IT Governance Institutionalisation at a Thai Bank. *International Journal of Technology Diffusion*, *7*(3), 33–58. doi:10.4018/IJTD.2016070104

Hajilari, A. B., Ghadaksaz, M., & Fasghandis, G. S. (2017). Assessing Organizational Readiness for Implementing ERP System Using Fuzzy Expert System Approach. *International Journal of Enterprise Information Systems*, *13*(1), 67–85. doi:10.4018/IJEIS.2017010105

Haldorai, A., Ramu, A., & Murugan, S. (2018). Social Aware Cognitive Radio Networks: Effectiveness of Social Networks as a Strategic Tool for Organizational Business Management. In H. Bansal, G. Shrivastava, G. Nguyen, & L. Stanciu (Eds.), *Social Network Analytics for Contemporary Business Organizations* (pp. 188–202). Hershey, PA: IGI Global. doi:10.4018/978-1-5225-5097-6.ch010

Hall, O. P. Jr. (2017). Social Media Driven Management Education. *International Journal of Knowledge-Based Organizations*, *7*(2), 43–59. doi:10.4018/IJKBO.2017040104

Hanifah, H., Halim, H. A., Ahmad, N. H., & Vafaei-Zadeh, A. (2017). Innovation Culture as a Mediator Between Specific Human Capital and Innovation Performance Among Bumiputera SMEs in Malaysia. In N. Ahmad, T. Ramayah, H. Halim, & S. Rahman (Eds.), *Handbook of Research on Small and Medium Enterprises in Developing Countries* (pp. 261–279). Hershey, PA: IGI Global. doi:10.4018/978-1-5225-2165-5.ch012

Hartlieb, S., & Silvius, G. (2017). Handling Uncertainty in Project Management and Business Development: Similarities and Differences. In Y. Raydugin (Ed.), *Handbook of Research on Leveraging Risk and Uncertainties for Effective Project Management* (pp. 337–362). Hershey, PA: IGI Global. doi:10.4018/978-1-5225-1790-0.ch016

Hass, K. B. (2017). Living on the Edge: Managing Project Complexity. In Y. Raydugin (Ed.), *Handbook of Research on Leveraging Risk and Uncertainties for Effective Project Management* (pp. 177–201). Hershey, PA: IGI Global. doi:10.4018/978-1-5225-1790-0.ch009

Hassan, A., & Privitera, D. S. (2016). Google AdSense as a Mobile Technology in Education. In J. Holland (Ed.), *Wearable Technology and Mobile Innovations for Next-Generation Education* (pp. 200–223). Hershey, PA: IGI Global. doi:10.4018/978-1-5225-0069-8.ch011

Hassan, A., & Rahimi, R. (2016). Consuming "Innovation" in Tourism: Augmented Reality as an Innovation Tool in Digital Tourism Marketing. In N. Pappas & I. Bregoli (Eds.), *Global Dynamics in Travel, Tourism, and Hospitality* (pp. 130–147). Hershey, PA: IGI Global. doi:10.4018/978-1-5225-0201-2.ch008

Hawking, P., & Carmine Sellitto, C. (2017). Developing an Effective Strategy for Organizational Business Intelligence. In M. Tavana (Ed.), *Enterprise Information Systems and the Digitalization of Business Functions* (pp. 222–237). Hershey, PA: IGI Global. doi:10.4018/978-1-5225-2382-6.ch010

Hawking, P., & Sellitto, C. (2017). A Fast-Moving Consumer Goods Company and Business Intelligence Strategy Development. *International Journal of Enterprise Information Systems*, *13*(2), 22–33. doi:10.4018/IJEIS.2017040102

Hawking, P., & Sellitto, C. (2017). Business Intelligence Strategy: Two Case Studies. *International Journal of Business Intelligence Research*, *8*(2), 17–30. doi:10.4018/IJBIR.2017070102

Haynes, J. D., Arockiasamy, S., Al Rashdi, M., & Al Rashdi, S. (2016). Business and E Business Strategies for Coopetition and Thematic Management as a Sustained Basis for Ethics and Social Responsibility in Emerging Markets. In M. Al-Shammari & H. Masri (Eds.), *Ethical and Social Perspectives on Global Business Interaction in Emerging Markets* (pp. 25–39). Hershey, PA: IGI Global. doi:10.4018/978-1-4666-9864-2.ch002

Hee, W. J., Jalleh, G., Lai, H., & Lin, C. (2017). E-Commerce and IT Projects: Evaluation and Management Issues in Australian and Taiwanese Hospitals. *International Journal of Public Health Management and Ethics*, *2*(1), 69–90. doi:10.4018/IJPHME.2017010104

Hernandez, A. A. (2018). Exploring the Factors to Green IT Adoption of SMEs in the Philippines. *Journal of Cases on Information Technology*, *20*(2), 49–66. doi:10.4018/JCIT.2018040104

Hernandez, A. A., & Ona, S. E. (2016). Green IT Adoption: Lessons from the Philippines Business Process Outsourcing Industry. *International Journal of Social Ecology and Sustainable Development*, *7*(1), 1–34. doi:10.4018/IJSESD.2016010101

Hollman, A., Bickford, S., & Hollman, T. (2017). Cyber InSecurity: A Post-Mortem Attempt to Assess Cyber Problems from IT and Business Management Perspectives. *Journal of Cases on Information Technology*, *19*(3), 42–70. doi:10.4018/JCIT.2017070104

Igbinakhase, I. (2017). Responsible and Sustainable Management Practices in Developing and Developed Business Environments. In Z. Fields (Ed.), *Collective Creativity for Responsible and Sustainable Business Practice* (pp. 180–207). Hershey, PA: IGI Global. doi:10.4018/978-1-5225-1823-5.ch010

Ilahi, L., Ghannouchi, S. A., & Martinho, R. (2016). A Business Process Management Approach to Home Healthcare Processes: On the Gap between Intention and Reality. In M. Cruz-Cunha, I. Miranda, R. Martinho, & R. Rijo (Eds.), *Encyclopedia of E-Health and Telemedicine* (pp. 439–457). Hershey, PA: IGI Global. doi:10.4018/978-1-4666-9978-6.ch035

Iwata, J. J., & Hoskins, R. G. (2017). Managing Indigenous Knowledge in Tanzania: A Business Perspective. In P. Jain & N. Mnjama (Eds.), *Managing Knowledge Resources and Records in Modern Organizations* (pp. 198–214). Hershey, PA: IGI Global. doi:10.4018/978-1-5225-1965-2.ch012

Jabeen, F., Ahmad, S. Z., & Alkaabi, S. (2016). The Internationalization Decision-Making of United Arab Emirates Family Businesses. In N. Zakaria, A. Abdul-Talib, & N. Osman (Eds.), *Handbook of Research on Impacts of International Business and Political Affairs on the Global Economy* (pp. 1–22). Hershey, PA: IGI Global. doi:10.4018/978-1-4666-9806-2.ch001

Jain, P. (2017). Ethical and Legal Issues in Knowledge Management Life-Cycle in Business. In P. Jain & N. Mnjama (Eds.), *Managing Knowledge Resources and Records in Modern Organizations* (pp. 82–101). Hershey, PA: IGI Global. doi:10.4018/978-1-5225-1965-2.ch006

Jamali, D., Abdallah, H., & Matar, F. (2016). Opportunities and Challenges for CSR Mainstreaming in Business Schools. *International Journal of Technology and Educational Marketing*, *6*(2), 1–29. doi:10.4018/IJTEM.2016070101

James, S., & Hauli, E. (2017). Holistic Management Education at Tanzanian Rural Development Planning Institute. In N. Baporikar (Ed.), *Management Education for Global Leadership* (pp. 112–136). Hershey, PA: IGI Global. doi:10.4018/978-1-5225-1013-0.ch006

Janošková, M., Csikósová, A., & Čulková, K. (2018). Measurement of Company Performance as Part of Its Strategic Management. In R. Leon (Ed.), *Managerial Strategies for Business Sustainability During Turbulent Times* (pp. 309–335). Hershey, PA: IGI Global. doi:10.4018/978-1-5225-2716-9.ch017

Jean-Vasile, A., & Alecu, A. (2017). Theoretical and Practical Approaches in Understanding the Influences of Cost-Productivity-Profit Trinomial in Contemporary Enterprises. In A. Jean Vasile & D. Nicolò (Eds.), *Sustainable Entrepreneurship and Investments in the Green Economy* (pp. 28–62). Hershey, PA: IGI Global. doi:10.4018/978-1-5225-2075-7.ch002

Jha, D. G. (2016). Preparing for Information Technology Driven Changes. In S. Tiwari & L. Nafees (Eds.), *Innovative Management Education Pedagogies for Preparing Next-Generation Leaders* (pp. 258–274). Hershey, PA: IGI Global. doi:10.4018/978-1-4666-9691-4.ch015

Joia, L. A., & Correia, J. C. (2018). CIO Competencies From the IT Professional Perspective: Insights From Brazil. *Journal of Global Information Management, 26*(2), 74–103. doi:10.4018/JGIM.2018040104

Juma, A., & Mzera, N. (2017). Knowledge Management and Records Management and Competitive Advantage in Business. In P. Jain & N. Mnjama (Eds.), *Managing Knowledge Resources and Records in Modern Organizations* (pp. 15–28). Hershey, PA: IGI Global. doi:10.4018/978-1-5225-1965-2.ch002

K., I., & A, V. (2018). Monitoring and Auditing in the Cloud. In K. Munir (Ed.), *Cloud Computing Technologies for Green Enterprises* (pp. 318-350). Hershey, PA: IGI Global. doi:10.4018/978-1-5225-3038-1.ch013

Kabra, G., Ghosh, V., & Ramesh, A. (2018). Enterprise Integrated Business Process Management and Business Intelligence Framework for Business Process Sustainability. In A. Paul, D. Bhattacharyya, & S. Anand (Eds.), *Green Initiatives for Business Sustainability and Value Creation* (pp. 228–238). Hershey, PA: IGI Global. doi:10.4018/978-1-5225-2662-9.ch010

Kaoud, M. (2017). Investigation of Customer Knowledge Management: A Case Study Research. *International Journal of Service Science, Management, Engineering, and Technology, 8*(2), 12–22. doi:10.4018/IJSSMET.2017040102

Kara, M. E., & Fırat, S. Ü. (2016). Sustainability, Risk, and Business Intelligence in Supply Chains. In M. Erdoğdu, T. Arun, & I. Ahmad (Eds.), *Handbook of Research on Green Economic Development Initiatives and Strategies* (pp. 501–538). Hershey, PA: IGI Global. doi:10.4018/978-1-5225-0440-5.ch022

Katuu, S. (2018). A Comparative Assessment of Enterprise Content Management Maturity Models. In N. Gwangwava & M. Mutingi (Eds.), *E-Manufacturing and E-Service Strategies in Contemporary Organizations* (pp. 93–118). Hershey, PA: IGI Global. doi:10.4018/978-1-5225-3628-4.ch005

Khan, M. A. (2016). MNEs Management Strategies in Developing Countries: Establishing the Context. In M. Khan (Ed.), *Multinational Enterprise Management Strategies in Developing Countries* (pp. 1–33). Hershey, PA: IGI Global. doi:10.4018/978-1-5225-0276-0.ch001

Khan, M. A. (2016). Operational Approaches in Organizational Structure: A Case for MNEs in Developing Countries. In M. Khan (Ed.), *Multinational Enterprise Management Strategies in Developing Countries* (pp. 129–151). Hershey, PA: IGI Global. doi:10.4018/978-1-5225-0276-0.ch007

Kinnunen, S., Ylä-Kujala, A., Marttonen-Arola, S., Kärri, T., & Baglee, D. (2018). Internet of Things in Asset Management: Insights from Industrial Professionals and Academia. *International Journal of Service Science, Management, Engineering, and Technology*, *9*(2), 104–119. doi:10.4018/IJSSMET.2018040105

Klein, A. Z., Sabino de Freitas, A., Machado, L., Freitas, J. C. Jr, Graziola, P. G. Jr, & Schlemmer, E. (2017). Virtual Worlds Applications for Management Education. In L. Tomei (Ed.), *Exploring the New Era of Technology-Infused Education* (pp. 279–299). Hershey, PA: IGI Global. doi:10.4018/978-1-5225-1709-2.ch017

Kożuch, B., & Jabłoński, A. (2017). Adopting the Concept of Business Models in Public Management. In M. Lewandowski & B. Kożuch (Eds.), *Public Sector Entrepreneurship and the Integration of Innovative Business Models* (pp. 10–46). Hershey, PA: IGI Global. doi:10.4018/978-1-5225-2215-7.ch002

Kumar, J., Adhikary, A., & Jha, A. (2017). Small Active Investors' Perceptions and Preferences Towards Tax Saving Mutual Fund Schemes in Eastern India: An Empirical Note. *International Journal of Asian Business and Information Management*, *8*(2), 35–45. doi:10.4018/IJABIM.2017040103

Lassoued, Y., Bouzguenda, L., & Mahmoud, T. (2016). Context-Aware Business Process Versions Management. *International Journal of e-Collaboration*, *12*(3), 7–33. doi:10.4018/IJeC.2016070102

Lavassani, K. M., & Movahedi, B. (2017). Applications Driven Information Systems: Beyond Networks toward Business Ecosystems. *International Journal of Innovation in the Digital Economy*, *8*(1), 61–75. doi:10.4018/IJIDE.2017010104

Lazzareschi, V. H., & Brito, M. S. (2017). Strategic Information Management: Proposal of Business Project Model. In G. Jamil, A. Soares, & C. Pessoa (Eds.), *Handbook of Research on Information Management for Effective Logistics and Supply Chains* (pp. 59–88). Hershey, PA: IGI Global. doi:10.4018/978-1-5225-0973-8.ch004

Lederer, M., Kurz, M., & Lazarov, P. (2017). Usage and Suitability of Methods for Strategic Business Process Initiatives: A Multi Case Study Research. *International Journal of Productivity Management and Assessment Technologies*, *5*(1), 40–51. doi:10.4018/IJPMAT.2017010103

Lee, I. (2017). A Social Enterprise Business Model and a Case Study of Pacific Community Ventures (PCV). In V. Potocan, M. Ünğan, & Z. Nedelko (Eds.), *Handbook of Research on Managerial Solutions in Non-Profit Organizations* (pp. 182–204). Hershey, PA: IGI Global. doi:10.4018/978-1-5225-0731-4.ch009

Lee, L. J., & Leu, J. (2016). Exploring the Effectiveness of IT Application and Value Method in the Innovation Performance of Enterprise. *International Journal of Enterprise Information Systems*, *12*(2), 47–65. doi:10.4018/IJEIS.2016040104

Lee, Y. (2016). Alignment Effect of Entrepreneurial Orientation and Marketing Orientation on Firm Performance. *International Journal of Customer Relationship Marketing and Management*, *7*(4), 58–69. doi:10.4018/IJCRMM.2016100104

Leon, L. A., Seal, K. C., Przasnyski, Z. H., & Wiedenman, I. (2017). Skills and Competencies Required for Jobs in Business Analytics: A Content Analysis of Job Advertisements Using Text Mining. *International Journal of Business Intelligence Research*, *8*(1), 1–25. doi:10.4018/IJBIR.2017010101

Leu, J., Lee, L. J., & Krischke, A. (2016). Value Engineering-Based Method for Implementing the ISO14001 System in the Green Supply Chains. *International Journal of Strategic Decision Sciences*, *7*(4), 1–20. doi:10.4018/IJSDS.2016100101

Levy, C. L., & Elias, N. I. (2017). SOHO Users' Perceptions of Reliability and Continuity of Cloud-Based Services. In M. Moore (Ed.), *Cybersecurity Breaches and Issues Surrounding Online Threat Protection* (pp. 248–287). Hershey, PA: IGI Global. doi:10.4018/978-1-5225-1941-6.ch011

Levy, M. (2018). Change Management Serving Knowledge Management and Organizational Development: Reflections and Review. In N. Baporikar (Ed.), *Global Practices in Knowledge Management for Societal and Organizational Development* (pp. 256–270). Hershey, PA: IGI Global. doi:10.4018/978-1-5225-3009-1.ch012

Lewandowski, M. (2017). Public Organizations and Business Model Innovation: The Role of Public Service Design. In M. Lewandowski & B. Kożuch (Eds.), *Public Sector Entrepreneurship and the Integration of Innovative Business Models* (pp. 47–72). Hershey, PA: IGI Global. doi:10.4018/978-1-5225-2215-7.ch003

Lhannaoui, H., Kabbaj, M. I., & Bakkoury, Z. (2017). A Survey of Risk-Aware Business Process Modelling. *International Journal of Risk and Contingency Management, 6*(3), 14–26. doi:10.4018/IJRCM.2017070102

Li, J., Sun, W., Jiang, W., Yang, H., & Zhang, L. (2017). How the Nature of Exogenous Shocks and Crises Impact Company Performance?: The Effects of Industry Characteristics. *International Journal of Risk and Contingency Management, 6*(4), 40–55. doi:10.4018/IJRCM.2017100103

Lu, C., & Liu, S. (2016). Cultural Tourism O2O Business Model Innovation-A Case Study of CTrip. *Journal of Electronic Commerce in Organizations, 14*(2), 16–31. doi:10.4018/JECO.2016040102

Machen, B., Hosseini, M. R., Wood, A., & Bakhshi, J. (2016). An Investigation into using SAP-PS as a Multidimensional Project Control System (MPCS). *International Journal of Enterprise Information Systems, 12*(2), 66–81. doi:10.4018/IJEIS.2016040105

Malega, P. (2017). Small and Medium Enterprises in the Slovak Republic: Status and Competitiveness of SMEs in the Global Markets and Possibilities of Optimization. In M. Vemić (Ed.), *Optimal Management Strategies in Small and Medium Enterprises* (pp. 102–124). Hershey, PA: IGI Global. doi:10.4018/978-1-5225-1949-2.ch006

Malewska, K. M. (2017). Intuition in Decision-Making on the Example of a Non-Profit Organization. In V. Potocan, M. Üngan, & Z. Nedelko (Eds.), *Handbook of Research on Managerial Solutions in Non-Profit Organizations* (pp. 378–399). Hershey, PA: IGI Global. doi:10.4018/978-1-5225-0731-4.ch018

Maroofi, F. (2017). Entrepreneurial Orientation and Organizational Learning Ability Analysis for Innovation and Firm Performance. In N. Baporikar (Ed.), *Innovation and Shifting Perspectives in Management Education* (pp. 144–165). Hershey, PA: IGI Global. doi:10.4018/978-1-5225-1019-2.ch007

Martins, P. V., & Zacarias, M. (2017). A Web-based Tool for Business Process Improvement. *International Journal of Web Portals, 9*(2), 68–84. doi:10.4018/IJWP.2017070104

Matthies, B., & Coners, A. (2017). Exploring the Conceptual Nature of e-Business Projects. *Journal of Electronic Commerce in Organizations, 15*(3), 33–63. doi:10.4018/JECO.2017070103

McKee, J. (2018). Architecture as a Tool to Solve Business Planning Problems. In M. Khosrow-Pour, D.B.A. (Ed.), Encyclopedia of Information Science and Technology, Fourth Edition (pp. 573-586). Hershey, PA: IGI Global. doi:10.4018/978-1-5225-2255-3.ch050

McMurray, A. J., Cross, J., & Caponecchia, C. (2018). The Risk Management Profession in Australia: Business Continuity Plan Practices. In N. Bajgoric (Ed.), *Always-On Enterprise Information Systems for Modern Organizations* (pp. 112–129). Hershey, PA: IGI Global. doi:10.4018/978-1-5225-3704-5.ch006

Meddah, I. H., & Belkadi, K. (2018). Mining Patterns Using Business Process Management. In R. Hamou (Ed.), *Handbook of Research on Biomimicry in Information Retrieval and Knowledge Management* (pp. 78–89). Hershey, PA: IGI Global. doi:10.4018/978-1-5225-3004-6.ch005

Mendes, L. (2017). TQM and Knowledge Management: An Integrated Approach Towards Tacit Knowledge Management. In D. Jaziri-Bouagina & G. Jamil (Eds.), *Handbook of Research on Tacit Knowledge Management for Organizational Success* (pp. 236–263). Hershey, PA: IGI Global. doi:10.4018/978-1-5225-2394-9.ch009

Mnjama, N. M. (2017). Preservation of Recorded Information in Public and Private Sector Organizations. In P. Jain & N. Mnjama (Eds.), *Managing Knowledge Resources and Records in Modern Organizations* (pp. 149–167). Hershey, PA: IGI Global. doi:10.4018/978-1-5225-1965-2.ch009

Mokoqama, M., & Fields, Z. (2017). Principles of Responsible Management Education (PRME): Call for Responsible Management Education. In Z. Fields (Ed.), *Collective Creativity for Responsible and Sustainable Business Practice* (pp. 229–241). Hershey, PA: IGI Global. doi:10.4018/978-1-5225-1823-5.ch012

Muniapan, B. (2017). Philosophy and Management: The Relevance of Vedanta in Management. In P. Ordóñez de Pablos (Ed.), *Managerial Strategies and Solutions for Business Success in Asia* (pp. 124–139). Hershey, PA: IGI Global. doi:10.4018/978-1-5225-1886-0.ch007

Muniapan, B., Gregory, M. L., & Ling, L. A. (2016). Marketing Education in Sarawak: Looking at It from the Employers' Viewpoint. In B. Smith & A. Porath (Eds.), *Global Perspectives on Contemporary Marketing Education* (pp. 112–130). Hershey, PA: IGI Global. doi:10.4018/978-1-4666-9784-3.ch008

Murad, S. E., & Dowaji, S. (2017). Using Value-Based Approach for Managing Cloud-Based Services. In A. Turuk, B. Sahoo, & S. Addya (Eds.), *Resource Management and Efficiency in Cloud Computing Environments* (pp. 33–60). Hershey, PA: IGI Global. doi:10.4018/978-1-5225-1721-4.ch002

Mutahar, A. M., Daud, N. M., Thurasamy, R., Isaac, O., & Abdulsalam, R. (2018). The Mediating of Perceived Usefulness and Perceived Ease of Use: The Case of Mobile Banking in Yemen. *International Journal of Technology Diffusion, 9*(2), 21–40. doi:10.4018/IJTD.2018040102

Naidoo, V. (2017). E-Learning and Management Education at African Universities. In N. Baporikar (Ed.), *Management Education for Global Leadership* (pp. 181–201). Hershey, PA: IGI Global. doi:10.4018/978-1-5225-1013-0.ch009

Naidoo, V., & Igbinakhase, I. (2018). Opportunities and Challenges of Knowledge Retention in SMEs. In N. Baporikar (Ed.), *Knowledge Integration Strategies for Entrepreneurship and Sustainability* (pp. 70–94). Hershey, PA: IGI Global. doi:10.4018/978-1-5225-5115-7.ch004

Nayak, S., & Prabhu, N. (2017). Paradigm Shift in Management Education: Need for a Cross Functional Perspective. In N. Baporikar (Ed.), *Management Education for Global Leadership* (pp. 241–255). Hershey, PA: IGI Global. doi:10.4018/978-1-5225-1013-0.ch012

Ndede-Amadi, A. A. (2016). Student Interest in the IS Specialization as Predictor of the Success Potential of New Information Systems Programmes within the Schools of Business in Kenyan Public Universities. *International Journal of Information Systems and Social Change*, *7*(2), 63–79. doi:10.4018/IJISSC.2016040104

Nedelko, Z., & Potocan, V. (2016). Management Practices for Processes Optimization: Case of Slovenia. In G. Alor-Hernández, C. Sánchez-Ramírez, & J. García-Alcaraz (Eds.), *Handbook of Research on Managerial Strategies for Achieving Optimal Performance in Industrial Processes* (pp. 545–561). Hershey, PA: IGI Global. doi:10.4018/978-1-5225-0130-5.ch025

Nedelko, Z., & Potocan, V. (2017). Management Solutions in Non-Profit Organizations: Case of Slovenia. In V. Potocan, M. Üngan, & Z. Nedelko (Eds.), *Handbook of Research on Managerial Solutions in Non-Profit Organizations* (pp. 1–22). Hershey, PA: IGI Global. doi:10.4018/978-1-5225-0731-4.ch001

Nedelko, Z., & Potocan, V. (2017). Priority of Management Tools Utilization among Managers: International Comparison. In V. Wang (Ed.), *Encyclopedia of Strategic Leadership and Management* (pp. 1083–1094). Hershey, PA: IGI Global. doi:10.4018/978-1-5225-1049-9.ch075

Nedelko, Z., Raudeliūnienė, J., & Črešnar, R. (2018). Knowledge Dynamics in Supply Chain Management. In N. Baporikar (Ed.), *Knowledge Integration Strategies for Entrepreneurship and Sustainability* (pp. 150–166). Hershey, PA: IGI Global. doi:10.4018/978-1-5225-5115-7.ch008

Nguyen, H. T., & Hipsher, S. A. (2018). Innovation and Creativity Used by Private Sector Firms in a Resources-Constrained Environment. In S. Hipsher (Ed.), *Examining the Private Sector's Role in Wealth Creation and Poverty Reduction* (pp. 219–238). Hershey, PA: IGI Global. doi:10.4018/978-1-5225-3117-3.ch010

Nycz, M., & Pólkowski, Z. (2016). Business Intelligence as a Modern IT Supporting Management of Local Government Units in Poland. *International Journal of Knowledge and Systems Science*, *7*(4), 1–18. doi:10.4018/IJKSS.2016100101

Obaji, N. O., Senin, A. A., & Olugu, M. U. (2016). Supportive Government Policy as a Mechanism for Business Incubation Performance in Nigeria. *International Journal of Information Systems and Social Change*, *7*(4), 52–66. doi:10.4018/IJISSC.2016100103

Obicci, P. A. (2017). Risk Sharing in a Partnership. In *Risk Management Strategies in Public-Private Partnerships* (pp. 115–152). Hershey, PA: IGI Global. doi:10.4018/978-1-5225-2503-5.ch004

Obidallah, W. J., & Raahemi, B. (2017). Managing Changes in Service Oriented Virtual Organizations: A Structural and Procedural Framework to Facilitate the Process of Change. *Journal of Electronic Commerce in Organizations*, *15*(1), 59–83. doi:10.4018/JECO.2017010104

Ojasalo, J., & Ojasalo, K. (2016). Service Logic Business Model Canvas for Lean Development of SMEs and Start-Ups. In N. Baporikar (Ed.), *Handbook of Research on Entrepreneurship in the Contemporary Knowledge-Based Global Economy* (pp. 217–243). Hershey, PA: IGI Global. doi:10.4018/978-1-4666-8798-1.ch010

Ojo, O. (2017). Impact of Innovation on the Entrepreneurial Success in Selected Business Enterprises in South-West Nigeria. *International Journal of Innovation in the Digital Economy*, *8*(2), 29–38. doi:10.4018/IJIDE.2017040103

Okdinawati, L., Simatupang, T. M., & Sunitiyoso, Y. (2017). Multi-Agent Reinforcement Learning for Value Co-Creation of Collaborative Transportation Management (CTM). *International Journal of Information Systems and Supply Chain Management*, *10*(3), 84–95. doi:10.4018/IJISSCM.2017070105

Ortner, E., Mevius, M., Wiedmann, P., & Kurz, F. (2016). Design of Interactional Decision Support Applications for E-Participation in Smart Cities. *International Journal of Electronic Government Research*, *12*(2), 18–38. doi:10.4018/IJEGR.2016040102

Pal, K. (2018). Building High Quality Big Data-Based Applications in Supply Chains. In A. Kumar & S. Saurav (Eds.), *Supply Chain Management Strategies and Risk Assessment in Retail Environments* (pp. 1–24). Hershey, PA: IGI Global. doi:10.4018/978-1-5225-3056-5.ch001

Palos-Sanchez, P. R., & Correia, M. B. (2018). Perspectives of the Adoption of Cloud Computing in the Tourism Sector. In J. Rodrigues, C. Ramos, P. Cardoso, & C. Henriques (Eds.), *Handbook of Research on Technological Developments for Cultural Heritage and eTourism Applications* (pp. 377–400). Hershey, PA: IGI Global. doi:10.4018/978-1-5225-2927-9.ch018

Parry, V. K., & Lind, M. L. (2016). Alignment of Business Strategy and Information Technology Considering Information Technology Governance, Project Portfolio Control, and Risk Management. *International Journal of Information Technology Project Management, 7*(4), 21–37. doi:10.4018/IJITPM.2016100102

Pashkova, N., Trujillo-Barrera, A., Apostolakis, G., Van Dijk, G., Drakos, P. D., & Baourakis, G. (2016). Business Management Models of Microfinance Institutions (MFIs) in Africa: A Study into Their Enabling Environments. *International Journal of Food and Beverage Manufacturing and Business Models, 1*(2), 63–82. doi:10.4018/IJFBMBM.2016070105

Patiño, B. E. (2017). New Generation Management by Convergence and Individual Identity: A Systemic and Human-Oriented Approach. In N. Baporikar (Ed.), *Innovation and Shifting Perspectives in Management Education* (pp. 119–143). Hershey, PA: IGI Global. doi:10.4018/978-1-5225-1019-2.ch006

Pawliczek, A., & Rössler, M. (2017). Knowledge of Management Tools and Systems in SMEs: Knowledge Transfer in Management. In A. Bencsik (Ed.), *Knowledge Management Initiatives and Strategies in Small and Medium Enterprises* (pp. 180–203). Hershey, PA: IGI Global. doi:10.4018/978-1-5225-1642-2.ch009

Pejic-Bach, M., Omazic, M. A., Aleksic, A., & Zoroja, J. (2018). Knowledge-Based Decision Making: A Multi-Case Analysis. In R. Leon (Ed.), *Managerial Strategies for Business Sustainability During Turbulent Times* (pp. 160–184). Hershey, PA: IGI Global. doi:10.4018/978-1-5225-2716-9.ch009

Perano, M., Hysa, X., & Calabrese, M. (2018). Strategic Planning, Cultural Context, and Business Continuity Management: Business Cases in the City of Shkoder. In A. Presenza & L. Sheehan (Eds.), *Geopolitics and Strategic Management in the Global Economy* (pp. 57–77). Hershey, PA: IGI Global. doi:10.4018/978-1-5225-2673-5.ch004

Pereira, R., Mira da Silva, M., & Lapão, L. V. (2017). IT Governance Maturity Patterns in Portuguese Healthcare. In S. De Haes & W. Van Grembergen (Eds.), *Strategic IT Governance and Alignment in Business Settings* (pp. 24–52). Hershey, PA: IGI Global. doi:10.4018/978-1-5225-0861-8.ch002

Perez-Uribe, R., & Ocampo-Guzman, D. (2016). Conflict within Colombian Family Owned SMEs: An Explosive Blend between Feelings and Business. In J. Saiz-Álvarez (Ed.), *Handbook of Research on Social Entrepreneurship and Solidarity Economics* (pp. 329–354). Hershey, PA: IGI Global. doi:10.4018/978-1-5225-0097-1.ch017

Pérez-Uribe, R. I., Torres, D. A., Jurado, S. P., & Prada, D. M. (2018). Cloud Tools for the Development of Project Management in SMEs. In R. Perez-Uribe, C. Salcedo-Perez, & D. Ocampo-Guzman (Eds.), *Handbook of Research on Intrapreneurship and Organizational Sustainability in SMEs* (pp. 95–120). Hershey, PA: IGI Global. doi:10.4018/978-1-5225-3543-0.ch005

Petrisor, I., & Cozmiuc, D. (2017). Global Supply Chain Management Organization at Siemens in the Advent of Industry 4.0. In L. Saglietto & C. Cezanne (Eds.), *Global Intermediation and Logistics Service Providers* (pp. 123–142). Hershey, PA: IGI Global. doi:10.4018/978-1-5225-2133-4.ch007

Pierce, J. M., Velliaris, D. M., & Edwards, J. (2017). A Living Case Study: A Journey Not a Destination. In N. Silton (Ed.), *Exploring the Benefits of Creativity in Education, Media, and the Arts* (pp. 158–178). Hershey, PA: IGI Global. doi:10.4018/978-1-5225-0504-4.ch008

Radosavljevic, M., & Andjelkovic, A. (2017). Multi-Criteria Decision Making Approach for Choosing Business Process for the Improvement: Upgrading of the Six Sigma Methodology. In J. Stanković, P. Delias, S. Marinković, & S. Rochhia (Eds.), *Tools and Techniques for Economic Decision Analysis* (pp. 225–247). Hershey, PA: IGI Global. doi:10.4018/978-1-5225-0959-2.ch011

Radovic, V. M. (2017). Corporate Sustainability and Responsibility and Disaster Risk Reduction: A Serbian Overview. In M. Camilleri (Ed.), *CSR 2.0 and the New Era of Corporate Citizenship* (pp. 147–164). Hershey, PA: IGI Global. doi:10.4018/978-1-5225-1842-6.ch008

Raghunath, K. M., Devi, S. L., & Patro, C. S. (2018). Impact of Risk Assessment Models on Risk Factors: A Holistic Outlook. In K. Strang, M. Korstanje, & N. Vajjhala (Eds.), *Research, Practices, and Innovations in Global Risk and Contingency Management* (pp. 134–153). Hershey, PA: IGI Global. doi:10.4018/978-1-5225-4754-9.ch008

Raman, A., & Goyal, D. P. (2017). Extending IMPLEMENT Framework for Enterprise Information Systems Implementation to Information System Innovation. In M. Tavana (Ed.), *Enterprise Information Systems and the Digitalization of Business Functions* (pp. 137–177). Hershey, PA: IGI Global. doi:10.4018/978-1-5225-2382-6.ch007

Rao, Y., & Zhang, Y. (2017). The Construction and Development of Academic Library Digital Special Subject Databases. In L. Ruan, Q. Zhu, & Y. Ye (Eds.), *Academic Library Development and Administration in China* (pp. 163–183). Hershey, PA: IGI Global. doi:10.4018/978-1-5225-0550-1.ch010

Ravasan, A. Z., Mohammadi, M. M., & Hamidi, H. (2018). An Investigation Into the Critical Success Factors of Implementing Information Technology Service Management Frameworks. In K. Jakobs (Ed.), *Corporate and Global Standardization Initiatives in Contemporary Society* (pp. 200–218). Hershey, PA: IGI Global. doi:10.4018/978-1-5225-5320-5.ch009

Renna, P., Izzo, C., & Romaniello, T. (2016). The Business Process Management Systems to Support Continuous Improvements. In W. Nuninger & J. Châtelet (Eds.), *Handbook of Research on Quality Assurance and Value Management in Higher Education* (pp. 237–256). Hershey, PA: IGI Global. doi:10.4018/978-1-5225-0024-7.ch009

Rezaie, S., Mirabedini, S. J., & Abtahi, A. (2018). Designing a Model for Implementation of Business Intelligence in the Banking Industry. *International Journal of Enterprise Information Systems*, *14*(1), 77–103. doi:10.4018/IJEIS.2018010105

Riccò, R. (2016). Diversity Management: Bringing Equality, Equity, and Inclusion in the Workplace. In J. Prescott (Ed.), *Handbook of Research on Race, Gender, and the Fight for Equality* (pp. 335–359). Hershey, PA: IGI Global. doi:10.4018/978-1-5225-0047-6.ch015

Romano, L., Grimaldi, R., & Colasuonno, F. S. (2017). Demand Management as a Success Factor in Project Portfolio Management. In L. Romano (Ed.), *Project Portfolio Management Strategies for Effective Organizational Operations* (pp. 202–219). Hershey, PA: IGI Global. doi:10.4018/978-1-5225-2151-8.ch008

Rostek, K. B. (2016). Risk Management: Role and Importance in Business Organization. In D. Jakóbczak (Ed.), *Analyzing Risk through Probabilistic Modeling in Operations Research* (pp. 149–178). Hershey, PA: IGI Global. doi:10.4018/978-1-4666-9458-3.ch007

Rouhani, S., & Savoji, S. R. (2016). A Success Assessment Model for BI Tools Implementation: An Empirical Study of Banking Industry. *International Journal of Business Intelligence Research*, *7*(1), 25–44. doi:10.4018/IJBIR.2016010103

Ruan, Z. (2016). A Corpus-Based Functional Analysis of Complex Nominal Groups in Written Business Discourse: The Case of "Business". *International Journal of Computer-Assisted Language Learning and Teaching*, *6*(2), 74–90. doi:10.4018/IJCALLT.2016040105

Ruhi, U. (2018). Towards an Interdisciplinary Socio-Technical Definition of Virtual Communities. In M. Khosrow-Pour, D.B.A. (Ed.), Encyclopedia of Information Science and Technology, Fourth Edition (pp. 4278-4295). Hershey, PA: IGI Global. doi:10.4018/978-1-5225-2255-3.ch371

Ryan, J., Doster, B., Daily, S., & Lewis, C. (2016). A Case Study Perspective for Balanced Perioperative Workflow Achievement through Data-Driven Process Improvement. *International Journal of Healthcare Information Systems and Informatics*, *11*(3), 19–41. doi:10.4018/IJHISI.2016070102

Safari, M. R., & Jiang, Q. (2018). The Theory and Practice of IT Governance Maturity and Strategies Alignment: Evidence From Banking Industry. *Journal of Global Information Management*, *26*(2), 127–146. doi:10.4018/JGIM.2018040106

Sahoo, J., Pati, B., & Mohanty, B. (2017). Knowledge Management as an Academic Discipline: An Assessment. In B. Gunjal (Ed.), *Managing Knowledge and Scholarly Assets in Academic Libraries* (pp. 99–126). Hershey, PA: IGI Global. doi:10.4018/978-1-5225-1741-2.ch005

Saini, D. (2017). Relevance of Teaching Values and Ethics in Management Education. In N. Baporikar (Ed.), *Management Education for Global Leadership* (pp. 90–111). Hershey, PA: IGI Global. doi:10.4018/978-1-5225-1013-0.ch005

Sambhanthan, A. (2017). Assessing and Benchmarking Sustainability in Organisations: An Integrated Conceptual Model. *International Journal of Systems and Service-Oriented Engineering*, *7*(4), 22–43. doi:10.4018/IJSSOE.2017100102

Sambhanthan, A., & Potdar, V. (2017). A Study of the Parameters Impacting Sustainability in Information Technology Organizations. *International Journal of Knowledge-Based Organizations*, *7*(3), 27–39. doi:10.4018/IJKBO.2017070103

Sánchez-Fernández, M. D., & Manríquez, M. R. (2018). The Entrepreneurial Spirit Based on Social Values: The Digital Generation. In P. Isaias & L. Carvalho (Eds.), *User Innovation and the Entrepreneurship Phenomenon in the Digital Economy* (pp. 173–193). Hershey, PA: IGI Global. doi:10.4018/978-1-5225-2826-5.ch009

Sanchez-Ruiz, L., & Blanco, B. (2017). Process Management for SMEs: Barriers, Enablers, and Benefits. In M. Vemić (Ed.), *Optimal Management Strategies in Small and Medium Enterprises* (pp. 293–319). Hershey, PA: IGI Global. doi:10.4018/978-1-5225-1949-2.ch014

Sanz, L. F., Gómez-Pérez, J., & Castillo-Martinez, A. (2018). Analysis of the European ICT Competence Frameworks. In V. Ahuja & S. Rathore (Eds.), *Multidisciplinary Perspectives on Human Capital and Information Technology Professionals* (pp. 225–245). Hershey, PA: IGI Global. doi:10.4018/978-1-5225-5297-0.ch012

Sarvepalli, A., & Godin, J. (2017). Business Process Management in the Classroom. *Journal of Cases on Information Technology*, *19*(2), 17–28. doi:10.4018/JCIT.2017040102

Satpathy, B., & Muniapan, B. (2016). Ancient Wisdom for Transformational Leadership and Its Insights from the Bhagavad-Gita. In U. Aung & P. Ordoñez de Pablos (Eds.), *Managerial Strategies and Practice in the Asian Business Sector* (pp. 1–10). Hershey, PA: IGI Global. doi:10.4018/978-1-4666-9758-4.ch001

Saygili, E. E., Ozturkoglu, Y., & Kocakulah, M. C. (2017). End Users' Perceptions of Critical Success Factors in ERP Applications. *International Journal of Enterprise Information Systems*, *13*(4), 58–75. doi:10.4018/IJEIS.2017100104

Saygili, E. E., & Saygili, A. T. (2017). Contemporary Issues in Enterprise Information Systems: A Critical Review of CSFs in ERP Implementations. In M. Tavana (Ed.), *Enterprise Information Systems and the Digitalization of Business Functions* (pp. 120–136). Hershey, PA: IGI Global. doi:10.4018/978-1-5225-2382-6.ch006

Seidenstricker, S., & Antonino, A. (2018). Business Model Innovation-Oriented Technology Management for Emergent Technologies. In M. Khosrow-Pour, D.B.A. (Ed.), Encyclopedia of Information Science and Technology, Fourth Edition (pp. 4560-4569). Hershey, PA: IGI Global. doi:10.4018/978-1-5225-2255-3.ch396

Senaratne, S., & Gunarathne, A. D. (2017). Excellence Perspective for Management Education from a Global Accountants' Hub in Asia. In N. Baporikar (Ed.), *Management Education for Global Leadership* (pp. 158–180). Hershey, PA: IGI Global. doi:10.4018/978-1-5225-1013-0.ch008

Sensuse, D. I., & Cahyaningsih, E. (2018). Knowledge Management Models: A Summative Review. *International Journal of Information Systems in the Service Sector*, *10*(1), 71–100. doi:10.4018/IJISSS.2018010105

Sensuse, D. I., Wibowo, W. C., & Cahyaningsih, E. (2016). Indonesian Government Knowledge Management Model: A Theoretical Model. *Information Resources Management Journal, 29*(1), 91–108. doi:10.4018/irmj.2016010106

Seth, M., Goyal, D., & Kiran, R. (2017). Diminution of Impediments in Implementation of Supply Chain Management Information System for Enhancing its Effectiveness in Indian Automobile Industry. *Journal of Global Information Management, 25*(3), 1–20. doi:10.4018/JGIM.2017070101

Seyal, A. H., & Rahman, M. N. (2017). Investigating Impact of Inter-Organizational Factors in Measuring ERP Systems Success: Bruneian Perspectives. In M. Tavana (Ed.), *Enterprise Information Systems and the Digitalization of Business Functions* (pp. 178–204). Hershey, PA: IGI Global. doi:10.4018/978-1-5225-2382-6.ch008

Shaikh, A. A., & Karjaluoto, H. (2016). On Some Misconceptions Concerning Digital Banking and Alternative Delivery Channels. *International Journal of E-Business Research, 12*(3), 1–16. doi:10.4018/IJEBR.2016070101

Shams, S. M. (2016). Stakeholder Relationship Management in Online Business and Competitive Value Propositions: Evidence from the Sports Industry. *International Journal of Online Marketing, 6*(2), 1–17. doi:10.4018/IJOM.2016040101

Shamsuzzoha, A. (2016). Management of Risk and Resilience within Collaborative Business Network. In R. Addo-Tenkorang, J. Kantola, P. Helo, & A. Shamsuzzoha (Eds.), *Supply Chain Strategies and the Engineer-to-Order Approach* (pp. 143–159). Hershey, PA: IGI Global. doi:10.4018/978-1-5225-0021-6.ch008

Shaqrah, A. A. (2018). Analyzing Business Intelligence Systems Based on 7s Model of McKinsey. *International Journal of Business Intelligence Research, 9*(1), 53–63. doi:10.4018/IJBIR.2018010104

Sharma, A. J. (2017). Enhancing Sustainability through Experiential Learning in Management Education. In N. Baporikar (Ed.), *Management Education for Global Leadership* (pp. 256–274). Hershey, PA: IGI Global. doi:10.4018/978-1-5225-1013-0.ch013

Shetty, K. P. (2017). Responsible Global Leadership: Ethical Challenges in Management Education. In N. Baporikar (Ed.), *Innovation and Shifting Perspectives in Management Education* (pp. 194–223). Hershey, PA: IGI Global. doi:10.4018/978-1-5225-1019-2.ch009

Sinthupundaja, J., & Kohda, Y. (2017). Effects of Corporate Social Responsibility and Creating Shared Value on Sustainability. *International Journal of Sustainable Entrepreneurship and Corporate Social Responsibility*, *2*(1), 27–38. doi:10.4018/IJSECSR.2017010103

Škarica, I., & Hrgović, A. V. (2018). Implementation of Total Quality Management Principles in Public Health Institutes in the Republic of Croatia. *International Journal of Productivity Management and Assessment Technologies*, *6*(1), 1–16. doi:10.4018/IJPMAT.2018010101

Smuts, H., Kotzé, P., Van der Merwe, A., & Loock, M. (2017). Framework for Managing Shared Knowledge in an Information Systems Outsourcing Context. *International Journal of Knowledge Management*, *13*(4), 1–30. doi:10.4018/IJKM.2017100101

Soares, E. R., & Zaidan, F. H. (2016). Information Architecture and Business Modeling in Modern Organizations of Information Technology: Professional Career Plan in Organizations IT. In G. Jamil, J. Poças Rascão, F. Ribeiro, & A. Malheiro da Silva (Eds.), *Handbook of Research on Information Architecture and Management in Modern Organizations* (pp. 439–457). Hershey, PA: IGI Global. doi:10.4018/978-1-4666-8637-3.ch020

Sousa, M. J., Cruz, R., Dias, I., & Caracol, C. (2017). Information Management Systems in the Supply Chain. In G. Jamil, A. Soares, & C. Pessoa (Eds.), *Handbook of Research on Information Management for Effective Logistics and Supply Chains* (pp. 469–485). Hershey, PA: IGI Global. doi:10.4018/978-1-5225-0973-8.ch025

Spremic, M., Turulja, L., & Bajgoric, N. (2018). Two Approaches in Assessing Business Continuity Management Attitudes in the Organizational Context. In N. Bajgoric (Ed.), *Always-On Enterprise Information Systems for Modern Organizations* (pp. 159–183). Hershey, PA: IGI Global. doi:10.4018/978-1-5225-3704-5.ch008

Steenkamp, A. L. (2018). Some Insights in Computer Science and Information Technology. In *Examining the Changing Role of Supervision in Doctoral Research Projects: Emerging Research and Opportunities* (pp. 113–133). Hershey, PA: IGI Global. doi:10.4018/978-1-5225-2610-0.ch005

Studdard, N., Dawson, M., Burton, S. L., Jackson, N., Leonard, B., Quisenberry, W., & Rahim, E. (2016). Nurturing Social Entrepreneurship and Building Social Entrepreneurial Self-Efficacy: Focusing on Primary and Secondary Schooling to Develop Future Social Entrepreneurs. In Z. Fields (Ed.), *Incorporating Business Models and Strategies into Social Entrepreneurship* (pp. 154–175). Hershey, PA: IGI Global. doi:10.4018/978-1-4666-8748-6.ch010

Sun, Z. (2016). A Framework for Developing Management Intelligent Systems. *International Journal of Systems and Service-Oriented Engineering*, *6*(1), 37–53. doi:10.4018/IJSSOE.2016010103

Swami, B., & Mphele, G. T. (2016). Problems Preventing Growth of Small Entrepreneurs: A Case Study of a Few Small Entrepreneurs in Botswana Sub-Urban Areas. In N. Baporikar (Ed.), *Handbook of Research on Entrepreneurship in the Contemporary Knowledge-Based Global Economy* (pp. 479–508). Hershey, PA: IGI Global. doi:10.4018/978-1-4666-8798-1.ch020

Tabach, A., & Croteau, A. (2017). Configurations of Information Technology Governance Practices and Business Unit Performance. *International Journal of IT/ Business Alignment and Governance*, *8*(2), 1–27. doi:10.4018/IJITBAG.2017070101

Talaue, G. M., & Iqbal, T. (2017). Assessment of e-Business Mode of Selected Private Universities in the Philippines and Pakistan. *International Journal of Online Marketing*, *7*(4), 63–77. doi:10.4018/IJOM.2017100105

Tam, G. C. (2017). Project Manager Sustainability Competence. In *Managerial Strategies and Green Solutions for Project Sustainability* (pp. 178–207). Hershey, PA: IGI Global. doi:10.4018/978-1-5225-2371-0.ch008

Tambo, T. (2018). Fashion Retail Innovation: About Context, Antecedents, and Outcome in Technological Change Projects. In I. Management Association (Ed.), Fashion and Textiles: Breakthroughs in Research and Practice (pp. 233-260). Hershey, PA: IGI Global. doi:10.4018/978-1-5225-3432-7.ch010

Tambo, T., & Mikkelsen, O. E. (2016). Fashion Supply Chain Optimization: Linking Make-to-Order Purchasing and B2B E-Commerce. In S. Joshi & R. Joshi (Eds.), *Designing and Implementing Global Supply Chain Management* (pp. 1–21). Hershey, PA: IGI Global. doi:10.4018/978-1-4666-9720-1.ch001

Tandon, K. (2016). Innovative Andragogy: The Paradigm Shift to Heutagogy. In S. Tiwari & L. Nafees (Eds.), *Innovative Management Education Pedagogies for Preparing Next-Generation Leaders* (pp. 238–257). Hershey, PA: IGI Global. doi:10.4018/978-1-4666-9691-4.ch014

Tantau, A. D., & Frăţilă, L. C. (2018). Information and Management System for Renewable Energy Business. In *Entrepreneurship and Business Development in the Renewable Energy Sector* (pp. 200–244). Hershey, PA: IGI Global. doi:10.4018/978-1-5225-3625-3.ch006

Teixeira, N., Pardal, P. N., & Rafael, B. G. (2018). Internationalization, Financial Performance, and Organizational Challenges: A Success Case in Portugal. In L. Carvalho (Ed.), *Handbook of Research on Entrepreneurial Ecosystems and Social Dynamics in a Globalized World* (pp. 379–423). Hershey, PA: IGI Global. doi:10.4018/978-1-5225-3525-6.ch017

Trad, A., & Kalpić, D. (2016). The E-Business Transformation Framework for E-Commerce Architecture-Modeling Projects. In I. Lee (Ed.), *Encyclopedia of E-Commerce Development, Implementation, and Management* (pp. 733–753). Hershey, PA: IGI Global. doi:10.4018/978-1-4666-9787-4.ch052

Trad, A., & Kalpić, D. (2016). The E-Business Transformation Framework for E-Commerce Control and Monitoring Pattern. In I. Lee (Ed.), *Encyclopedia of E-Commerce Development, Implementation, and Management* (pp. 754–777). Hershey, PA: IGI Global. doi:10.4018/978-1-4666-9787-4.ch053

Trad, A., & Kalpić, D. (2018). The Business Transformation Framework, Agile Project and Change Management. In M. Khosrow-Pour, D.B.A. (Ed.), Encyclopedia of Information Science and Technology, Fourth Edition (pp. 620-635). Hershey, PA: IGI Global. doi:10.4018/978-1-5225-2255-3.ch054

Trad, A., & Kalpić, D. (2018). The Business Transformation and Enterprise Architecture Framework: The Financial Engineering E-Risk Management and E-Law Integration. In B. Sergi, F. Fidanoski, M. Ziolo, & V. Naumovski (Eds.), *Regaining Global Stability After the Financial Crisis* (pp. 46–65). Hershey, PA: IGI Global. doi:10.4018/978-1-5225-4026-7.ch003

Turulja, L., & Bajgoric, N. (2018). Business Continuity and Information Systems: A Systematic Literature Review. In N. Bajgoric (Ed.), *Always-On Enterprise Information Systems for Modern Organizations* (pp. 60–87). Hershey, PA: IGI Global. doi:10.4018/978-1-5225-3704-5.ch004

van Wessel, R. M., de Vries, H. J., & Ribbers, P. M. (2016). Business Benefits through Company IT Standardization. In K. Jakobs (Ed.), *Effective Standardization Management in Corporate Settings* (pp. 34–53). Hershey, PA: IGI Global. doi:10.4018/978-1-4666-9737-9.ch003

Vargas-Hernández, J. G. (2017). Professional Integrity in Business Management Education. In N. Baporikar (Ed.), *Management Education for Global Leadership* (pp. 70–89). Hershey, PA: IGI Global. doi:10.4018/978-1-5225-1013-0.ch004

Vasista, T. G., & AlAbdullatif, A. M. (2017). Role of Electronic Customer Relationship Management in Demand Chain Management: A Predictive Analytic Approach. *International Journal of Information Systems and Supply Chain Management, 10*(1), 53–67. doi:10.4018/IJISSCM.2017010104

Vergidis, K. (2016). Rediscovering Business Processes: Definitions, Patterns, and Modelling Approaches. In P. Papajorgji, F. Pinet, A. Guimarães, & J. Papathanasiou (Eds.), *Automated Enterprise Systems for Maximizing Business Performance* (pp. 97–122). Hershey, PA: IGI Global. doi:10.4018/978-1-4666-8841-4.ch007

Vieru, D., & Bourdeau, S. (2017). Survival in the Digital Era: A Digital Competence-Based Multi-Case Study in the Canadian SME Clothing Industry. *International Journal of Social and Organizational Dynamics in IT, 6*(1), 17–34. doi:10.4018/IJSODIT.2017010102

Vijayan, G., & Kamarulzaman, N. H. (2017). An Introduction to Sustainable Supply Chain Management and Business Implications. In M. Khan, M. Hussain, & M. Ajmal (Eds.), *Green Supply Chain Management for Sustainable Business Practice* (pp. 27–50). Hershey, PA: IGI Global. doi:10.4018/978-1-5225-0635-5.ch002

Vlachvei, A., & Notta, O. (2017). Firm Competitiveness: Theories, Evidence, and Measurement. In A. Vlachvei, O. Notta, K. Karantininis, & N. Tsounis (Eds.), *Factors Affecting Firm Competitiveness and Performance in the Modern Business World* (pp. 1–42). Hershey, PA: IGI Global. doi:10.4018/978-1-5225-0843-4.ch001

von Rosing, M., Fullington, N., & Walker, J. (2016). Using the Business Ontology and Enterprise Standards to Transform Three Leading Organizations. *International Journal of Conceptual Structures and Smart Applications, 4*(1), 71–99. doi:10.4018/IJCSSA.2016010104

von Rosing, M., & von Scheel, H. (2016). Using the Business Ontology to Develop Enterprise Standards. *International Journal of Conceptual Structures and Smart Applications, 4*(1), 48–70. doi:10.4018/IJCSSA.2016010103

Walczak, S. (2016). Artificial Neural Networks and other AI Applications for Business Management Decision Support. *International Journal of Sociotechnology and Knowledge Development, 8*(4), 1–20. doi:10.4018/IJSKD.2016100101

Wamba, S. F., Akter, S., Kang, H., Bhattacharya, M., & Upal, M. (2016). The Primer of Social Media Analytics. *Journal of Organizational and End User Computing, 28*(2), 1–12. doi:10.4018/JOEUC.2016040101

Wang, C., Schofield, M., Li, X., & Ou, X. (2017). Do Chinese Students in Public and Private Higher Education Institutes Perform at Different Level in One of the Leadership Skills: Critical Thinking?: An Exploratory Comparison. In V. Wang (Ed.), *Encyclopedia of Strategic Leadership and Management* (pp. 160–181). Hershey, PA: IGI Global. doi:10.4018/978-1-5225-1049-9.ch013

Wang, F., Raisinghani, M. S., Mora, M., & Wang, X. (2016). Strategic E-Business Management through a Balanced Scored Card Approach. In I. Lee (Ed.), *Encyclopedia of E-Commerce Development, Implementation, and Management* (pp. 361–386). Hershey, PA: IGI Global. doi:10.4018/978-1-4666-9787-4.ch027

Wang, J. (2017). Multi-Agent based Production Management Decision System Modelling for the Textile Enterprise. *Journal of Global Information Management, 25*(4), 1–15. doi:10.4018/JGIM.2017100101

Wiedemann, A., & Gewald, H. (2017). Examining Cross-Domain Alignment: The Correlation of Business Strategy, IT Management, and IT Business Value. *International Journal of IT/Business Alignment and Governance*, *8*(1), 17–31. doi:10.4018/IJITBAG.2017010102

Wolf, R., & Thiel, M. (2018). Advancing Global Business Ethics in China: Reducing Poverty Through Human and Social Welfare. In S. Hipsher (Ed.), *Examining the Private Sector's Role in Wealth Creation and Poverty Reduction* (pp. 67–84). Hershey, PA: IGI Global. doi:10.4018/978-1-5225-3117-3.ch004

Wu, J., Ding, F., Xu, M., Mo, Z., & Jin, A. (2016). Investigating the Determinants of Decision-Making on Adoption of Public Cloud Computing in E-government. *Journal of Global Information Management*, *24*(3), 71–89. doi:10.4018/JGIM.2016070104

Xu, L., & de Vrieze, P. (2016). Building Situational Applications for Virtual Enterprises. In I. Lee (Ed.), *Encyclopedia of E-Commerce Development, Implementation, and Management* (pp. 715–724). Hershey, PA: IGI Global. doi:10.4018/978-1-4666-9787-4.ch050

Yablonsky, S. (2018). Innovation Platforms: Data and Analytics Platforms. In *Multi-Sided Platforms (MSPs) and Sharing Strategies in the Digital Economy: Emerging Research and Opportunities* (pp. 72–95). Hershey, PA: IGI Global. doi:10.4018/978-1-5225-5457-8.ch003

Yusoff, A., Ahmad, N. H., & Halim, H. A. (2017). Agropreneurship among Gen Y in Malaysia: The Role of Academic Institutions. In N. Ahmad, T. Ramayah, H. Halim, & S. Rahman (Eds.), *Handbook of Research on Small and Medium Enterprises in Developing Countries* (pp. 23–47). Hershey, PA: IGI Global. doi:10.4018/978-1-5225-2165-5.ch002

Zanin, F., Comuzzi, E., & Costantini, A. (2018). The Effect of Business Strategy and Stock Market Listing on the Use of Risk Assessment Tools. In *Management Control Systems in Complex Settings: Emerging Research and Opportunities* (pp. 145–168). Hershey, PA: IGI Global. doi:10.4018/978-1-5225-3987-2.ch007

Zgheib, P. W. (2017). Corporate Innovation and Intrapreneurship in the Middle East. In P. Zgheib (Ed.), *Entrepreneurship and Business Innovation in the Middle East* (pp. 37–56). Hershey, PA: IGI Global. doi:10.4018/978-1-5225-2066-5.ch003

About the Contributors

Noor Aniza Abdul Rahim is a food technologist in the Division of Food Safety and Quality, Ministry of Health Malaysia since December 2012. She is currently pursuing her postgraduate degree in Food Safety at Universiti Putra Malaysia. She received her first degree in Food Technology (2010) from Universiti Malaysia Terengganu.

Muhammad Haseeb Ahmad currently working as Assistant Professor in Department of Food Science, Government College University Faisalabad, Pakistan. Before this, he completed his doctorate from University of Hohenheim, Stuttgart Germany in 2016.He is expert in development of rapid and non invasive methodologies in food applications. He presented his research work in various international conferences. He is the author of 25 research articles 8 book chapters and 5 review articles.

Rabia Shabir Ahmad has a strong academic, teaching, and research background, with a number of credits and honors received. During her academic career, she secured a merit scholarship from the University of Agriculture, Faisalabad for her Bachelor's and Master's degrees, and received a PhD Indigenous Scholarship during her PhD study. She played an important role in the establishment and development of the Institute of Home and Food Science at GC University, Faisalabad, and developed and taught courses to graduate and post-graduate level students, both in food science and technology, and human nutrition and dietetics. During her stay at the Government College University Faisalabad, she won various HEC-funded research projects and was also honored to receive the highest funded project under the NRPU scheme from HEC. Along with her teaching and research supervising responsibilities, she has also been working as the reviewer of several journals, and has published numerous research papers in high-impact international and national journals. She is active in developing laboratories and laying the strong foundation of research at the Institute of Home and Food Sciences.

Muhammad Faizan Afzal completed his BS (Hons.) in Human Nutrition and Dietetics from Government College University, Faisalabad. After this he has got MS Food Science and Technology from Government College University, Faisalabad under the supervision of Dr. Muhammad Haseeb Ahmad and currently a PhD scholar in the same department.

Carolina K. de Souza received the B.E.degree in food engineering from the Federal University of Santa Catarina, Brazil, in 2003, specialization in bromatology from the University of Ferrara, Italy, and the Ph.D. degree in chemistry and toxicology of foods from the University of Perugia, Italy, in 2008. She was a Postdoctoral Researcher of Biochemical Engineering with the Federal University of Santa Catarina, in 2013. She is currently a Professor with the Courses of Food Engineering, Chemical Engineering, and Nutrition, and the Graduate Program in Chemical Engineering (PPGEQ), Regional University of Blumenau. She experience in food science with emphasis on food chemistry and functional foods. She operates in product development research lines, food preservation, and food safety.

M. Bilal Hussain is currently working as a Co-ordinator and Lecturer in Department of Food Science and Technology, Government college University, Faisalabad, Pakistan.

Muhammad Imran is Assistant Professor in the Department of Food Science, Faculty of Life Sciences, Government College University, Faisalabad–Pakistan since September 03, 2012. Dr. Muhammad Imran completed his Doctorate (Food Technology) from National Institute of Food Science and Technology, University of Agriculture, Faisalabad. He won the Indigenous and IRSIP (Department of Food Science and Human Nutrition, Michigan State University, East Lansing, USA) Fellowships for completion of doctorate research funded by HEC, Islamabad, Pakistan. Dr. Muhammad Imran has expertise in extrusion technology, microencapsulation, lipids chemistry, sensory evaluation and food process engineering. Until today, Dr. Muhammad Imran has authored 70 publications (International & National) in various Impact Journals of Scientific repute and written 10 Book Chapters as principal author and co-author. Dr. Muhammad Imran has attended several International Conferences (held nationally and locally) as Invited and Keynote Speaker and also participated as organizing member to conduct International Conferences, Seminar and Workshops etc. Dr. Muhammad Imran is the Life Time Member of PSFST.

Bushra Ishfaq is the doctoral student of Food Technology at National Institute of Food Science and Technology, University of Agriculture, Faisalabad.

Simmi Jain received the B.Sc. degree from Jai Narain Vyas University, Jodhpur, Rajasthan, India, in 1999 with specialization in Food and Nutrition and Ph.D. in 2006. She has work experience of more than 19 years at various Institutions and Colleges. She is currently Assistant Professor and Head with Department of Food Technology and Management, School of Food Science, M. O. P. Vaishnav College for Women (Autonomous), Chennai, Tamil Nadu, India. Her areas of research and core areas of expertise include Beverage Technology, Baking Technology, Product Development, Sensory Science, Food Processing and Preservation. Dr. Simmi Jain has18 research papers in National and International Journals in her name and is on the Editorial Board for International and National Journals.

Lekhraj Katariya is working as Assistant Director (Tech) at Export Inspection Agency- Delhi (HO) an organisation under Export Inspection Council, Ministry of Commerce & Industry, Govt. of India, New Delhi since September 2017. Before joining this organisation having almost 10 Years of working experience as Assistant Plant Protection Officer (Plant Pathology) and Scientific Assistant at Directorate of Plant Protection, Quarantine & Storage, NH-IV, Faridabad under Ministry of Agriculture & Farmers Welfare, Govt. of India during the tenure working at different states like Locust Warning Organisation, Bikaner, Rajasthan, Regional Plant Quarantine Station, Mumbai and Central Integrated Pest Management Centre, Aizawl, Mizoram (NE). Qualified M.SC. (Agri) in Plant Pathology from Swami Keshwanand Rajasthan Agricultural University, Bikaner Qualified 4 times NET (Plant Pathology) from Agricultural Scientists Recruitment Board (ICAR), New Delhi. He has published 04 Research Papers, 01 Short communication published in NAAS Rated Journal - Indian Journal of Mycology and Plant Pathology, Udaipur and 40 populer articles published in various monthly magazines.

Muhammad Kamran Khan completed my doctoral studies from University of Avignon, France in 2010 and won two post-doctoral fellowships: one from "University of French Guiana, France (2011)" and the second one "Endeavour Research Fellowship Award – Australia (2014)". I have joined the Department of Food Science, Government College University Faisalabad, Pakistan on 13-03-2012 on the post of Assistant Professor under IPFP (Interim Placement for Fresh PhDs) and on 19-09-2012 on the same post under Tenure Track System (TTS). During my professional career at GCUF, various research and review articles have been published in peer-reviewed journals. The article published in "Journal of Food Composition and Analysis" remained among the top 3 downloaded articles of 2014. Another noteworthy achievement is the approval of project entitled "Impact of fruit processing modes on extracted juice quality: conventional vs. Innovative technologies" with worth of 6.2 million PKR (~ $ 60,000) funded by Higher Education Commission of

Pakistan. Then visit to Curtin University, Perth, Australia on "Endeavour Research Fellowship Award-2014" was a great success and helped to develop collaborative publications between two institutions. Being Pak-France Alumni, I actively prospered the educational wing of "the Embassy of France in Pakistan" and have been selected as "France Alumni Ambassador-2016". The professional experience gave me various opportunities to represent the university on different platforms at national as well as international level.

Usman Khan is the doctoral student of Food Technology at National Institute of Food Science and Technology, University of Agriculture, Faisalabad.

Kiran Meghwal is M. Sc. Specialization in Organic chemistry, 2011; Qualified N.E.T. and GATE in 2012; received Ph. D. Degree in Polymer specialization from Mohanlal sukhadia university udaipur in 2018; Received B. Ed degree in general chemistry from M. L. S. university, udaipur in 2020; Currently working as Lecturer, with the courses of Industrial chemistry, polymer chemistry, medicinal chemistry and chemical engineering in M. L. S. University, udaipur(Raj.)

Murlidhar Meghwal received his B. Tech. degree (Agricultural Engineering) in 2008 from College of Agricultural Engineering Bapatla, Acharya N. G. RANGA Agricultural University, Hyderabad, India; his M. Tech. degree (Dairy and Food Engineering) in 2010and PhD degree (Food Process Engineering) in 2014 from Indian Institute of Technology Kharagpur, West Bengal, India. He worked for one-year as research associate at INDUS Kolkata, (rice parboiling, milling and processing industry) for development of a quicker and industrial level parboiling system for paddy and rice milling. Currently, Dr. Meghwalis working on: Developing inexpensive, disposable and biodegradable food containers using agricultural wastes; Quality improvement, and Freeze drying of milk. At present, he is actively involved in research and teaching at the Food Science and Technology Division, Jain University Bangalore, India. He has written one book and many research publications in food engineering. He has attended many national and international seminars and conferences. He is reviewer and member of editorial boards of reputed journals.

Fathmath Nuzla finished her undergraduate degree in BSc. Biochemistry, genetics and biotechnology (2013) and got her post-graduate diploma in Public Health in 2017. She completed her MSc in Food Safety and Quality Assurance from University of Putra Malaysia in 2020 and currently working at Maldives Food and Drug Authority in Scientific Officer post as section head overseeing food import units in the Maldives.

Genevie Eleanor Ruby graduated with Master of Education (Biology) from UPSI, before continued her PhD at Faculty of Food Science and Technology, UPM in the field of Food Management. To date, she has published in several academic research journals in the field of food safety among food handlers. She also interested in studies related to food microbiology. Currently, she works as a lecturer at Sarawak Matriculation College.

Mahalakshmi S. completed B.Tech in Food Process Engineering at Tamil Nadu Agricultural University, Coimbatore. Currently pursuing M.Sc Food Technology and Management at M.O.P Vaishnav College for Women, Chennai.

Amna Sahar is working as Associate Professor at the Department of Food Engineering, University of Agriculture Faisalabad. She has expertise in food technology and food engineering.

Aysha Sameen is working as Associate Professor at the National Institute of Food Science and Technology, University of Agriculture, Faisalabad. She has expertise in Food Technology and dairy technology.

Maimunah Sanny is an Associate Professor at the faculty of Food Science and Technology, University Putra Malaysia (UPM). She has been researching various mitigation strategies for reducing heat-generated toxicants (acrylamide, polyaromatic hydrocarbon (PAH), and heterocyclic amines (HCAs)) in food since 2008. She is currently researching the magnitude and significance of migration of chemical contaminants (phthalate and styrene) from plastic into the contacting food. She is also assessing the performance of Food Safety Management System (FSMS) of selected food industries in Malaysia and understanding the effects of assessment results on food safety objectives.

Farwa Tariq is a doctoral student of Human Nutrition and Dietetics at the National Institute of Food Science and Technology, University of Agriculture, Faisalabad.

Tayyaba Tariq is a doctoral student of Human Nutrition and Dietetics at the National Institute of Food Science and Technology, University of Agriculture, Faisalabad.

Marwa Waheed is a PhD scholar (Food Science & Technology), Government College University Faisalabad, Pakistan.

Nur Hanani Zainal Abedin is an associate professor in the Department of Food Technology, Faculty of Food Science and Technology at Universiti Putra Malaysia (UPM), Malaysia. She earned a Doctorate in Food and Nutritional Science from University College Cork (UCC), Ireland. She is currently involved in national and international research projects. Her research interest is on food packaging, including biodegradable packaging, active and intelligent packaging, edible films and coatings, modified atmosphere packaging (MAP) and shelf-life study. She has published some articles in peer-reviewed international journals and book chapters and has presented her research findings at many international conferences with significant publications. She is also an assessor for few international grants and a reviewer for many high impact journals in food packaging. Nur Hanani is an associate editor of the Journal of Food Packaging and Shelf Life and an editorial board member of the Journal of Heliyon published by Elsevier.

Index

www.ingramcontent.com/pod-product-compliance
Lightning Source LLC
Chambersburg PA
CBHW081124180926
51611CB00004B/49